Nonconvex Optimization and Its Applications

Volume 88

Shashi Kant Mishra · Giorgio Giorgi

Invexity and Optimization

Springer

Shashi Kant Mishra
Reader
Department of Mathematics
Faculty of Science
Banaras Hindu University
Varanasi–221005
India

Giorgio Giorgi
Universita Pavia Dipto. Ricerche Aziendali
Sez. Matematica Generale
Via S. Felice, 5
27100 Pavia
Italy
ggiorgi@eco.unipv.it

ISBN 978-3-540-78561-3 e-ISBN 978-3-540-78562-0

Nonconvex Optimization and Its Applications ISSN 1571-568X

Library of Congress Control Number: 2008923063

Cover design: WMX Design GmbH, Heidelberg

Printed on acid-free paper

9 8 7 6 5 4 3 2 1

springer.com

The first author dedicates this book to his son Rohan Mishra
The second author dedicates this book to his mother Olga

Preface

Generalized convexity and generalized monotonicity are the core of many important subjects in the context of various research fields such as mathematics, economics, management science, engineering and other applied sciences. After the introduction of quasi-convex functions by de Finetti in 1949, many other authors have defined and studied several types of generalized convex functions and their applications in the context of scalar and vector optimization problems, calculus of variations and optimal control theory and financial and economic decision models.

In many cases, generalized convex functions preserve some of the valuable properties of convex functions. One of the important generalizations of convex functions is invex functions, a notion originally introduced for differentiable functions $f : X \to R$, X an open set of R^n, for which there exists some function $\eta : X \times X \to R^n$ such that $f(x) - f(y) \geq \eta(x, y)^T \nabla f(u), \forall x, u \in X$. Invex functions have the property that all stationary points are global minimizers and, since their introduction in 1981, have been used in many applications.

The interest in these topics is continuous, as shown by eight specific international meetings (the next one is scheduled in Kaohsiung, July 21–25, 2008, along with 2nd Summer School for Generalized convexity from July 15–19, 2008 at the Department of Applied Mathematics, The National Sun Yat-sen University, Kaohsiung, Taiwan) held to date (Vancouver in 1980; Canton (USA) in 1986; Pisa in 1988; Pecs (Hungary) in 1992; Luminy (France) in 1996; Hanoi (Vietnam) in 2001; Varese (Italy) in 2005) and by the foundation of the Scientific Committee of the Working Group on Generalized Convexity, the group sponsored by the Mathematical Programming Society.

This book deals with invex functions and their applications in nonlinear scalar and vector optimization problems, nonsmooth optimization problems, fractional and quadratic programming problems and continuous-time optimization problems. This book provides a comprehensive discussion on invex functions and their applications, based on the research work carried out over the past several decades.

Pantnagar, India,
Pavia, Italy,
November, 2007

Shashi Kant Mishra
Giorgio Giorgi

Contents

1

Introduction

The convexity of sets and the convexity or concavity of functions have been the object of many investigations during the past century. This is mainly due to the development of the theory of mathematical programming, both linear and nonlinear, which is closely tied with convex analysis. Optimality conditions, duality and related algorithms were mainly established for classes of problems involving the optimization of convex objective functions over convex feasible regions. Such assumptions were very convenient, due to the basic properties of convex (or concave) functions concerning optimality conditions. However, not all practical problems, when formulated as mathematical problems, fulfill the requirements of convexity (or concavity). Fortunately, such problems were often found to have some characteristics in common with convex problems and these properties could be exploited to establish theoretical results or develop algorithms. In the second half of the past century various generalizations of convex functions have been introduced. We mention here the early work by de Finetti [54], Fenchel [65], Arrow and Enthoven [5], Mangasarian [142], Ponstein [203] and Karamardian [109]. Usually such generalizations were introduced by a particular problem in economics, management science or optimization theory. In 1980 the first International Conference on generalized convexity/concavity and related fields was held in Vancouver (Canada) and since then, similar international symposia have been organized every year. So, at present we dispose of the proceedings of such conferences, published by Academic Press [221], Analytic Publishing [222], Springer Verlag [25, 52, 80, 129, 130], and Kluwer Academic Publishers [51]. Moreover, a monograph on generalized convexity was published by Plenum Publishing Corporation in 1988 (see [10]) and Handbook of Generalized Convexity and Generalized Monotonicity was published by Springer in 2005 (see, [81]). A useful survey is provided by Pini and Singh [202]. The Working Group on Generalized Convexity (WGGC) was founded during the 15th International Symposium on Mathematical Programming in Ann Arbor (Michigan, USA), August 1994. It is a working group of researchers who carry on their interests in generalized convexity, generalized monotonicity and related fields.

The present monograph is concerned with one particular class of generalized convex functions, the invex functions, which were initially employed by Hanson [83] and named by Craven [43]. There are many valuable contributions to this class of functions, contributions appeared so far mainly in journals and in proceedings of conferences. Perhaps this is the first monograph concerned entirely with invex and related functions. In order to place this generalization in perspective, we give here, for the reader's convenience, the basic notions on convexity and generalized convexity (in an n-dimensional real space R^n).

Definition 1.1. *A set $C \subseteq R^n$ is convex if for every $x, u \in C$ and $\lambda \in [0,1]$ we have*

$$\lambda x + (1-\lambda)u \in C.$$

Definition 1.2. *A function f defined on a convex set $C \subseteq R^n$ is convex (CX) on C if $\forall x, u \in C$ and $\lambda \in [0,1]$ we have*

$$f(\lambda x + (1-\lambda)u) \leq \lambda f(x) + (1-\lambda)f(u). \tag{1.1}$$

f is strictly convex (SCX) if the inequality in (1.1) is strict, $\forall x \neq u$ and $\lambda \in (0,1)$.

A function $f : C \to R, C \subseteq R^n$ convex set, is concave (strictly concave) if and only if $-f$ is convex (strictly convex).

An equivalent condition for the convexity of a function $f : C \to R$ that its epigraph, defined by

$$epi f = \{(x,\alpha) \in C \times R : f(x) \leq \alpha\}$$

is a convex set in R^{n+1}. For concave functions it is considered the hypograph:

$$\text{hypo} f = \{(x,\alpha) \in C \times R : f(x) \geq \alpha\}.$$

If f is Frechet differentiable on the open convex set $C \subseteq R^n$, then f is convex on C if and only if

$$f(x) - f(u) \geq (x-u)^T \nabla f(u), \quad \forall x, u \in C,$$

where $\nabla f(u)$ is the gradient of f at $u \in C$ (see, e.g., [143]).

Convexity at a point can also be defined: f is said to be convex at $u \in C$, if $\forall x \in C$ and $\lambda \in [0,1]$, we have

$$f(\lambda x + (1-\lambda)u) \leq \lambda f(x) + (1-\lambda)f(u).$$

If in addition, f is differentiable at u, then f is convex at u if and only if

$$f(x) - f(u) \geq (x-u)^T \nabla f(u), \quad \forall x \in C.$$

Another characterization of differentiable convex functions, generalizing the monotonicity property of the first-order derivative of a single variable convex function, is possible (see, e.g., [143]): if f is differentiable on the open convex set $C \subseteq R^n$, then f is convex on C if and only if

$$(x-u)^T\left[\nabla f(x) - \nabla f(u)\right] \geq 0, \quad \forall x, u \in C.$$

Note that both characterizations of differentiable convex functions involve conditions on two points. If the function under consideration is twice continuously differentiable, there is a useful characterization by second-order derivatives at only one point [65]: let f be twice continuously differentiable (i.e., of class C^2) on the open convex set $C \subseteq R^n$. Then f is convex on C if and only if its Hessian matrix $\nabla^2 f(x)$ is positive semidefinite, for each $x \in C$.

This characterization cannot be strictly extended to strictly convex functions by requiring that the Hessian matrix $\nabla^2 f(x)$ to be positive definite $\forall x \in C$. This is only a sufficient condition for strict convexity, but not a necessary one: consider, e.g., the strictly convex function $f(x) = x^4$ on R. Clearly the second-order derivative vanishes at the origin.

Some basic properties of convex functions are as follows. Let $f : C \to R$ be convex on the convex set $C \subseteq R^n$. Then:

1. The lower level sets $L(\alpha) = \{x \in C : f(x) \leq \alpha\}$ are convex sets in R^n for each $\alpha \in R$.
2. The maximum of f along any line segment occurs at an end point.
3. Every local minimum of f is a global minimum.
4. If f is differentiable (on the open convex set C), then every stationary point is a global minimizer; i.e., $\nabla f(\bar{x}) = 0 \Rightarrow f(\bar{x}) \leq f(x), \forall x \in C$.

As it appears from the previous properties, convex (and concave) functions play an important role in optimization theory (besides the role they play in economics, management science, statistics, econometrics, etc.). More precisely we can consider the following optimization problem under constraints (nonlinear programming problem):
(P)

$$\text{Minimize} f(x)$$

$$\text{Subject to } g_i(x) \leq 0, \quad i = 1, \ldots, m,$$

where $f : C \to R$, $g_i : C \to R, i = 1, \ldots, m$, and $C \subset R^n$. Such a problem is called a convex program if all functions f, $g_i, i = 1, \ldots, m$, are convex on the convex set C. Convex programs have many useful properties, summarized below.

Given a convex program (P), we have:

1. The set of feasible solutions is convex; the set of optimal solutions is convex.
2. Any local minimum is a global minimum.
3. The Karush–Kuhn–Tucker optimality conditions are sufficient for a (global) minimum. As it is well known, assuming that the functions f and $g_i, i = 1, \ldots, m$, are differentiable, if the nonlinear program (P) has an optimal

solution $x^* \in C$ and a constraint qualification is satisfied, there must exist a vector $\lambda \in R^m$ such that

$$\nabla f(x^*) + \sum_{i=1}^{m} \lambda_i \nabla g_i(x^*) = 0$$

$$\lambda^T g(x^*) = 0$$

$$\lambda \geq 0.$$

(Karush–Kuhn–Tucker conditions).

4. Dual programs (D) of (P) and duality relations between (D) and (P) can be established. For example we can consider the following dual problem of (P), given by Wolfe [247]: (D)

$$\text{Maximize} f(u) + \lambda^T g(u)$$

$$\text{Subject to } \nabla f(u) + \lambda^T \nabla g(u) = 0$$

$$\lambda \geq 0.$$

Wolfe [247] showed that, if (P) is a convex program, every feasible solution of the dual had an objective value less than or equal to the objective value of every feasible solution of the primal. Such a result is known as weak duality. Furthermore, under a constraint qualification, strong duality was established, so that if x^* is optimal for the primal, then there exists some $\lambda^* \in R^m$ such that (x^*, λ^*) is optimal for the dual.
5. A minimum of (P) is unique if f is strictly convex.

At this point question arises whether there are non-convex functions which share some of the useful properties of convex functions. For example if we take a monotone transform $h(f(x))$ of a convex (concave) function $f(x)$, where h is increasing, we see that $h(f(x))$ is often not convex (not concave), however, e.g., its lower level sets (upper level sets) are still convex. Indeed, property (1) is a direct consequence of the convexity of the sets $L_f(\alpha)$. This convexity is not an exclusive feature of convex functions: consider, e.g., the function $f(x) = x^3, x \in R$, which is not convex nor concave, but has convex lower level sets. This fact motivated de Finetti [54] to introduce the important class of quasi-convex functions, class which strictly contains the class of convex functions.

Definition 1.3. *A function f defined on a convex set C of R^n is quasi-convex (QCX) on C if its lower level sets*

$$L_f(\alpha) = \{x \in C : f(x) \leq \alpha\}$$

are convex sets for every $\alpha \in R$. f is quasi-concave if and only if $-f$ is quasi-convex, i.e., its upper level sets

$$U_f(\alpha) = \{x \in C : f(x) \geq \alpha\}$$

are convex for every $\alpha \in R$.

We can show that if f is quasi-convex on a convex set $C \subseteq R^n$ if and only if for every $x, u \in C$ and $\lambda \in [0,1]$

$$f(\lambda x + (1-\lambda)u) \leq \max\{f(x), f(u)\}$$

or, equivalently:

$$x, u \in C, \lambda \in [0,1], f(x) \leq f(u) \\ \Rightarrow f(\lambda x + (1-\lambda)u) \leq f(u).$$

Turning to differentiable functions, we have the following characterization due to Arrow and Enthoven [5]:

Theorem 1.4. *Let f be differentiable on the open convex set $C \subseteq R^n$. Then f is quasi-convex on C if and only if*

$$x, u \in C, f(x) \leq f(u) \Rightarrow (x-u)^T \nabla f(u) \leq 0.$$

For twice continuously differentiable functions the characterization of quasi-convexity is more complicated than the case of convex functions. We have the following result, due to Crouzeix [48] and Crouzeix and Ferland [50].

Theorem 1.5. *Let $f : C \to R$ be twice continuously differentiable on the open convex set $C \subseteq R^n$. Then f is quasi-convex on C if and only if*

1. *$x \in C, y \in R^n$,*

$$y^T \nabla f(x) = 0 \Rightarrow y^T \nabla^2 f(x) y \geq 0;$$

   and
2. *Whenever $\nabla f(x) = 0$, then $\forall y \in R^n$, the function $\varphi_{x,y}(t) = f(x+ty)$ is quasi-convex on the interval $I_{x,y} = \{t \in R : x + ty \in C\}$.*

We have to note that, contrary to convex functions, quasi-convex functions can have local minima that are not global and that its stationary points are not necessarily global minimum points: Consider again, e.g., the quasi-convex function $f(x) = x^3, x \in R$, for which $f'(0) = 0$, but for which $x^* = 0$ is an efficient point.

In order to obtain some of the properties of convex functions, we have to isolate certain subclasses of quasi-convex functions. We consider the following ones.

Definition 1.6. *A function f defined on a convex set $C \subseteq R^n$ is strictly quasi-convex (SQCX) on C if, for every $x, u \in C, x \neq u$, and $\lambda \in (0,1)$:*

$$f(\lambda x + (1-\lambda)) < \max\{f(x), f(u)\}.$$

f is semi-strictly quasi-convex (SSQCX) if $f(x) \neq f(u)$, instead of $x \neq u$ is assumed above.

It is immediate to prove that f is SQCX if and only if

$$x, u \in C, x \neq u, \lambda \in (0,1), f(x) \leq f(u) \\ \Rightarrow f(\lambda x + (1-\lambda)u) < f(u)$$

and that f is SSQCX if and only if

$$x, u \in C, \lambda \in (0,1), f(x) < f(u) \\ \Rightarrow f(\lambda x + (1-\lambda)u) < f(u).$$

Strictly quasi-convex functions are also called "strongly quasi-convex" in Avriel [8], "unnamed convex" in Ponstein [203] and "X-convex" in Thompson and Parke [235]. Strictly quasi-convex functions attain global minima over their domain at no more than one point, and every local minimum is global.

The family of semi-strictly quasi-convex functions is between the families of quasi-convex and strictly quasi-convex functions, in case of lower semi-continuity [109]. Every local minimum of a continuous semi-strictly quasi-convex function is also global, but contrary to strictly quasi-convex functions, the minimum can be attained at more than one point (see, e.g., [143]). Semi-strictly quasi-convex functions are also called "strictly quasi-convex" by Mangasarian [143], Ponstein [203], Thompson and Parke [235] and Avriel [8], "explicitly quasi-convex" by Martos [146] and "functionally convex" by Hanson [82].

Two other special classes of generalized convex functions are here introduced. We have already remarked that a stationary point of a (differentiable) quasi-convex function is not necessarily a minimum point (recall the example $f(x) = x^3, x \in R$). For this reason Mangasarian [142] introduced the class of pseudo-convex functions, which is a class of differentiable functions, wider than the class of differentiable convex functions.

Definition 1.7. *A differentiable function on the open set $C \subseteq R^n$ is pseudo-convex (PCX) on C if, for every $x, u \in C$*

$$(x-u)^T \nabla f(u) \geq 0 \Rightarrow f(x) \geq f(u),$$

or, equivalently

$$f(x) < f(u) \Rightarrow (x-u)^T \nabla f(u) < 0.$$

f is strictly pseudo-convex (SPCX) if, for every $x, u \in C, x \neq u$,

$$(x-u)^T \nabla f(u) \geq 0 \Rightarrow f(x) > f(u).$$

For a pseudo-convex function a stationary point $\bar{u}$ ($\nabla f(\bar{u}) = 0$) is obviously a global minimum point. A function is pseudo-convex, but not strictly pseudo-convex if f is constant over a line segment at the minimal level of f.

Ortega and Rheinboldt [191] have introduced pseudo-convexity also for non-differentiable functions: A real-valued function f defined on an open set

$C \subseteq R^n$ is said to be pseudo-convex on C if for every $x, u \in C$ and $\lambda \in (0,1)$ we have

$$\begin{aligned} f(x) &< f(u) \\ &\Rightarrow f(\lambda x + (1-\lambda)u) \le f(u) + (\lambda - 1)\lambda k(x,u), \end{aligned}$$

where $k(x,u)$ is a positive number, depending, in general on x and u. In case of differentiability, the above definition collapses to the one already given (see, e.g., [10]). For the twice differentiable case we have the following characterization (see, e.g., [58]).

Theorem 1.8. *Let f be twice continuously differentiable on the open convex set $C \subseteq R^n$. Then f is pseudo-convex (strictly pseudo-convex) if and only if*

1. $x \in C$, $\|v\| = 1$,

$$v^T \nabla f(x) = 0 \Rightarrow v^T \nabla^2 f(x) v > 0;$$

2. $v^T \nabla^2 f(x) = 0$ and $F(t) = f(x + tv)$ attains a local minimum (a strict local minimum) at $t = 0$.

Now we consider again the optimization problem
(P)

$$\text{Minimize } f(x)$$

$$\text{Subject to } g_i(x) \le 0, \quad i = 1, \dots, m,$$

$$x \in C \subseteq R^n.$$

This problem is called a quasi-convex program, if all functions f, g_i are quasi-convex on the convex set C. Given a quasi-convex program, the following properties hold:

1. The set of feasible solutions is convex and the set of optimal solutions is convex.
2. Any local minimum is a global minimum, if f is semi-strictly quasi-convex.
3. The Karush–Kuhn–Tucker optimality conditions are sufficient for a global minimum (i.e., a solution for (P)), if $f(x)$ is pseudo-convex.
4. Dual program (D) of (P) and duality relations between (D) and (P) can be established (see, e.g., [49, 197]).
5. A minimum of (P) is unique if f is strictly quasi-convex.

Hanson [83] observed that the sufficiency of the Karush–Kuhn–Tucker conditions and weak duality could be achieved by replacing the linear term $(x - u)$, appearing in the definition of convexity for differentiable functions, by an arbitrary vector-valued function, usually denoted $\eta(x, u)$ and also called "kernel" provided the same kernel was employed for each f and g_i. Craven [43] noted that functions satisfying the inequality

$$f(x) - f(u) \ge \eta(x,u)^T \nabla f(u),$$

could be generated by applying a differentiable injective domain transformation to a convex function and thus coined the term "invex" a contraction of "invariant convex."

This new generalization of convexity inspired a large number of papers dealing mainly with analysis and applications of this functional class in optimization theory and related fields. See the bibliographical reference at the end of this monograph. The purpose of this volume is to present an overview (together with some new results) of invexity theory and the extent to which it has replaced convexity and generalized convexity in various types of optimization problems, static and dynamic, with single-valued objective functions and with vector-valued objective functions (Pareto or multi-objective programming problems).

Chapter 2 features a discussion on the definition and meaning of invexity, under the original differentiability assumption, including conditions for several functions to be invex with respect to a common kernel. A section is concerned with the analysis of restricted invexity and pointwise invexity. A comparison of invexity with other classes of generalized convex functions is discussed. We also consider the relationships of invexity with domain and range transformations. Finally the question on the continuity of the kernel function is examined.

Chapter 3 is concerned with those functions are called η-pseudolinear (f and $-f$ both pseudo-invex). Another section of this chapter is concerned with the links between invexity and generalized monotonicity.

Chapter 4 is concerned with various notions of invexity for non-differentiable functions; in particular, it is considered the relevance of the Lipschitz case, in order to make use of Clarke's theory on generalized directional derivatives and subdifferentials.

The role of invexity in standard nonlinear programming is the subject of Chap. 5. After a discussion of the use of invexity for necessary and sufficient optimality conditions and a discussion on a sufficient condition for invexity and on η-pseudolinear program, we examine the relevance of invexity to duality theory. The two last sections of this chapter are concerned, respectively, to second and higher order invexity and to saddle points, optimality and duality for not necessarily differentiable invex functions.

Chapter 6 is concerned about the multiobjective programming problems involving invex and generalized invex functions. We have presented optimality conditions for multiobjective programming problems. Wolfe type, Mond–Weir type and general Mond–Weir type duality are presented for multiobjective programming problem. Nonsmooth case of multiobjective programming problems are also discussed. Nonsmooth multiobjective programming problems are discussed on abstract space as well. Multiobjective composite programming problems under invexity is presented. Saddle points and symmetric duality relations are also discussed in this chapter.

Chapter 7 is concerned with the optimization problems in infinite dimensional normal spaces. Two types of problems fitting into this scheme are variational and control problems. An early result on variational problems is addressed by Friedrichs [66]. Hanson [82] observed that variational and control problems are continuous analogues of finite dimensional nonlinear programs. Since, then the fields of nonlinear programming and the calculus of variations have to some extent, merged together within optimization theory, enhancing the potential for continued research in both. In Chap. 7, we have considered scalar variational problems and invexity. We discussed optimality and duality results for scalar problem in Sect. 7.1, multi-objective variational problems involving invexity is studied in Sect. 7.2. Control problems with single-objective is considered in Sect. 7.3. Finally, in Sect. 7.4, we have discussed multi-objective control problems with invex functions.

Finally, Chap. 8 is concerned with applications of invexity to some special functions and problems. More precisely, we consider the case of quadratic functions, the case of fractional functions and fractional programming problems, the case of class of non-differentiable programs, non-differentiable symmetric dual problems, symmetric duality for multi-objective fractional variational problems and symmetric duality for non-differentiable fractional variational problems.

2

Invex Functions (The Smooth Case)

2.1 Introduction

Usually, generalized convex functions have been introduced in order to weaken as much as possible the convexity requirements for results related to optimization theory (in particular, optimality conditions and duality results), to optimal control problems, to variational inequalities, etc. For instance, this is the motivation for employing pseudo-convex and quasi-convex functions in [142, 143]; [228] use convexlike functions to give a very general condition for minimax problems on compact sets. Some approaches to generate new classes of generalized convex functions have been to select a property of convex functions which is to be retained and then forming the wider class of functions having this property: both pseudo-convexity and quasi-convexity can be assigned to this perspective. Other generalizations have been obtained through altering the expressions in the definition of convexity, such as the arcwise convex functions in [8] and [9], the (h, ϕ)-convex function in [17], the (α, λ)-convex functions in [27], the semilocally generalized convex functions in [113], etc.

The reasons for Hanson's conception of invex functions [83] may have stemmed from any of these motivating forces, although in that paper Hanson dealt only with the relationships of invex functions to the Kuhn–Tucker conditions and Wolfe duality. More precisely, Hanson [83] noted that the usual convexity (or pseudo-convexity or quasi-convexity) requirements, appearing in the sufficient Kuhn–Tucker conditions for a mathematical programming problems, can be further weakened. Indeed, in the related proofs of the said conditions, there is no explicit dependence of the linear term $(x - y)$, appearing in the definition of differentiable convex, pseudo-convex and quasi-convex functions. This linear term was therefore substituted by an arbitrary vector-valued function, usually denoted by η and sometimes called "kernel function."

2.2 Invex Functions: Definitions and Properties

Definition 2.1. *Assume $X \subseteq R^n$ is an open set. The differentiable function $f : X \to R$ is invex if there exists a vector function $\eta : X \times X \to R^n$ such that*

$$f(x) - f(y) \geq \eta(x,y)^T \nabla f(y), \quad \forall x, y \in X. \tag{2.1}$$

It is obvious that the particular case of (differentiable) convex function is obtained from (2.1) by choosing $\eta(x,y) = x - y$. The term "invex" is due to Craven [43] and is an abbreviation of "invariant convex," since it is possible to create an invex function with the following method:
Let $g : R^n \to R$ be differentiable and convex and $\Phi : R^r \to R^n (r \geq n)$ be differentiable with $\nabla\Phi$ of rank n. Then $f = g \circ \Phi$ is invex, $\forall x, y \in R^r$, we have

$$f(x) - f(y) = g(\Phi(x)) - g(\Phi(y)) \geq (\Phi(x) - \Phi(y))^T \nabla g(\Phi(y)).$$

As $\nabla f(y) = \nabla\Phi(y)\nabla g(\Phi(y))$ and $\nabla\Phi(y)$ is of rank n, the equation $(\Phi(x) - \Phi(y))^T \nabla g(\Phi(y)) = \eta(x,y)^T \nabla f(y)$ has a solution $\eta(x,y) \in R^r$. Hence, $f(x) - f(y) \geq \eta(x,y)^T \nabla f(y), \forall x, y \in R^r$ for some $\eta : R^r \times R^r \to R^r$.

This characterization of invexity is closely related to (h, F)-convexity, a generalization of convexity based on the use of generalized means (see, e.g., [146,169]). The class of (h, F)-convex functions, with h, h^{-1} and F differentiable, from a subclass of invex functions. It was stated earlier that invexity was used by Hanson [83] to obtain sufficient optimality conditions (in terms of Kuhn–Tucker conditions) for a nonlinear programming problem. This is possible, an invex function shares with convex function the property that every stationary point is a global minimum point. Craven and Glover [45] and Ben-Israel and Mond [18] established the basic relationship between this property and the function η of Definition 2.1.

Theorem 2.2. *Let $f : X \to R$ be differentiable. Then f is invex if and only if every stationary point is a global minimizer.*

Proof. Necessity: Let f be invex and assume $\bar{x} \in X$ with $\nabla f(\bar{x}) = 0$. Then $f(x) - f(\bar{x}) \geq 0, \forall x \in X$, so $\bar{x}$ is a global minimizer of f over X.
Sufficiency: Assume that every stationary point is a global minimizer. If $\nabla f(y) = 0$, let $\eta(x,y) = 0$. If $\nabla f(y) \neq 0$, let

$$\eta(x,y) = \frac{[f(x) - f(y)]\nabla f(y)}{\nabla f(y)^T \nabla f(y)}.$$

Then f is invex with respect to η.

□

This is, of course, not the only possible choice of η. Indeed, if $\nabla f(y) = 0$, then $\eta(x,y)$ may be chosen arbitrarily, and if $\nabla f(y) \neq 0$, then

$$\eta(x,y) \in \left\{ \frac{[f(x) - f(y)]\nabla f(y)}{\nabla f(y)^T \nabla f()y} + v : v^T \nabla f(y) \leq 0 \right\},$$

a half-space in R^n.

This importance of functions with the stationary points as global minimizers had been recognized also by Zang et al. [255], who however, did not pursue any further analysis and applications.

Let us denote by $L_f(\alpha)$ the lower α-level set of a function $f : X \to R$, i.e., the set $L_f(\alpha) = \{x : x \in X, f(x) \leq \alpha\}, \forall \alpha \in R$. Zang et al. [255] characterized by means of the sets $L_f(\alpha)$ the functions whose stationary points are global minimizers, i.e., the class of invex functions.

Definition 2.3. *If $L_f(\alpha)$ is non-empty, then it is said to be strictly lower semi-continuous if, for every $x \in L_f(\alpha)$ and sequence $\{\alpha_i\}$, with $\alpha_i \to \alpha, L_f(\alpha_i)$ non-empty, there exist $k \in N$, a sequence $\{x^i\}$, with $x^i \to x$ and $\beta(x) \in R, \beta(x) > 0$, such that*

$$x^i \in L_f[\alpha_i - \beta(x) \left\|x^i - x\right\|], \quad i = k, k+1, \ldots.$$

The authors proved the following result.

Theorem 2.4. *A function $f : X \to R$, differentiable on the open set $X \subseteq R^n$, is invex if and only if $L_f(\alpha)$ is strictly lower semi-continuous, for every α such that $L_f(\alpha) \neq \Phi$.*

Proof. See Zang et al. [255].

□

Another characterization of invex functions stemming from Theorem 2.2, can be obtained through the conjugation operation. Let $f : X \to R$, $X \subseteq R^n$; given $\xi \in R^n$, we consider the collection of all affine functions $\xi^T x - \alpha$, with slope ξ, that minorize $f(x)$, i.e., $\xi^T - \alpha \leq f(x), \forall x \in X$. This collection, if non-empty, gives rise to the smallest α^* for which the above relation holds. If there is no affine function with slope ξ minorizing $f(x)$, we agree to set $\alpha^* = +\infty$. In any case $\alpha^* = f^*(\xi) = \sup_x\{\xi^T x - f(x)\}$ is precisely what is called the conjugate function of f (see [211]). By reiterating the operation $f \to f^*$ on X, we get the biconjugate of $f(x)$, defined by

$$f^{**} = \sup_{\xi}\{\xi^T x - f^*(\xi)\}.$$

It can be proved (see [91]) the following result.

Theorem 2.5. *Let $f : X \to R$ be differentiable on the open set $X \subseteq R^n$. Then $x^0 \in X$ is a (global) minimum point of f on X if and only if: (i) $\nabla f(x^0)$, and (ii) $f^{**}(x^0) = f(x^0)$. In such a case f^{**} is differentiable at x^0 and $\nabla f^{**}(x^0) = 0$.*

Proof. See Hiriart-Urruty [91].

□

Thus Theorem 2.5 gives another characterization of an invex function: it is a C differentiable function whose value at stationary points equals the value of its biconjugate.

Hanson and Rueda [89] sufficient conditions for invexity of a function are established through the use of linear programming. We shall revert to this question when we shall treat the applications of invexity to nonlinear programming problems. From Theorem 2.2 we get immediately that if f has no stationary points, then f is invex. Furthermore, Theorem 2.2 will be useful to state some relationships between invex functions and other classes of generalized convex functions. Some nice properties of convex functions are however lost in the invex case. In fact, unlike convex (or pseudo-convex) case, the restriction of an invex function on a not open set does not maintain the local/global property. Let us consider the following example.

Example 2.6. Let $f(x,y) = y(x^2-1)^2$, considered on the closed set $S = \left\{(x,y) \in R^2 : x \geq -\frac{1}{2}, y \geq 1\right\}$. Every stationary point of f on S is a global minimum point of f on S and therefore f is invex on S. The point $(-\frac{1}{2}, 1)$ is a local minimum point of f on S, with

$$f(-\frac{1}{2},1) = \frac{9}{16} > f(1,y) = f(-1,y) = 0.$$

The points $(1,y)$, $(-1,y), y \geq 1$, are the global minimizers for f on S.

If f is invex on an open set $X \subseteq R^n$, contrary to what asserted in Pini [201], it is not true that the set $A = \{x \in X, \nabla f(x) = 0\}$ is a convex set (as for convex functions). Let us consider the following example.

Example 2.7. Let $f(x,y) = y(x^2-1)^2$, defined on the open set $S = \left\{(x,y) \in R^2 : x \in R, y > 0\right\}$. The set of all its stationary points coincides with the set of all its minimum points (i.e., f on S). This set is given by $\{(1,y) : y > 0\} \bigcup \{(-1,y) : y > 0\}$, which is not a convex set in R^2.

As a consequence, for an invex function the set of all minimum points (the set of all stationary points if f is defined on an open set) is not necessarily a convex set. Ben-Israel and Mond [18] observed that there is an analogue of Theorem 2.2 for pseudo-convex functions.

Theorem 2.8. *A differentiable function on the open set $X \subseteq R^n$ is pseudo-convex on X if and only if*

$$(x-y)^T \nabla f(y) = 0 \Rightarrow f(y) \leq f(y + t(x-y)), \quad \forall t > 0. \tag{2.2}$$

Proof. Necessity: Obvious from the definition of pseudo-convexity. Here (2.2) holds for all real t.

Sufficiency: Suppose f is not pseudo-convex; that is, there exists (x,y) such that $(x-y)^T \nabla f(y) \geq 0$ and $f(x) < f(y)$. If $(x-y)^T \nabla f(y) = 0$, then (2.2) is contradicted. If $(x-y)^T \nabla f(y) > 0$, then there exists v which maximizes f on the line segment from y to x. Thus $\nabla f(v) = 0$ and therefore $(x-y)^T \nabla f(v) = 0$ and

$$f(v) \geq f(y) > f(x) = f(v + 1(x-v)),$$

contradicting (2.2).

□

We note that the class of functions differentiable on an open set X and all invex with respect to the same $\eta(x,y)$, is closed under addition on any domain contained in X, unlike the classes of quasi-convex and pseudo-convex functions which do not retain this property of convex functions. However, the class of functions invex on an open set X, but not necessarily with respect to the same $\eta(x,y)$, need not be closed under addition. For instance (see, [178,224]), consider $f_1 : R \to R$ and $f_2 : R \to R$ defined by $f_1(x) = 1 - e^{-(x+5)^2}$. Both f_1 and f_2 are invex, but $f_1 + f_2$ has a stationary point at $\bar{x} = 0$ which is not a global minimizer. In fact, for a given $\eta(x,y)$, the set of functions invex with respect to $\eta(x,y)$, form a convex cone; that is, the set is closed under addition and positive scalar multiplication. Therefore, we can state the following result.

Theorem 2.9. *Let $f_1, f_2, \ldots, f_m : X \to R$ all invex on the open set $X \subseteq R^n$, with respect to the same function $\eta(x,y) : X \times X \to R^n$. Then:*

1. *For each $\alpha \in R, \alpha > 0$, the function $\alpha f_i, i = 1, \ldots, m$, is invex with respect to the same η.*
2. *The linear combination of $f_1, f_2, \ldots, f_m$, with nonnegative coefficients is invex with respect to the same η.*

Following Smart [224] and Mond and Smart [179], a natural question is now the following: given two (or more) invex functions, how do we know if they are invex with respect to a common η. It is convenient to first prove a result characterizing functions for which no common η exists.

Lemma 2.10. *Let $f : X \to R, g : X \to R$ be invex. There does not exist a common η, with respect to which f and g are both invex if and only if there exists $x, y \in X, \lambda > 0$ such that $\nabla f(y) = -\lambda \nabla g(y)$ and $f(x) - f(y) + \lambda(g(x) - g(y)) < 0$.*

Proof. (a) Sufficiency: Assume there exist $x, y \in X, \lambda > 0$ such that $\nabla f(y) = -\lambda \nabla g(y)$ and $f(x) - f(y) + \lambda(g(x) - g(y)) < 0$. We wish to show that the system

$$f(x) - f(y) \geq \eta(x,y)^T \nabla f(y)$$
$$g(x) - g(y) \geq \eta(x,y)^T \nabla g(y)$$

has no solution $\eta(x,y) \in R^n$. Assume such an $\eta(x,y)$ exists. Now, as $\lambda > 0$, $g(x) - g(y) \geq \eta(x,y)^T \nabla g(y) \Rightarrow \lambda[g(x) - g(y)] \geq \lambda \eta(x,y)^T \nabla g(y)$. Therefore,

$$\begin{aligned} f(x) - f(y) + \lambda(g(x) - g(y)) &\geq \eta(x,y)^T \nabla f(y) + \lambda \eta(x,y)^T \nabla g(y) \\ &= \eta(x,y)^T [\nabla f(y) + \lambda \nabla g(y)] \\ &= 0, \end{aligned}$$

which contradicts $f(x) - f(y) + \lambda(g(x) - g(y)) < 0$. Hence, no common function $\eta(x,y)$ exists.

(b) Necessity: Assume no common function $\eta(x,y)$ exists. Then there exists $x, y \in X$ such that the system

$$f(x) - f(y) \geq \eta(x,y)^T \nabla f(y)$$

$$g(x) - g(y) \geq \eta(x,y)^T \nabla g(y)$$

has no solution $\eta(x,y)^T \in R^n$.

Rewrite the system as $A\eta(x,y) \leq C$, where $A = \begin{pmatrix} \nabla f(y)^T \\ \nabla g(y)^T \end{pmatrix}, C = \begin{pmatrix} f(x) - f(y) \\ g(x) - g(y) \end{pmatrix}$.

By Gale's Theorem of the alternative for linear inequalities (see, e.g., [143]), there exists $y \in R^2, y = (y_1, y_2)^T$, such that $A^T y = 0, C^T y = -1, y \geq 0$, that is,

$$\nabla f(y)y_1 + \nabla g(y)y_2 = 0,$$

$$[f(x) - f(y)y_1] + [g(x) - g(y)y_2] = -1,$$

$$y_1 \geq 0, \quad y_2 \geq 0.$$

Now, if $y_1 = 0$, then $\nabla g(y)y_2 = 0, [g(x) - g(y)]y_2 = -1, y_2 \geq 0$, which implies that $\nabla g(y) = 0$ and $g(x) - g(y) < 0$, which contradicts the invexity of g. Hence, $y_1 \geq 0$. Similarly, $y_2 > 0$. Thus,

$$\nabla f(y) = -\frac{y_2}{y_1}\nabla g(y) = -\lambda \nabla g(y), \quad \text{where} \quad \lambda = \frac{y_2}{y_1} > 0$$

and

$$f(x) - f(y) + \frac{y_2}{y_1}[g(x) - g(y)] = -1,$$

that is,

$$f(x) - f(y) + \lambda[g(x) - g(y)] < 0.$$

□

The negation of the Lemma 2.10 yields the next result.

Theorem 2.11. *Let $f : X \to R, g : X \to R$ be invex. A common η, with respect to which both f and g are invex, exists if and only if $\forall x, y \in X$ either*

1. $\nabla f(y) \neq \lambda \nabla g(y)$ for any $\lambda > 0$ or
2. $\nabla f(y) = -\lambda \nabla g(y)$ for some $\lambda > 0$ and

$$f(x) - f(y) \geq -\lambda[g(x) - g(y)].$$

Using Theorem 2.11, it is possible to give a more useful characterization of invex functions with respect to a common η.

Theorem 2.12. *Let $f : X \to R, g : X \to R$ be invex. A common η, with respect to which both f and g are invex, exists if and only if $f + \lambda g$ is invex for all $\lambda > 0$.*

Proof. (a) Necessity: this follows since the set of functions invex with respect to η is a convex cone.
(b) Sufficiency: assume $f+\lambda g$ is invex, for all $\lambda > 0$. Then, whenever $\nabla f(y) = -\lambda \nabla g(y)$ for some $\lambda > 0$, we have

$$f(x) + \lambda g(x) \geq f(y) + \lambda g(y), \quad \forall x \in X,$$

by invexity of $f + \lambda g$. That is,

$$\nabla f(y) = -\lambda \nabla g(y) \Rightarrow f(x) - f(y) \geq -\lambda [g(x) - g(y)], \quad \forall x \in X.$$

By Theorem 2.11, a common η exists.

□

Theorem 2.12 generalizes to any finite number of functions, and is useful for the requirements of invexity in sufficiency and duality results in optimization.

Corollary 2.13. *Let $f : X \to R, g_1, g_2, \ldots, g_m : X \to$ be invex. A common η with respect to which $f, g_1, g_2, \ldots, g_m$ are invex, exists if and only if $f + \lambda_1 g_1 + \lambda_2 g_2 + \cdots + \lambda_m g_m$ is invex for all $\lambda_1 > 0, \lambda_2 > 0, \ldots, \lambda_m > 0$.*

Proof. By induction; the case $m = 1$ is proved in Theorem 2.12. Assume the statement is true for some $k \in N$. Now $f, g_1, g_2, \ldots, g_{k+1}$ have a common η if and only if $f, g_1, g_2, \ldots, g_k$ have a common η with respect to which g_{k+1} is also invex. Now $f, g_1, g_2, \ldots, g_k$ have a common η if and only if $f + \lambda_1 g_1 + \lambda_2 g_2 + \cdots + \lambda_k g_k$ is invex for all $\lambda_1 > 0, \lambda_2 > 0, \ldots, \lambda_k > 0$. Therefore, $f, g_1, g_2, \ldots, g_{k+1}$ have a common η if and only if $f + \lambda_1 g_1 + \lambda_2 g_2 + \cdots + \lambda_k g_k$ is invex with respect to same η independent of $\lambda_1 > 0, \lambda_2 > 0, \ldots, \lambda_k > 0$, and g_{k+1} is invex with respect to the same η. But $f + \lambda_1 g_1 + \lambda_2 g_2 + \cdots + \lambda_k g_k$ and g_{k+1} have a common η if and only if $f + \lambda_1 g_1 + \lambda_2 g_2 + \cdots + \lambda_{k+1} g_{k+1}$ is invex for all $\lambda_{k+1} > 0$. Therefore, $f, g_1, g_2, \ldots, g_{k+1}$ have a common η if and only if $f + \lambda_1 g_1 + \lambda_2 g_2 + \cdots + \lambda_{k+1} g_{k+1}$ is invex for all $\lambda_1 > 0, \lambda_2 > 0, \ldots, \lambda_{k+1} > 0$.

□

Since it is assumed in Corollary 2.13 that f is invex, the necessary and sufficient condition could also be expressed as: $f + \lambda_1 g_1 + \lambda_2 g_2 + \cdots + \lambda_m g_m$ is invex for all $\lambda_1 \geq 0, \lambda_2 \geq 0, \ldots, \lambda_m \geq 0$.

Like convex functions, invex functions with respect to a certain η are transformed into invex functions with respect to the same η, by a suitable class of monotone functions.

Theorem 2.14. *Let $\psi : R \to R$ be a monotone increasing differentiable convex function. If f is invex on X with respect to η, then the composite function $\psi \circ f$ is invex with respect to the same η.*

Proof. By the fact that $\psi(x + h) \geq \psi(x) + \psi'(x)h, \forall x, h \in R$, we get

$$\begin{aligned} \psi(f(x)) &\geq \psi(f(y)) + \nabla f(y)\eta(x, y) \\ &\geq \psi(f(y)) + \psi'(f(y))\nabla f(y)\eta(x, y) \\ &= \psi(f(y)) + \nabla(f \circ \psi)(y)\eta(x, y). \end{aligned}$$

□

Further generalizations of invexity are possible; indeed Hanson [83] introduced also the following classes of generalized convex functions.

Definition 2.15. *The differentiable function $f : X \to R$ is pseudo-invex if there exists $\eta : X \times X \to R^n$ such that for all $x, y \in X$,*

$$\eta(x,y)^T \nabla f(y) \geq 0 \Rightarrow f(x) - f(y) \geq 0;$$

f is quasi-invex if there exists $\eta : X \times X \to R^n$ such that $\forall x, y \in X$,

$$f(x) - f(y) \leq 0 \Rightarrow \eta(x,y)^T \nabla f(y) \leq 0.$$

We point out that if we do not specify the function η in the definition of quasi-invexity, it turns out that every function f is quasi-invex: it is sufficient to take η identically equal to zero. Definitions 2.1 and 2.15 can be further weakened if we consider, as in Kaul and Kaur [114], pointwise characterization at a point $x^0 \in X$. In this respect we say that a differentiable function $f : X \to R$ is invex at $x^0 \in X$, if there exists $\eta(x, x^0)$ such that $\forall x \in X$,

$$f(x) - f(x^0) \geq \eta(x,x^0)^T \nabla f(x^0).$$

f is pseudo-invex at $x^0 \in X$, if there exists $\eta(x, x^0)$ such that $\forall x \in X$,

$$\eta(x,x^0)^T \nabla f(x^0) \geq 0 \Rightarrow f(x) - f(x^0) \geq 0.$$

f is quasi-invex at $x^0 \in X$, if there exists $\eta(x, x^0)$ such that $\forall x \in X$,

$$f(x) - f(x^0) \leq 0 \Rightarrow \eta(x,x^0)^T \nabla f(x^0) \leq 0.$$

Craven [43] introduced further relaxations: the local invexity at a point and the invexity with respect to a cone.

Definition 2.16. *The differentiable function $f : X \to R, X \subseteq R^n, X$ open, is said to be locally invex at $x^0 \in X$, if there exist a function $\eta(x, x^0)$ and a positive scalar δ such that*

$$f(x) - f(x^0) \geq \eta(x,x^0)^T \nabla f(x^0), \quad \forall x \in X, \left\| x - x^0 \right\| < \delta.$$

Definition 2.17. *Let $f : X \to R^k$ be a differentiable vector-valued function; f is invex with respect to the cone K in R^k if*

$$f(x) - f(y) - \nabla f(y)\eta(x,y) \in K.$$

If K is polyhedral convex cone and $q^j, j = 1, \ldots, l$, denote the generating vectors of the dual cone K^* such that

$$K = \left\{ x \in R^k : q^j x \geq 0, j = 1, \ldots, l \right\},$$

the Definition 2.17 is nothing but the invexity with respect to η, Craven [43] has given a characterization of local invexity with respect to a cone. Assume

$f : R^n \to R^k$ and $\eta : R^n \times R^n \to R^n$ are functions of class C^2. Given, y, we write the Taylor expression of $\eta(\cdot, y)$ and $f(\cdot)$ up to quadratic terms as follows

$$\eta(x,y) = \eta(y,y) + A(x-y) + \frac{1}{2}(x-y)^T Q_0 (x-y) + O(\|x-y\|^2)$$

$$f(x) = f(y) + B(x-y) + \frac{1}{2}(x-y)^T M_0 (x-y) + O(\|x-y\|^2),$$

where A, B, Q and M_0 have the obvious significance. Then the following holds.

Theorem 2.18. *Let $f : R^n \to R^k$ be a function of class C^2; denote by K a closed convex cone in R^k such that $K \bigcap (-K) = 0$. If f is locally invex at y, with respect to η and with respect to the cone K and $\eta : R^n \times R^n \to R^n$ is a function of class C^2, for which $\eta(x,x) = 0$, then, after substitution of a term in the null space of B, η has the form*

$$\eta(x,y) = x - y + \frac{1}{2}(x-y)^T Q_0 (x-y) + O(\|x-y\|^2), \tag{2.3}$$

where $M_0 - BQ_0$ is K-semidefinite. Conversely, if η has the form (2.3), and if $M_0 - BQ_0$ is K-positive definite, then f is locally invex at y, with respect to η and K.

Proof. See Craven [43].

□

Note that if f is a function defined on R and the cone K is the interval $[0, +\infty]$, the positive semidefiniteness of $M_0 - BQ_0$ is nothing but the condition

$$f''(y) - f'(y)\eta(y,y) \geq 0.$$

The conditions of Theorem 2.18 are however, from a computational point of view, difficult to apply. In Sect. 3, we shall see other sufficient conditions for invexity in nonlinear programming, through the use of linear programming. Further generalizations of invex functions can be obtained through notions similar to the ones utilized by Vial [239] to define strong and weak convex functions. On these lines Jeyakumar [100, 103] defined the following class of generalized invex functions.

Definition 2.19. *A differentiable function $f : X \to R, X \subseteq R^n$, is called ρ-invex with respect to the vector-valued function η and θ, if there exists some real number ρ such that, for every $x, y \in X$*

$$f(x) - f(y) \geq \eta(x,y)^T \nabla f(y) + \rho(\|\theta(x,y)\|^2).$$

If $\rho > 0$, then f is called strongly ρ-invex. If $\rho = 0$, we obviously get the usual definition of invexity and if $\rho < 0$, then f is called weakly ρ-invex.

It is clear that strongly ρ-invexity $\Rightarrow$ invexity $\Rightarrow$ weakly ρ-invexity. Rueda [213] points out that, under the assumption that $\|\nabla f(x)\| \neq 0$, Definition 2.19 is equivalent to invexity. Indeed, define

$$\eta_1(x,y) = \eta_x(x,y) + \rho(\|\theta(x,y)\|^2)\frac{\nabla f(x)}{[\nabla f(x)]^T \nabla f(x)}.$$

Thus, f is invex with respect to η_1.

Definition 2.20. *A differentiable function $f : X \to R$, is called ρ-pseudo-invex with respect to the vector-valued functions η and θ, if there exists some real number ρ such that, for every $x, y \in X$*

$$\eta(x,y)^T \nabla f(y) \geq -\rho(\|\theta(x,y)\|^2) \Rightarrow f(x) \geq f(y).$$

Definition 2.21. *A differentiable function $f : X \to R$, is called ρ-quasi-invex with respect to the vector-valued functions η and θ, if there exists some real number ρ such that, for every $x, y \in X$*

$$f(x) \leq f(y) \Rightarrow \eta(x,y)^T \nabla f(y) \leq -\rho(\|\theta(x,y)\|^2).$$

Pointwise definitions follow easily. The above definitions can be used to obtain general optimality and duality results for a nonlinear programming problem.

2.3 Restricted Invexity and Pointwise Invexity

The results characterizing invex functions as the class of functions for which stationary points are global minimizers, may be viewed as a special case of a more general theorem, due to Smart [224]; see also Mond and Smart [179], Molho and Schaible [166] and Chandra et al. [33].

For given $x, y \in R^n$, let $m(x,y)$ be a point in R^n and $\Lambda(x,y)$ a cone of R^n with vertex at $0 \in \Lambda$. Let $\Lambda^*(x,y)$ be the (positive) polar cone of $\Lambda(x,y)$, i.e.,

$$\Lambda^*(x,y) = \left\{ v \in R^n : v^T t \geq 0, \forall t \in \Lambda(x,y) \right\}.$$

Theorem 2.22. *Let $f : X \subseteq R^n \to R$ be differentiable. A necessary and sufficient condition for f to be invex with respect to $\eta : X \times X \to R^n$, subject to the restriction $\eta(x,y) \in m(x,y) + \Lambda(x,y), \forall x, y \in X$, is the following:*

$$\nabla f(y) \in \Lambda^*(x,y) \Rightarrow f(x) - f(y) - m(x,y)^T \nabla f(y) \geq 0.$$

Proof. Necessity: Assume f is invex with respect to $\eta(x,y) \in m(x,y) + \Lambda(x,y)$. Then $f(x) - f(y) \geq \eta(x,y)^T \nabla f(y) = (m(x,y) + t(x,y)^T)\nabla f(y)$, for some $t(x,y) \in \Lambda(x,y)$. Thus

$$\nabla f(y) \in \Lambda^*(x,y) \Rightarrow f(x) - f(y) \geq m(x,y)^T \nabla f(y), \quad \forall x \in X.$$

Sufficiency: assume that

$$\nabla f(y) \in \Lambda^*(x,y) \Rightarrow f(x) - f(y) - m(x,y)^T \nabla f(y) \geq 0.$$

Case (a). $\nabla f(y) \in \Lambda^*(x,y)$. Then take $\eta(x,y) = m(x,y)$. Since $0 \in \Lambda(x,y)$, we have $\eta(x,y) \in m(x,y) + \Lambda(x,y)$.
Case (b). $\nabla f(y) \notin \Lambda^*(x,y)$. Then there exists $t_1(x,y) \in \Lambda(x,y)$ such that $t_1(x,y)^T \nabla f(y) < 0$. If $f(x) - f(y) - m(x,y)^T \nabla f(y) \geq 0$, take $\eta(x,y) = m(x,y)$. On the other hand, if $f(x) - f(y) - m(x,y)^T \nabla f(y) < 0$, take $\eta(x,y) = m(x,y) + t_2(x,y)$, where

$$t_2(x,y) = \frac{f(x) - f(y) - m(x,y)^T \nabla f(y)}{t_1(x,y) \nabla f(y)} t_1(x,y).$$

Then

$$\begin{aligned} f(x) - f(y) - \eta(x,y)^T \nabla f(y) &= f(x) - f(y) - m(x,y)^T \nabla f(y) \\ &\quad - t_2(x,y)^T \nabla f(y) \\ &= f(x) - f(y) - m(x,y)^T \nabla f(y) \\ &\quad - \left(\frac{f(x) - f(y) - m(x,y)^T \nabla f(y)}{t_1(x,y)^T \nabla f(y)} \right)^T \\ &\quad \times t_1(x,y)^T \nabla f(y) \\ &= 0. \end{aligned}$$

Since

$$\frac{f(x) - f(y) - m(x,y)^T \nabla f(y)}{t_1(x,y)^T \nabla f(y)} > 0$$

and $\Lambda(x,y)$ is a cone, we have $t_2(x,y) \in \Lambda(x,y)$. Hence f is invex with respect to η, subject to the restriction $\eta(x,y) \in m(x,y) + \Lambda(x,y)$.

□

Let us apply the above results to some special cases:

(a) For convexity, take $m(x,y) = x - y$ and $\Lambda(x,y) = \{0\}$. The necessary and sufficient condition is $f(x) - f(y) \geq (x-y)^T \nabla f(y), \forall x, y \in X$.
(b) For arbitrary invexity, take $m(x,y)$ arbitrary, $\Lambda(x,y) = R^n$, so the necessary and sufficient condition is

$$y \in X, \nabla f(y) = 0 \Rightarrow f(x) - f(y) \geq 0, \quad \forall x \in X.$$

(c) For invexity with $\eta(x,y) \geq x - y$, take $m(x,y) = x - y, \Lambda(x,y) = R^n_+$. The condition is

$$y \in X, \nabla f(y) \geq 0 \Rightarrow f(x) - f(y) \geq (x-y)^T \nabla f(y), \quad \forall x \in X.$$

(d) For invexity with $\eta(x, y) + y \geq 0$, take $m(x, y) = -y, \Lambda(x, y) = R^n_+$. The condition is

$$y \in X, \nabla f(y) \geq 0 \Rightarrow f(x) - f(y) \geq -y^T \nabla f(y), \quad \forall x \in X.$$

We remark that if there does not exist a $y \in X$ such that $\nabla f(y) \geq 0$, then it is immediate that f is invex with respect to the desired η in both cases (c) and (d). Consider the following (non-convex) example: $f : R^2 \to R$, $f(x_1, x_2) = -x_2(x_1^2 + 1) + g(x_1)$, where $g : R \to R$ is any differentiable function. As $\nabla f(x_1, x_2) = (-2x_1x_2 + \nabla g(x_1); -(x_1^2 + 1)^T)$, there is no $(y_1, y_2) \in R^2$ such $\nabla f(y_1, y_2) \geq 0$, so f is invex with respect to some η_1 with $\eta_1(x, y) \geq x - y$, and also with respect to some η_2 with $\eta_2(x, y) + y \geq 0$.

A further special case of Theorem 2.22 concerns quadratic functions; we postpone the analysis of this case to Chap. 8, due to its importance in mathematical programming. We have already given the definitions, due to Kaul and Kaur [114], of invexity at a point x^0. We now make some other considerations on this case, under the assumption of twice differentiability of the functions. Let us therefore consider invex functions that are twice continuous differentiable. If $\nabla f(x^0) = 0$ for some $x^0 \in X$, a necessary condition for (global) invexity is that the Hessian matrix $\nabla^2 f(x^0)$ of f at x^0 is positive semidefinite. Indeed, if $\nabla f(x^0) = 0$ and f is invex, then x^0 is a point of global minimum. Therefore, $\nabla^2 f(x^0)$ is positive semidefinite.

2.4 Invexity and Other Generalizations of Convexity

In this section, we examine the main relationships between invexity definitions and other forms of generalized convexity. Obviously, for any assertion on a generalized convexity concept there is a generalized concavity counterpart. For invexity, the "incavity" is defined in a natural way by replacing $\geq$ with $\leq$.

First of all we note that:

(I) A differentiable convex function is also invex (take $\eta(x, y) = x - y$) but the converse is not true. Take, for example, the function $f(x) = \log x, x \in R$, which has no stationary points and is therefore invex. Obviously $f(x) = \log(x), x \in R$, is not convex (it is strictly concave) on its domain.

(II) A differentiable pseudo-convex function is also pseudo-invex, but not conversely. This property will be best precised in Theorem 2.25.

(III) A differentiable quasi-convex function is also quasi-invex, but not conversely (recall that every differentiable function is trivially quasi-invex).

For the reader's convenience we recall the basic definitions and properties of quasi-convex and pseudo-convex functions.

Definition 2.23 (Mangasarian [143], Avriel et al. [10]). *The function $f : X \to R$ is said to be quasi-convex on the convex set $X \subseteq R^n$ if for each $x, y \in X$ such that $f(x) - f(y) \leq 0$ and for each $\lambda \in [0,1]$, we have $f(\lambda x + (1-\lambda)y) \leq f(y)$.*

It is well known that f is quasi-convex on C if and only if the lower level sets $L_f(\alpha)$ are convex sets in R^n for each $\alpha \in R$. In case f is differentiable on the open convex set X, then f is quasi-convex on X if and only if $x, y \in X, f(x) - f(y) \leq 0 \Rightarrow (x-y)^T \nabla f(y) \leq 0$; or equivalently, $x, y \in X, (x-y)^T \nabla f(y) > 0 \Rightarrow f(x) - f(y) > 0$.

Definition 2.24. *The function $f : X \to R$, differentiable on the open set $X \subseteq R^n$, is pseudo-convex on X if*

$$x, y \in X, (x-y)^T \nabla f(y) \geq 0 \Rightarrow f(x) - f(y) \geq 0;$$

or equivalently,

$$x, y \in X, f(x) - f(y) < 0 \Rightarrow (x-y)^T \nabla f(y) < 0.$$

Furthermore, we say that f is strictly pseudo-convex on X if

$$x, y \in X, f(x) - f(y) \leq 0 \Rightarrow (x-y)^T \nabla f(y) < 0,$$

and we say that f is strongly pseudo-convex on X if f is pseudo-convex and satisfies the following conditions: For every $x^0 \in X$ and for every $v \in R^n, \|v\| = 1$, such that $v^T \nabla f(x^0) = 0$, there exist positive ϵ and α such that

$$f(x^0 + tv) \geq f(x^0) + \frac{1}{2}\alpha t^2,$$

for every $t \in R, 0 \leq t \leq \epsilon$.

(IV) Every invex function is also pseudo-invex for the same function η, but not conversely (see [114]). We have already remarked that a (differentiable) function without stationary points is invex, thanks to Theorem 2.2. Moreover, it results that the class of invex and pseudo-invex functions are coincident. This is not in contrast with property (IV), which is established with respect to the same η. We may note that some authors (see, e.g., Hanson and Mond [87], Kim [118] still consider pseudo-invexity as a generalization of invexity. We can therefore assert the following property:
(V) Every pseudo-convex function is invex; every pseudo-invex function is quasi-invex, but not conversely. For what concerns property (II) or its equivalent statement expressed by the first part of property (V), we have the following results, due to Pini [201].

Theorem 2.25. *The class of pseudo-convex functions on $X \subseteq R^n$ is strictly included in the class of invex functions if $n > 1$; if $n = 1$ the two classes coincide.*

Instead of following the proof of Pini [201], it is more useful to prove the following lemma [178, 224]).

Lemma 2.26. *Let $f : X \to R$, where X is an interval (open, half-open or closed) in R. If f is invex on X then it is also quasi-convex on X.*

Proof. We show that for every $\alpha \in R$, the lower level sets $L_f(\alpha)$ are convex. Assume to contrary that there exists $\alpha \in R$ such that $L_f(\alpha)$ is not convex. Then $L_f(\alpha)$ is the union of more than one disjoint intervals in X. Consider any two such intervals, I_1 and I_2, which are consecutive. Without loss of generality, $x_1 \in I_1$ and $x_2 \in I_2 \Rightarrow x_1 < x_2$. By continuity of f, I_1 must be closed on the right and I_2 must be closed on the left.

That is, there exists $\bar{x}_1 \in I_1$ such that $x_1 \leq \bar{x}_1, \forall x_1 \in I_1$ and $f(\bar{x}_1) = \alpha$; and there exists $\bar{x}_2 \in I_2$ such that $x_2 \geq \bar{x}_2, \forall x_2 \in I_2$ and $f(\bar{x}_2) = \alpha$. By assumption, $f(\alpha) > \alpha, \forall x \in (\bar{x}_1, \bar{x}_2)$. Since f is differentiable, then by the Mean Value Theorem, there exists $\bar{x} \in (\bar{x}_1, \bar{x}_2)$ such that $\nabla f(\bar{x}) = 0$. As $f(\bar{x}) > \alpha$, then $\bar{x}$ is not a global minimizer, which contradicts f being invex. □

The converse of Lemma 2.26 does not hold: take, e.g., the function $f : X \to R$, $f(x) = x^3$, which is quasi-convex (quasi-concave) on R, but not invex, since $\bar{x} = 0$ is a stationary point which is not global minimizer. Moreover, Lemma 2.26 does not hold when $X \subseteq R^n$ with $n > 1$. Consider the following example: $f : R^2 \to R, f(x_1, x_2) = 1 + x_1^2 - e^{-x_2^2}$. The function f has one stationary point, namely $x^* = (0, 0)$, and x^* is a global minimizer of f, so f is invex. However, f is not quasi-convex; take, e.g., $x = (1.12, 2.32940995)$ and $y = (1.31, 1.64704975)$. Now, $f(x) \leq f(y)$, but $(x - y)^T \nabla f(y) > 0$.

Another example is given by Ben-Israel and Mond [18]: The function $f : R^2 \to R, f(x_1, x_2) = x_1^3 + x_1 - 10x_2^3 - x_2$ is invex, since there are no stationary points. Taking $y = (0, 0), x_1 = 2, x_2 = 1$, gives $f(x) < f(y) < 0$ but $(x - y)^T \nabla f(y) > 0$, so f is not quasi-convex.

Another result useful to detect the relationships between the different classes of functions here considered is the following Theorem, due to Crouzeix and Ferland [50] and Giorgi [69]. See also Smart [224] and Mond and Smart [178].

Theorem 2.27. *Let f be differentiable quasi-convex function on the open convex set $X \subseteq R^n$. Then f is pseudo-convex on X if and only if f has a global minimum point at $x \in X$, whenever $\nabla f(x) = 0$.*

Theorem 2.27 asserts, in other words, that, under the assumption of quasi-convexity, invexity and pseudo-convexity coincides. So for an invex function not to be pseudo-convex, it must also not be quasi-convex. Taking this result into account, together with Lemma 2.26 and the related remarks, the proof of Theorem 2.25 is immediate.

Proof (of Theorem 2.27 Giorgi [69]). The necessary part of the theorem follows from the definition of pseudo-convex functions. As for sufficiency, let

$x^0 \in X, \nabla f(x^0) = 0 \Rightarrow x^0$ is a global minimum point of $f(x)$ on X, i.e., $(x - x^0)^T \nabla f(x^0) = 0 \Rightarrow f(x) \geq f(x^0), \forall x \in X$. It is obvious that $f(x)$ is then pseudo-convex at x^0 with respect to X. Let us now prove that: $f(x)$ quasi-convex on $X; x^0 \in X, \nabla f(x^0) \neq 0$ implies $f(x)$ pseudo-convex at x^0, i.e., $(x - x^0)^T \nabla f(x^0) \geq f(x) \geq f(x^0), \forall x \in X$. Let us consider a point $x^1 \in X$, such that

$$(x^1 - x^0)^T \nabla f(x^0) \geq 0 \tag{2.4}$$

but for which it is

$$f(x^1) < f(x^0). \tag{2.5}$$

Thus x^1 belongs to the nonvoid set

$$X_0 = \{x : x \in X, f(x) \leq f(x^0)\}$$

whose elements, thanks to the quasi-convexity of $f(x)$, verify the relation

$$x \in X_0 \Rightarrow (x - x^0)^T \nabla f(x^0) \leq 0. \tag{2.6}$$

Let us now consider the sets, both non-void,

$$W = \{x : x \in X, (x - x^0)^T \nabla f(x^0) \geq 0\}, \quad \text{and} \quad X_{00} = X_0 \cap W.$$

the following implication obviously holds:

$$x \in X_{00} \Rightarrow x \in H_0 = \{x : x \in X, (x - x^0)^T \nabla f(x^0) = 0\}.$$

It is therefore, evident that X_{00} is included in the hyperplane (recall that $\nabla f(x^0) \neq 0$) $H = \{x : x \in R^n, (x - x^0)^T \nabla f(x^0) = 0\}$, a hyperplane supporting X_0 covering to (2.6). Relation (2.4) and (2.5) point out that x^1 belongs to W and X_0 and hence to X_{00}, H_0 and H. Moreover, (2.5) says that x^1 lies in the interior of X_0; therefore x^1 at the same time belongs to the interior of a set and to a hyperplane supporting the same set, which is absurd. So relation (2.5) is false and (2.4) implies $f(x^1) \geq f(x^0)$.

□

We remark that the previous result states that a quasi-convex function $f(x)$ is thus pseudo-convex at every point $x \in X$ whenever $\nabla f(x) \neq 0$. Consequently we note that those sufficient conditions to test the quasi-convexity of a function in a convex set X where $\nabla f(x) \neq 0, \forall x \in X$, really locate the class of pseudo-convex functions. This is for example, the case of determinantal conditions for twice continuously differentiable functions, established by Arrow and Enthoven [5].

We can therefore add to the previous result, the following ones:

(VI) The classes of invex and pseudo-invex functions coincide.
(VII) The classes of quasi-convex and invex functions have only a partial overlapping.

We consider again pseudo-invex and quasi-invex functions. For what concerns pseudo-invex functions, we already know that if we do not impose further specifications on the choice of the kernel function η, this class coincides with the class of invex functions. However, if we consider the properties of these two classes of functions (invex and pseudo-invex) with respect to a specific function η, these properties are not the same. For example, unlike invex functions, the sum of pseudo-invex functions with respect to the same η is not pseudo-invex, with respect to that η. Consider, e.g., the following functions: $f(x) = \log x$ and $g(x) = -2x^2$ both defined on $X = \{x \in R : x > 0\}$. Both functions are pseudo-invex for $\eta(x,y) = x - y$. Indeed, $f(x) = \log x$ is strictly increasing function, being $f'(x) = \frac{1}{x} > 0, \forall x \in X$; therefore, $\eta(x,y)f'(x) \geq 0 \Leftrightarrow \eta(x,y) \geq 0$. Thus $\eta(x,y) = x - y \geq 0 \Leftrightarrow x \geq y \Rightarrow f(x) \geq f(y)$. So f is pseudo-invex with respect to $\eta(x,y) = x - y$.

The function g is strictly decreasing on X, as $g'(x) = -4x < 0, \forall x \in X$. We have $\eta(x,y)g'(y) \geq 0 \Leftrightarrow \eta(x,y) \leq 0; \eta(x,y) = x - y \leq 0 \Leftrightarrow x \leq y \Rightarrow g(x) \geq g(y)$, so g is pseudo-invex with respect to $\eta(x,y) = x - y$. The sum $z = f + g$ is $z = \log x - 2x^2, x > 0$. We have $z' = \frac{1}{x} - 4x = \frac{1-4x^2}{x}$. Thus $z' \geq 0 \Leftrightarrow 1 - 4x^2 \geq 0 \Rightarrow x \leq \frac{1}{2}$. Therefore $z(x)$ has a maximum point at $x = \frac{1}{2}$, so it is not pseudo-invex.

As for what concerns quasi-invex functions, we know that the class of pseudo-invex functions (i.e., invex functions) is strictly contained in the class of quasi-invex functions.

However, if we consider a pseudo-invex function f with respect to a certain function η, it is no longer true that f is also quasi-invex with respect to the same η. The converse also holds. Consider the following example.

Example 2.28. Let $f(x) = x^2 - 2x$ defined on R and

$$\eta(x,y) = \begin{cases} -1, & \forall (x,y) = (2,0) \\ 1, & \forall (x,y) = (x,1) \\ \frac{(x-y)(x+y-2)}{2(y-1)}, & \forall (x,y) \neq (x,1). \end{cases}$$

Let us verify that f is pseudo-invex with respect to $\eta(x,y)$; we have $f'(y) = 2y - 2$. If $(x - y) \neq (2,0)$ and $(x,y) \neq (x,1)$, then

$$\begin{aligned} \eta(x,y)f'(y) &= (x-y)(x+y-2) \\ &= x^2 - 2x - (y^2 - 2y) \geq 0. \end{aligned}$$

$$\Rightarrow x^2 - 2x \geq y^2 - 2y \Leftrightarrow f(x) \geq f(y).$$

If $(x,y) = (x,1)$, then $\eta(x,1)f'(1) = 0$ and $f(x) \geq f(1)$, being $x^2 - 2x \geq -1 \Leftrightarrow (x-1)^2 \geq 0, \forall x \in R$. If $(x,y) = (2,0)$, then $\eta(2,0)f'(0) = 2$ and

$f(2) = f(0) = 0$. So, f is not quasi-invex with respect to the same η; indeed if we choose $x = 2$ and $y = 0$, we have $f(x) \le f(y)$, but $\eta(2,0)f'(0) = 2 > 0$.

To verify that a quasi-invex function f with respect to a certain η, may not be pseudo-invex with respect to the same η, consider the function $f(x) = \tan x, x \in (-\frac{\pi}{2}, \frac{\pi}{2})$ which is quasi-convex and also therefore quasi-invex with respect to $\eta(x,y) = x - y$ but not pseudo-convex. Choose, e.g., $x = -\frac{\pi}{4}, y = 0$; we have $f(\frac{\pi}{4}) = 1 < f(0) = 0$, but $(x-y)f'(y) = 0$. So f is not pseudo-invex with respect to $\eta(x,y) = x - y$ (and it is not invex with respect to the same η).

We recall again that, if no specification is made on the choice of η, the class of quasi-invex functions coincides with the class of differentiable functions.

Similar to pseudo-invex functions, the sum of quasi-invex functions with respect to the same functions η, need not be quasi-invex with respect to that η. For example:

Example 2.29. Consider the functions $f(x) = \arctan(x)$ and $g(x) = -x^2$, both defined on $X = \{x \in R : x \ge 0\}$. Both functions are quasi-convex and therefore quasi-invex with respect to $\eta(x,y) = x - y$. The sum $z = \arctan(x) - x^2$ is not quasi-convex on X : Choose $x = 0$ and $y = 0.8$. We have $z(x) = 0 < z(y) = 0.03$. Therefore, we should have $z(x) \le 0.03$ for every $x \in \left(0, \frac{8}{10}\right)$. But if we consider $x = 0.5$, we have $z(x) = 0.21 > 0.03$. So $z(x)$ is not quasi-invex with respect to $\eta(x,y) = x - y$.

We now give following results from Pini [201] which ensure that an invex function is pseudo-convex or quasi-convex.

Theorem 2.30. *Assume that $X \subseteq R^n$ is an open convex set and $f : X \to R$ is an invex function, with respect to η. If*

$$(x-y)^T \nabla f(y) \le \eta(x,y)^T \nabla f(y), \quad \forall x, y \in X, \tag{2.7}$$

such that $f(x) < f(y)$, then f is pseudo-convex. If

$$(x-y)^T \nabla f(y) < \eta(x,y)^T \nabla f(y), \quad \forall x, y \in X \tag{2.8}$$

such that $f(x) \le f(y)$, then f is strictly pseudo-convex.

Proof. If $x, y \in X$ and $f(x) < f(y)$, by the hypothesis of invexity and (2.7), we get

$$\begin{aligned}(x-y)^T \nabla f(y) &= [(x-y) - \eta(x,y)]^T \nabla f(y) + \eta(x,y) \nabla f(y) \\ &\le [(x-y) - \eta(x,y)]^T \nabla f(y) + f(x) - f(y) \\ &< [(x-y) - \eta(x,y)]^T \nabla f(y) < 0.\end{aligned}$$

If $x, y \in X$ and $f(x) \le f(y)$, then (2.8) implies that

$$\begin{aligned}(x-y)^T \nabla f(y) &= [(x-y) - \eta(x,y)]^T \nabla f(y) + \eta(x,y) \nabla f(y) \\ &\le [(x-y) - \eta(x,y)]^T \nabla f(y) + f(x) - f(y) \\ &\le [(x-y) - \eta(x,y)]^T \nabla f(y) < 0.\end{aligned}$$

□

Theorem 2.31. *Assume that for every $y \in R^n$ the function $x \to \eta(x,y)$ is differentiable at the point $x = y, \eta(x,y) = 0$ and $\eta_x(y,y) = 1$. If $f : X \to R$ is invex with respect to η and*

$$f(x) < f(y) \Rightarrow \eta(x,y)^T \nabla f(y) \leq (x-y)^T \nabla f(y)$$

and

$$v^T \nabla f(y) = 0 \Rightarrow (v^T \eta_{xx}(y,y)v) \nabla f(y) > 0,$$

then f is strongly pseudo-convex.

Proof. Choose $x^0 \in X$ and $v \in R^n$, with $\|v\| = 1$ such that $v^T \nabla f(x^0) = 0$. Since f is invex, we have

$$f(x^0 + tv) - f(x^0) \geq [\eta(x^0 + tv, x^0) - tv]^T \nabla f(x^0).$$

Since

$$\frac{d}{dt}[\eta(x^0 + tv, x^0) - tv]^T \nabla f(x^0)_{t=0} = 0,$$

it is sufficient to prove that

$$\frac{d^2}{dt^2}[\eta(x^0 + tv, x^0) - tv]^T \nabla f(x^0)_{t=0} > 0;$$

this is equivalent to

$$\nabla f(x^0)[v^T \eta_{xx}(x^0, x^0)v] > 0,$$

which is indeed true by assumption.

□

Theorem 2.32. *Let $f : X \to R$ be invex on the open convex set $X \subseteq R^n$, with respect to the kernel function η. If $(x-y)^T \nabla f(y) > 0 \Rightarrow \eta(x,y)^T \nabla f(y) \geq (x-y)^T \nabla f(y)$, for every $x, y \in X$, then f is quasi-convex on X.*

Proof. We estimate the difference $f(x) - f(y)$ whenever $(x-y)^T \nabla f(y) > 0$. We readily get

$$\begin{aligned} f(x) - f(y) &\geq \eta(x,y)^T \nabla f(y) \\ &= [\eta(x,y) - (x-y)]^T \nabla f(y) + (x-y)^T \nabla f(y) \\ &> [\eta(x,y) - (x-y)]^T \nabla f(y) > 0. \end{aligned}$$

□

Recall now the following definitions (see [10]).

Definition 2.33. *Let f be a function defined on the convex set $X \subseteq R^n$. We say that f is semi-strictly quasi-convex on X if*

$$f(x) < f(y) \Rightarrow f(\lambda x + (1-\lambda)y) < f(y), \quad \forall x, y \in X, x \neq y, \lambda \in (0,1).$$

Following Pini [201], we can give a sufficient condition for semi-strictly quasi-convexity.

Theorem 2.34. *Suppose that $f : X \to R$ is invex with respect to η, and that for every $x^0 \in X$ and $v \in R^n, \|v\| = 1$, such that $v\nabla f(x^0) = 0$, one of the two following conditions hold:*

$$\eta(x^0 + tv, x^0)^T \nabla f(x^0) > 0, \quad t \in [-a, b] \tag{2.9}$$

or

$$\eta(x^0 + tv, x^0)^T \nabla f(x^0) \geq 0, \quad t \in (-a, b),$$

$$\eta(x^0 - av, x^0)^T \nabla f(x^0) > 0, \quad \eta(x^0 + bv, x^0)^T \nabla f(x^0) > 0, \tag{2.10}$$

for some suitable $a, b > 0$. Then f is semi-strictly quasi-convex.

Proof. By Theorem 3.34 of Avriel et al. [10], it is sufficient to show that if $v\nabla f(x^0) = 0$, then the function $F(t) = f(x^0 + tv)$ does not admit a one-sided semi-strict local minimum at $t = 0$. Since f is an invex function, we have that

$$f(x^0 + tv) - f(x^0) \geq \eta(x^0 + tv, x^0)^T \nabla f(x^0),$$

that is,

$$F(0) \leq F(t) - \eta(x^0 + tv, x^0)^T \nabla f(x^0).$$

From (2.9), (2.10) it follows that $F(0) < F(-a), F(0) < F(b)$ and $F(0) \leq F(t), \forall t \in (-a, b)$. The thesis follows from the definition of one-sided semi-strict local maximum.

□

2.5 Domain and Range Transformations: The Hanson–Mond Functions

We follow here the approach of Smart [224], Mond and Smart [178] and Rueda [213]. These authors analyze in particular the article of Horst [94] dealing with non-convex nonlinear programs which may be transformed into convex programs via domain and/or range transformations in order to employ algorithms developed for convex programs. Convex range transformable functions, or F-convex functions, were first introduced by De Finetti [54].

Definition 2.35. *Let $f : X \to R, X$ a convex set in R^n. f is said to be convex range transformable or F-convex, if there exists a continuous, strictly monotone increasing function $F : range(f) \to R$, such that $F \circ f$ is convex on X. That is:*

$$F[f(\lambda x + (1 - \lambda)y)] \leq \lambda F[f(x)] + (1 - \lambda)F[f(x)], \tag{2.11}$$

$\forall x, y \in X, \forall \lambda \in [0, 1]$.

p-Convex functions (or power convex functions) and r-convex functions (see, [6–8,94,127,146,169]) are included in the class of convex range transformable functions. Concerning this subject we recall that the rth-generalized mean of $f(x)$ and $f(y)$, with $f(x)$ and $f(y)$ real and positive, defined as follows:

$$\begin{aligned} M_r\left(f(x), f(y), \lambda\right) &= M_r(f,\lambda) \\ &= \left[\lambda(f(x))^r + (1-\lambda)(f(y))^r\right]^{\frac{\lambda}{r}}, \end{aligned} \tag{2.12}$$

if $r \neq 0, \lambda \in [0,1]$.

It is possible to generalize (2.12) to the following cases:

$$M_0(f,\lambda) = \lim_{r\to 0} M_r(f,\lambda) = [f(y)]^{\lambda} \cdot [f(x)]^{1-\lambda}$$

$$M_{+\infty}(f,\lambda) = \lim_{r\to+\infty} (f,\lambda) = \max\left[f(x), f(y)\right].$$

$$M_{-\infty}(f,\lambda) = \lim_{r\to-\infty} (f,\lambda) = \min\left[f(x), f(y)\right].$$

Definition 2.36. *The function $f(x) > 0$ defined on the convex set $X \subseteq R^n$ is p-convex on X if there exists $p \geq 1$ such that $F \circ f = f^p$ is convex on X, i.e.,*

$$f(\lambda x + (1-\lambda)y) \leq M_p(f,\lambda), \quad \forall x, y \in X, \ \ \forall \lambda \in [0,1].$$

The previous inequality gives the usual definition of convexity for $p = 1$. If $1 < p < +\infty$, p-convexity is a special case of pseudo-invexity, i.e., of invexity. Indeed, if f is p-convex, then f^p is convex by definition and therefore it is invex. Since pseudo-invexity is equivalent to invexity for real functions, then there exists η such that

$$\eta(x,y)^T(\nabla f(y))^p \geq 0 \Rightarrow (f(x))^p \geq (f(y))^p.$$

Hence $f(x) \geq f(y)$, which proves that f is pseudo-invex. Note that invex functions need not be p-convex.

Example 2.37. Let $f : \left(0, \frac{\pi}{2}\right) \to R$ be defined by $f(x) = \sin x$. Then f is invex but it is not p-convex as can be seen by taking $y = \frac{\pi}{4}, x = \frac{\pi}{3}, p = 2$ and $\lambda = \frac{1}{2}$. In order to get rid of the restriction $f(x) > 0$, Avriel [6], Martos [146] and others proposed the following definition.

Definition 2.38. *The function $f : X \to R$ is r-convex on the convex set $X \subseteq R^n$, if for all $r, \lambda, -\infty \leq r \leq +\infty, 0 \leq \lambda \leq 1$, satisfies*

$$f(\lambda x + (1-\lambda)y) \leq \log M_r(e^{(f(x))}, e^{(f(y))}, \lambda).$$

Avriel [6] has proved that $f(x)$ is r-convex, with $r \neq 0$, if and only if the function $e^{rf(x)}$ is convex for $r > 0$ and concave $r < 0$. For $r > 0$, this is just the definition of r-convexity given by Horst [94]:

f is said to be r-convex if there exists $r > 0$ such that $F \circ f = e^{rf}$ is convex. For $0 < r < \infty$, we shall show that r-convexity is a special case of pseudo-invexity. If f is r-convex it follows that e^{rf} is convex and therefore it is invex (differentiability is assumed).

From

$$e^{rf(x)} - e^{rf(y)} \geq \eta(x,y)^T e^{rf(y)\nabla f(y)} \nabla f(y)$$

it follows that

$$e^{r(f(x)-f(y))} - 1 \geq \eta(x,y)^T.$$

Assume $\eta(x,y)^T \nabla f(y) \geq 0$. From the inequality above $e^{r(f(x)-f(y))} \geq 1$, which implies $r(f(x) - f(y)) \geq 0$. Since $r > 0$, it follows that $f(x) \geq f(y)$, which proves that f is pseudo-invex. From the previous example it follows that invex functions need not be r-convex. More generally, convex range transformable functions are quasi-convex [54, 94]. If in addition, a differentiable function f is convex range transformable with respect to a differentiable F, then f is invex. This may be seen by noting that, $\forall x, y \in X$,

$$F \circ f(x) - F \circ f(y) \geq (x-y)^T \nabla (F \circ f)(y) = (x-y)^T \nabla F(f(y)) \nabla f(y),$$

by convexity of $F \circ f$ and the by chain rule.

If $\nabla f(y) = 0$, then $F \circ f(x) \geq F \circ f(y), \forall x \in X$. By monotonicity of F, this implies that $f(x) \geq f(y), \forall x \in X$, so f is invex. By Theorem 2.27, f must also be pseudo-convex. Thus the class of differentiable convex range transformable (F-convex)functions, with F differentiable, form a strict sub-class of the invex functions.

A more general classification is obtained by incorporating a domain transformation [94].

Definition 2.39. *Let $f : X \to R, X \subseteq R^n, X$ convex. f is said to be (h, F)-convex if there exists a continuous one-to-one mapping $h : X \to h(X) \subseteq R^n$, and a continuous strictly monotone increasing function F : range$(f) \to R$ such that $h(X)$ is a convex set and $F \circ f \circ h^{-1}$ is a convex function on $h(X)$, i.e., $\forall x, y \in X$, and $\lambda \in [0,1]$, we have*

$$f\left[h^{-1}(\lambda h(x) + (1-\lambda) h(y))\right] \leq F^{-1}\left[\lambda F(f(x)) + (1-\lambda) F(f(y))\right].$$

Horst [94] has shown that (h, F)-convex functions need not be quasi-convex; the purpose of the domain transformation h is to obtain a quasi-convex function which is F-convex. Assuming that h and F are differentiable with ∇h of full rank, so that h^{-1} is differentiable, (h, F)-convexity implies invexity. This follows, since $\forall x, y \in X$,

$$\begin{aligned}(F \circ f \circ h^{-1})(x) &- (F \circ f \circ h^{-1})(y) \\ &\geq \nabla (F \circ f \circ h^{-1})(y) \\ &= \nabla F(f \circ h^{-1})(y) \cdot \nabla f(h^{-1}(y)) \nabla h^{-1}(y).\end{aligned}$$

If $\nabla f(x^*) = 0$, then as h is onto,there exists $y \in h(X)$ such that $h(x^*) = y$ and $h^{-1}(y) = x^*$. Therefore, $(F \circ f \circ h^{-1})(x) - (F \circ f)(x^*) \geq 0, \forall x \in h(X)$. As F is monotonic increasing, then $(f \circ h^{-1})(x) \geq f(x^*), \forall x \in h(X)$. Since h is onto, $f(z) \geq f(x^*), \forall z \in X$. Hence, every stationary point of f yields a global minimum on X, so f is invex on X.

Rueda [213] has shown that invex functions need not be (h, F)-convex. For further considerations on the relationships between invexity and (h, F)-convexity, see Smart [224] and Mond and Smart [178]. (h, F)-convex functions are actually a special case of the arcwise convex functions described in Avriel [8] and Avriel and Zang [9]. We can consider any continuous path from x to y instead of the straight line between x and y. Let $p_{x,y}(\lambda)$, where $p_{x,y}(0) = x$ and $p_{x,y}(1) = y$, represents a continuous path from x to y in R^n such that $f(p_{x,y}(\lambda)), 0 \leq \lambda \leq 1$, is defined. Let h be a continuous strictly increasing scalar function that implies $f(x)$ and $f(y)$ in its domain. Then f is said to be arcwise convex or (p, θ)-convex if

$$f(p_{x,y}(\lambda)) \leq h^{-1}\left[\lambda h(f(x)) + (1-\lambda) h(f(y))\right],$$

for all x, y in the domain of f, $0 \leq \lambda \leq 1$. For (h, F)-convexity these paths (or arcs) are h-mean value functions given by

$$p_{x,y}(\lambda) = h^{-1}\left[\lambda h(x) + (1-\lambda) h(y)\right].$$

Rueda [213] has shown that an arcwise convex function, with path and range transformation assumed to be differentiable, is pseudo-invex, and hence invex, but the converse does not hold.

We now briefly treat the so-called Hanson–Mond functions. Hanson and Mond [86] introduced a generalization of convexity based on sublinear functionals, intending to generalize both convex and invex functions. However, this class of functions is in fact the class of invex functions.

Definition 2.40. *The functional $F : D \to R, D \subseteq R^n$ is said to be sublinear if*

(i) $F(a + b) \leq F(a) + F(b), \forall a, b \in D$,
(ii) $F(\alpha x) \leq \alpha F(x), \forall x \in D, \forall \alpha \geq 0$ such that $x \in D, \alpha x \in D$.

Note that (ii) implies $F(0) = 0$.

Definition 2.41 (Hanson and Mond [86]). *Let $f : X \to R$ be differentiable; f is said to be a Hanson–Mond function if there exists a sublinear functional $F(x, y; \cdot) : X \times X \times R^n \to R$ such that $\forall x, y \in X$,*

$$f(x) - f(y) \geq F(x, y; \nabla f(y)).$$

These functions are also called F-convex functions (see, e.g., [20, 32, 77, 185, 204]).

Invex functions are Hanson–Mond functions, since if f is invex with respect to η, we can define F in Definition 2.41 by $F(x, y; a) = \eta(x, y)^T a$. But, note also that if f is a Hanson–Mond function and $\nabla f(y) = 0$, then since $F(x, y; a) = 0, \forall x \in X, y$ is a global minimizer of f, so f is invex. Therefore, the Hanson–Mond functions correspond to the invex functions. Craven and Glover [45] proved the equivalence between the two said classes. Caprari [26] proved this equivalence also with regards to the Lipschitzian case and also other type of equivalence involving pseudo-Hanson–Mond functions and quasi-Hanson–Mond functions. In spite of this, there is still a lot of papers dealing with (generalized) Hanson–Mond functions, with the conviction that these classes are true generalizations of the corresponding classes of invex functions.

2.6 On the Continuity of the Kernel Function

The continuity of the kernel of invex functions was studied by Smart [224, 225]. Here we follow his analysis. Usually, in the main applications of invexity (mathematical programming, variational and control problems, etc.) there are no restrictions on the analytical properties on the kernel function η, such as continuity or differentiability, etc. However, there are some type of problems where assumptions about the kernel η need to be made. Smart [225] describes two examples where continuity of η must be imposed.

In Parida et al. [195] a variational-like inequality problem is examined and applied to an invex mathematical program with the condition that η be continuous (in fact, continuity of η is included in the definition of invexity in [195]). The variational-like inequality problem considered is as follows:

Given a closed convex set K of R^n, and two continuous maps $F : K \to R^n$ and $\eta : K \times K \to R^n$, find $\bar{x} \in K$ such that

$$F(\bar{x})^T \eta(x, \bar{x}) \geq 0, \quad \forall x \in K.$$

For the applications of this problem to mathematical programming, they assume f is a continuously differentiable real-valued function on K, invex with respect to η and take $F = \nabla f$. Consider the program (PSK) Min $f(x)$, Subject to $x \in K$.

Parida et al. [195] show that if $\bar{x}$ solves the variational-like inequality problem, then $\bar{x}$ is an optimal solution of the program (PSK). The existence of a solution to the variational-like inequality problem depends on the continuity of η, allowing the Kakutani fixed-point theorem to be invoked.

Secondly, Ponstein [203] established six equivalent definitions of quasi-convexity, of which two apply to differentiable functions. The problem is to know whether the equivalence for these two can be extended to quasi-invexity. In fact, this equivalence is possible under a continuity property of the kernel. First, we recall Ponstein's results: assume $f : X \to R$ differentiable on the open convex set $X \subseteq R^n$. Then f is quasi-convex on X if either

$$f(x^2) \leq f(x^1) \Rightarrow (x^2 - x^1)^T \nabla f(x^1) \leq 0, \tag{2.13}$$

or equivalently

$$f(x^2) < f(x^1) \Rightarrow (x^2 - x^1)^T \nabla f(x^1) \leq 0. \tag{2.14}$$

We recall the definition of a quasi-invex function:

$$f(x^2) \leq f(x^1) \Rightarrow \eta(x^2, x^1)^T \nabla f(x^1) \leq 0. \tag{2.15}$$

Smart [225] gives a condition on η to guarantee that (2.15) is equivalent to

$$f(x^2) < f(x^1) \Rightarrow \eta(x^2, x^1)^T \nabla f(x^1) \leq 0. \tag{2.16}$$

Note that this result subsumes the results of Ponstein, taking $\eta(x^2, x^1) = x^2 - x^1$.

Theorem 2.42. *If the function f satisfies $\eta(x^2, \cdot)$ continuous at x^1 whenever $f(x^2) = f(x^1)$ and f is continuously differentiable, then conditions (2.15) and (2.16) are equivalent.*

Proof. Clearly, if (2.15) holds then (2.16) holds. Conversely, if (2.16) holds we need only establish that $f(x^2) = f(x^1) \rightarrow \eta(x^2, x^1)\nabla f(x^1) \leq 0$. Assume there exist $x^1, x^2 \in X$ (not necessarily distinct) such that $f(x^2) = f(x^1)$ and $\eta(X^2, x^1)\nabla f(x^1) > 0$. Then, by continuity of f, there exists $\bar{\lambda} > 0$, such that $\forall \lambda < \bar{\lambda}, \lambda \neq 0$, we have

$$f(x^1 + \lambda\eta(x^2, x^1)) > f(x^1) = f(x^2).$$

By (2.16), this gives

$$\eta(x^2, x^1 + \lambda\eta(x^2, x^1))^T \nabla f(x^1 + \lambda\eta(x^2, x^1)) \leq 0.$$

Taking limits as $\lambda \downarrow 0$, we obtain by continuity of $\eta(x^2, \cdot)$ and ∇f that $\eta(x^2, x^1)^T \nabla f(x^1) \leq 0$, a contradiction. Thus, if (2.16) holds then (2.15) holds. □

Now, given $f : X \rightarrow R (X \subseteq R^n)$, differentiable and invex, we know that f is invex with respect to $\eta : X \times X \rightarrow R^n$ if for every $x, y \in X$

$$\eta(x, y) = \left\{ \frac{(f(x) - f(y))\nabla f(y)}{\nabla f(y)^T \nabla f(y)} + v; v^T \nabla f(y) \leq 0 \right\},$$

where $\nabla f(y) \neq 0$. Under what conditions on f can a continuous η be chosen subject to the above constraint? For a given $f : X \rightarrow R$, one choice of η is given in the proof of Theorem 2.2:

$$\eta(x, y) = \begin{cases} \frac{(f(x) - f(y))\nabla f(y)}{\nabla f(y)^T \nabla f(y)}, & \text{if } \nabla f(y) \neq 0 \\ 0, & \text{if } \nabla f(y) = 0 \end{cases}$$

For $f : R \to R, f(x) = x^2$, this choice of η gives:

$$\eta(x,y) = \begin{cases} \dfrac{x^2 - y}{2y}, & y \neq 0 \\ 0, & y = 0. \end{cases}$$

Thus, for fixed $x \in R, x \neq 0, \lim_{y\to 0} \eta(x,y) = \lim_{y\to 0} \frac{x^2-y}{2y}$ which does not exist, so $\eta(x,y)$ is not continuous at 0 for any $x \in R - \{0\}$. An alternative choice of η is

$$\eta(x,y) = \begin{cases} 0, & f(y) \geq f(y) \\ \dfrac{(f(x) - f(y))\nabla f(y)}{\nabla f(y)^T \nabla f(y)}, & f(y) < f(y). \end{cases}$$

This η is formed by choosing v so that $v^T \nabla f(y) = f(y) - f(x)$ whenever $f(y) \leq f(x)$ with $\nabla f(y) \neq 0$, choosing $v = 0$ whenever $f(y) > f(x)$, and putting $\eta(x,y) = 0$ when $\nabla f(y) = 0$.

In the simple example above, we obtain

$$\eta(x,y) = \begin{cases} \frac{x^2-y^2}{2y}, & \text{if } |y| > |x| \\ 0, & \text{if } |y| \leq |x| \end{cases}$$

which is continuous in y for each $x \in R$ and furthermore, is continuous on R^2. The following theorem due to Smart [224,225], gives a sufficient condition for the continuity of the most recent choice of η.

Theorem 2.43. *Let $f : X \to R$ be continuously differentiable and invex. The function $\eta : X \times X \to R^n$ with respect to which f is invex, defined by*

$$\eta(x,y) = \begin{cases} 0, & \text{if } f(x) \geq f(y) \\ \frac{(f(x)-f(y))\nabla f(y)}{\nabla f(y)^T \nabla f(y)}, & \text{if } f(x) < f(y) \end{cases}$$

is continuous if, given y such that $\nabla f(y) = 0$, then for any sequence $\{y^n\}, y^n \to y, \nabla f(y^n) \neq 0$, we have

$$\lim_{n\to\infty} \frac{|f(y) - f(y^n)|}{\|\nabla f(y^n)\|} = 0,$$

where $\|\cdot\|$ is the usual Euclidean norm.

Proof. Let $(x,y) \in X \times X$ and assume $\{x^n\}$ and $\{y^n\}$ are sequences such that $(x^n, y^n) \in X \times X$ $x^n \to x$ and $y^n \to y$. We want to show that $\lim_{n\to\infty} \eta(x^n, y^n) = \eta(x,y)$. Three separate cases must be considered:
(a) From the definition of η, we have

$$\eta(x,y) = \frac{(f(x) - f(y))\nabla f(y)}{\nabla f(y)^T \nabla f(y)}.$$

By continuity of f, there exists an $N \in \aleph$ such that $\forall n \in N$, $f(x^n) < f(y^n)$. Therefore, for $n \geq N$,

$$\eta(x^n, y^n) = \frac{(f(x^n) - f(y^n))\nabla f(y^n)}{\nabla f(y^n)^T \nabla f(y^n)}.$$

by continuity of ∇f, $\lim_{n\to\infty} \eta(x^n, y^n) = \eta(x, y)$.
(b) By hypothesis, $\eta(x, y) = 0$. Again, by continuity of f there exists an $N \in \aleph$ such that $\forall n \in N$, $f(x^n) > f(y^n)$, and thus $\eta(x^n, y^n) = 0$. Therefore, $\lim_{n\to\infty} \eta(x^n, y^n) = \eta(x, y)$.
(c1) by continuity of f and $\nabla f, \forall \epsilon > 0$ there exists an $N \in \aleph$ such that $\forall n \geq N$,

$$|f(x^n) - f(x)| < \frac{\epsilon}{2}, |f(y^n) - f(y)| < \frac{\epsilon}{2}$$

and $f(y^n) \neq 0$.
Now, for $n \geq N$, if $f(x^n) \geq f(y^n)$, then $\eta(x^n, y^n) = 0$ and if $f(x^n) < f(y^n)$, then

$$\eta(x^n, y^n) = \frac{(f(x^n) - f(y^n))\nabla f(y^n)}{\nabla f(y^n) \nabla f(y^n)}.$$

We also have $|f(x^n) - f(y^n)| < \epsilon$. Hence, for $f(x^n) < f(y^n)$,

$$\begin{aligned}
\|\eta(x^n, y^n\|) &= \left\| \frac{(f(x^n) - f(y^n))\nabla f(y^n)}{\nabla f(y^n)^T \nabla f(y^n)} \right\| \\
&= \frac{\|(f(x^n) - f(y^n))\nabla f(y^n)\|}{\|\nabla f(y^n)\|^2} \\
&= \frac{|(f(x^n) - f(y^n))| \cdot \|\nabla f(y^n)\|}{\|\nabla f(y^n)\|^2} \\
&< \frac{\epsilon}{\|\nabla f(y^n)\|}.
\end{aligned}$$

As this holds $\forall \epsilon > 0$ and ∇f continuous, then $\lim_{n\to\infty} \|\eta(x^n, y^n)\| = 0$, so that $\lim_{n\to\infty} \eta(x^n, y^n) = 0 = \eta(x, y)$.
(c2) If $f(x^n) \geq f(y^n)$, then $\eta(x^n, y^n) = 0$. If $f(x^n) < f(y^n)$, then

$$\eta(x^n, y^n) = \frac{(f(x^n) - f(y^n))\nabla f(y^n)}{\nabla f(y^n)^T \nabla f(y^n)}$$

and so

$$\|\eta(x^n, y^n)\| = \frac{|f(x^n) - f(y^n)|}{\|\nabla f(y^n)\|}.$$

Note that $\nabla f(y) = 0$ and $f(x) = f(y)$ implies that x and y are global minimizers, so that when $f(x^n) < f(y^n)$, we have $f(y) = f(x) \leq f(x^n) < f(y^n)$. This gives

$$|f(y) - f(y^n)| \geq |f(x^n) - f(y^n)|$$

and hence

$$\|\eta(x^n, y^n)\| \le \frac{|f(x^n) - f(y^n)|}{\|\nabla f(y^n)\|}.$$

Now if there exists an $N \in \aleph$ such that $\forall n \ge N, f(x^n) \ge f(y^n)$, then we immediately have $\lim_{n\to\infty} \eta(x^n, y^n) = 0 = \eta(x, y)$. Otherwise, there exists a sub-sequence $\{y_i^n\}$ or $\{y^n\}$ such that $y_i^n \to y, f(x_i^n) < f(y_i^n)$, and $\nabla f(y_i^n) \neq 0$. By the hypothesis of the theorem

$$\lim_{n\to\infty} \|\eta(x_i^n, y_i^n)\| \le \lim_{n_i\to\infty} \frac{|f(y) - f(y_i^n)|}{\|\nabla f(y_i^n)\|} = 0.$$

Therefore, $\lim_{n\to\infty} \eta(x^n, y^n) = 0 = \eta(x, y)$.

□

The next result gives a simple second-order sufficient condition for the limit property of Theorem 2.43 to be satisfied.

Theorem 2.44. *Let $f : X \to R$ be invex and assume $\nabla f(y) = 0$. If f is twice continuously differentiable in some open neighborhood of y and $\nabla^2 f(y)$ is positive definite, then for any sequence $y^n, y^n \in X, y^n \to y, \nabla f(y^n) \neq 0$, we have*

$$\lim_{n\to\infty} \frac{|f(y) - f(y^n)|}{\|\nabla f(y^n)\|} = 0.$$

Proof. As f is twice continuously differentiable in some open neighborhood of y, and y is a global and therefore local minimizer with $\nabla^2 f(y)$ positive definite, then by continuity of $\nabla^2 f$, there exists some $\epsilon > 0$ such that $\forall x \in N(y, \epsilon)$ (the open ball of radius ϵ centered at y), f is twice continuously differentiable at x and $\nabla^2 f(x)$ is positive semi-definite. Now, consider $x \in N(y, \epsilon), x \neq y$, and define $g : [0, 1] \to R$ by $g(t) = f(y + t(x - y))$; g is twice differentiable, and its derivatives are given by

$$g' = (x - y)^T \nabla f(y + t(x - y)), \quad g''(t) = (x - y)^T \nabla^2 f(y + t(x - y))(x - y).$$

Let $t \in [0, 1]$. By the Mean Value Theorem, there exists $\xi \in [0, t]$ such that

$$g'(\xi) = \frac{g(t) - g(0)}{t},$$

that is, $g(t) - g(0) = tg'(\xi)$. But, as $\nabla f(x)$ is positive semi-definite on $\aleph(y, \epsilon)$, then $g'' \ge 0$ on $[0, 1]$. Hence g' is a non-decreasing function, so $g'(\xi) \le g'(t)$. Therefore, $g(t) - g(0) \le tg'(t)$. In particular, $g(1) - g(0) \le g'(1)$; that is, $f(x) - f(y) \le (x - y)^T \nabla f(x)$. Since the invexity of f implies that $f(x) \ge f(y)$, then by Cauchy–Schwarz inequality,

$$|f(x) - f(y)| \le \left|(x - y)^T \nabla f(x)\right| \le \|(x - y)\| \cdot \|\nabla f(x)\|.$$

Thus, if $\nabla f(x) \neq 0$, then

$$\frac{|f(x) - f(y)|}{\|\nabla f(x)\|} \le \|x - y\|.$$

Now, for any sequence $\{y^n\}, y^n \in X, y^n \to y, \nabla f(y^n) \neq$, there exists $N \in \aleph$ such that $\forall n \geq N$, we have $y^n \in N(y, \epsilon)$ and consequently

$$\frac{|f(y) - f(y^n)|}{\|\nabla f(y^n)\|} \leq \|y^n - y\| .$$

Therefore, by the squeeze principle,

$$\lim_{n\to\infty} \frac{|f(y) - f(y^n)|}{\|\nabla f(y^n)\|} = 0.$$

□

The limit property of Theorem 2.44 does not hold for all continuously differentiable invex functions. In the following example, due to Smart [224,225], the property does not hold. Furthermore, for invex functions of one variable if there exists $\bar{x} \in X$ such that $\bar{x}$ is a strict minimum and $\lim_{x\to\bar{x}} \frac{f(x)-f(\bar{x})}{f'(x)} \neq 0$, then there is no continuous η with respect to which f is invex [225].

Example 2.45. Define $f : R \to R$ by

$$f(x) = \begin{cases} 0, & x = 0 \\ \dfrac{n^2+n+1}{n+1}x^2 + \dfrac{-2n^2-2n-1}{(n+1)^2}x + \dfrac{4n^2+5n+2}{4(n+1)^3}, & \\ \qquad \dfrac{1}{n+1} \leq x \leq \dfrac{2n+1}{2n(n+1)}, & n = 1, 2, \ldots \\ \dfrac{1-n^2}{n}x^2 + \dfrac{2n^2-1}{n^2}x + \dfrac{-4n^2+n+1}{4n^3}, \ \dfrac{2n+1}{2n(n+1)} \leq x \leq \dfrac{1}{n}, & n = 1, 2, \ldots \\ x - \dfrac{1}{2}, & x \geq 1 \\ f(-x), & x < 0. \end{cases}$$

It is very easy to check that f is continuously differentiable, with $f'(y) = 0$ if and only if $y = 0$, which is a global minimizer. Consider the sequence y^n with $y^n = \frac{1}{n}, n = 1, 2, \ldots$. We have

$$f(y^n) = \frac{n+1}{4n^3} \quad \text{and} \quad f'(y^n) = \frac{1}{n^2}.$$

Therefore,

$$\lim_{n\to\infty} \frac{|f(y) - f(y^n)|}{|f'(y^n)|} = \lim_{n\to\infty} \frac{n^2(n+1)}{4n^3} = \frac{1}{4}.$$

Therefore, for this example, there is no choice of η which is continuous.

3

η-Pseudolinearity: Invexity and Generalized Monotonicity

3.1 η-Pseudolinearity

Chew and Choo [36] considered pseudolinear functions, i.e., functions that are both pseudo-convex and pseudo-concave. They showed that a function is pseudolinear if and only if there exists a positive functional $p\ (x, y) \in R$ such that

$$f(x) = f(y) + [p(x, y)(x - y)]^T \nabla f(y).$$

It is clear that every pseudolinear function is invex; however the converse is not true. As Kaul and Kaur [114] called the invex functions "η-convex functions" and pseudo-invex functions "η-pseudo-convex functions" it is quite natural to call an η-pseudolinear function a function f such that f and $-f$ are pseudo-invex with respect to the same η. There is a sizable literature on pseudolinear functions; see for example Chew and Choo [36], Bianchi and Schaible [23], Bianchi et al. [22], Jeyakumar and Yang [107], Kaul et al. [115], Komlosi [128], Kortanek and Evans [131], Kruk and Wolkowicz [132], Martos [146], Mishra [153], Rapcsak [207]. η-Pseudolinear functions have been introduced by Rueda [213] and studied in a more detailed way by Ansari et al. [3].

Definition 3.1. *A differentiable function f defined on an open set $X \subseteq R^n$ is called η-pseudolinear if f and $-f$ are pseudo-invex with respect to the same η.*

Obviously every pseudolinear function is η-pseudolinear with $\eta(x, y) = (x - y)$ but the converse is not true (see a counter example in [3]). The next two definitions will be considered again in the next chapter.

Definition 3.2 (Mohan and Neogy [165]). *For a given $\eta : K \times K \to R^n$ a non-empty set $K \subseteq R^n$ is called η-convex or simply invex, if for each $x, y \in K$, $0 \leq \lambda \leq 1$, $y + \lambda\eta(x, y) \in K$.*

Definition 3.3. *The function $\eta : K \times K \to R^n$ defined on the invex set $K \subseteq R^n$ satisfies Condition C [165], if for every $x, y \in K$: $\eta(y, y + \eta(x, y)) = -\lambda\eta(x, y)$ and $\eta(x, y + \eta(x, y)) = (1 - \lambda)\eta(x, y)$ for all $\lambda \in [0, 1]$.*

The following proposition is a necessary condition for f to be η-pseudolinear.

Theorem 3.4. *Let f be a differentiable function defined on an open set $X \subseteq R^n$ and K be an invex subset of X such that $\eta : K \times K \to R^n$ satisfies Condition C. Suppose that f is η-pseudolinear on K. Then for all $x, y \in K$, $\eta(x,y)^T \nabla f(y) = 0$ if and only if $f(x) = f(y)$.*

Proof. Suppose that f is η-pseudolinear on K. Then for all $x, y \in K$, we have

$$\eta(x,y)^T \nabla f(y) \geq 0 \Rightarrow f(x) \geq f(y)$$

and

$$\eta(x,y)^T \nabla f(y) \leq 0 \Rightarrow f(x) \leq f(y).$$

Combining these two inequalities, we obtain

$$\eta(x,y)^T \nabla f(y) = 0 \Rightarrow f(x) = f(y), \quad \forall\, x, y \in K.$$

Now we prove that $f(x) = f(y)$ implies $\eta(x,y)^T \nabla f(y) = 0$ for all $x, y \in K$. For that, we show that for any $x, y \in K$ such that $f(x) = f(y)$ implies that $f(y + \lambda\eta(x,y)) = f(y)$, $\forall\, \lambda \in (0,1)$. If $f(y + \lambda\eta(x,y)) > f(y)$ then by the definition of pseudo-invexity of f with respect to η, we have

$$\eta(y,z)^T \nabla f(z) < 0, \tag{3.1}$$

where $z = y + \lambda\eta(x,y)$.

We show that $\eta(y,z) = -\frac{\lambda}{1-\lambda}\eta(x,z)$. From Condition C, we have

$$\eta(y,z) = \eta(y, y + \eta(x,y)) = -\lambda\eta(x,y) = -\frac{\lambda}{1-\lambda}\eta(x,z).$$

Therefore, from (3.1), we obtain $-\frac{\lambda}{1-\lambda}\eta(x,z)^T \nabla f(z) < 0$ and hence $\eta(x,z)^T \nabla f(z) > 0$. By pseudo-invexity of f, with respect to η, we have

$$f(x) \geq f(z).$$

This contradicts the assumption that $f(z) > f(y) = f(x)$.

Similarly, we can also show that $f(y + \lambda\eta(x,y) < f(y))$, $\forall\, \lambda \in (0,1)$ leads to a contradiction, using pseudo-invexity of $-f$. This proves the claim that $f(y + \lambda\eta(x,y) = f(y))$, $\forall\, \lambda \in (0,1)$. Thus

$$\eta(x,y)^T \nabla f(y) = \lim_{\lambda \to 0+} \frac{f(y + \lambda\eta(x,y)) - f(y)}{\lambda} = 0.$$

□

Ansari et al. [3] gives an example where the converse of above theorem does not hold, that is if for all $x, y \in K$, $\eta(x,y)^T \nabla f(y) = 0$ if and only if $f(x) = f(y)$, then f need not be η-pseudolinear. The following result due to Rueda [213] and Ansari et al. [3] generalizes the corresponding result of Chew and Choo [36] quoted at the beginning of this section.

Theorem 3.5. *Let f be a differentiable function defined on an open set $X \subseteq R^n$ and K an invex subset of X with respect to η. Then f is η-pseudolinear on K if and only if there exists a function p defined on $K \times K$ such that $p(x,y) > 0$ and*

$$f(x) = f(y) + (p(x,y)\eta(x,y))^T \nabla f(y), \quad \forall x, y \in K.$$

Proof. Let f be an η-pseudolinear function. We have to construct a function p on $K \times K$ such that $p(x,y) > 0$ and $f(x) = f(y) + (p(x,y)\eta(x,y))^T \nabla f(y)$, $\forall x, y \in K$. If $\eta(x,y))^T \nabla f(y) = 0$, $\forall x, y \in K$. Then we define $p(x,y) = 1$. In this case we have $f(x) = f(y)$, thanks to Theorem 3.4. On the other hand, if $\eta(x,y))^T \nabla f(y) \neq 0$, then we define

$$p(x,y) = \frac{f(x) - f(y)}{\eta(x,y))^T \nabla f(y)}.$$

We have to show that $p(x,y) > 0$. Suppose that $f(x) > f(y)$. Then by pseudo-invexity of $-f$ we have $\eta(x,y))^T \nabla f(y) > 0$. Hence $p(x,y) > 0$. Similarly, if $f(x) < f(y)$ then we have $\eta(x,y))^T \nabla f(y) < 0$ by pseudo-invexity of f. Therefore, $p(x,y) > 0$. To prove the converse, we first show that f is pseudo-invex with respect to η, i.e., for any $x, y \in K$, $\eta(x,y))^T \nabla f(y) \geq 0 \Rightarrow f(x) \geq f(y)$. If $\eta(x,y))^T \nabla f(y) \geq 0$, then we have

$$f(x) - f(y) = (p(x,y)\eta(x,y))^T \nabla f(y) \geq 0.$$

Thus $f(x) \geq f(y)$. Likewise, we can prove that $-f$ is pseudoinvex with respect to η. Hence f is η-pseudolinear.

□

A second-order necessary condition for η-pseudolinearity is given in the next result due to Rueda [213] and which generalizes a similar result of Chew and Choo [36].

Theorem 3.6. *If $f : X \to R$ and $-f$ pseudo-invex with respect to the same η (f is η-pseudolinear) and f is twice continuously differentiable, then there exists $\alpha \in (0,1)$ such that*

$$\frac{1}{2}(x-y)^T \nabla^2 f(y + \alpha(x-y))(x-y) = (\eta(x,y) - (x-y))^T \nabla f(y).$$

Proof. The Taylor's expansion of f in terms of $x - y$ gives, up to quadratic terms

$$f(x) = f(y) + (x-y)^T \nabla f(y) + \frac{1}{2}(x-y)^T \nabla^2 f(y + \alpha(x-y))(x-y)$$

for some $\alpha, 0 < \alpha < 1$. Since f and $-f$ are invex with respect to the same η, then

$$f(x) - f(y) \geq \eta(x,y)^T \nabla f(y)$$

and

$$f(x) - f(y) \leq \eta(x,y)^T \nabla f(y).$$

It follows that

$$\frac{1}{2}(x-y)^T \nabla^2 f(y+\alpha(x-y))(x-y) \geq (\eta(x,y)-(x-y))^T \nabla f(y)$$

and

$$\frac{1}{2}(x-y)^T \nabla^2 f(y+\alpha(x-y))(x-y) \leq (\eta(x,y)-(x-y))^T \nabla f(y).$$

□

Other insights on η-pseudolinearity will be given in Chap. 5, when studying the characterizations of the solution sets of mathematical programming problems.

3.2 Invexity and Generalized Monotonicity

Several kinds of generalized monotone maps were introduced by various authors, mainly by Karamardian [110], Karamardian and Schaible [111], Karamardian et al. [112], Hadjisavvas and Schaible [78, 79]. See also the book edited by Hadjisavvas et al. [81].

These vector-valued functions (or maps) play a role in complementarity problems and variational inequality problems and are related to generalized convex functions. We assume that F denotes a map $F : C \to R^n$, where $C \subseteq R^n$. In the special case of a gradient map $F = \nabla f$, f denotes a differentiable function $f : C \to R$, where C is open and convex. The notion of monotone map $F : R^n \to R^n$ is a natural generalization of an increasing (non-decreasing) real-valued function of one variable.

Definition 3.7. *F is monotone (M) on C if for every pair of distinct points $x, y \in C$, we have*

$$(x-y)^T(F(x)-F(y)) \geq 0.$$

F is strictly monotone (SM) on C if for every pair of distinct points $x, y \in C$, we have

$$(x-y)^T(F(x)-F(y)) > 0.$$

The following proposition due to Minty [150] is well known. See also Mangasarian [143] and Avriel et al. [10].

Theorem 3.8. *Let $f : C \to R$ be differentiable on the open convex set $C \subseteq R^n$. Then f is convex (resp. strictly convex) on C if and only if ∇f is monotone (resp. strictly monotone) on C.*

This theorem opens the door to the study of generalized monotonicity and its relationships with generalized convexity. Karamardian [110] introduced the following definition.

Definition 3.9. *F is pseudo-monotone (PM) on C if $\forall x, y \in C$:*

$$(x-y)^T F(y) \geq 0 \Rightarrow (x-y)^T F(x) \geq 0.$$

Obviously a monotone map is pseudo-monotone but the converse is not true. Consider, e.g., the function

$$F(x) = \frac{1}{1+x}, \quad C = \{x \in R : x \geq 0\}.$$

It can be proved (see [111]) that F is pseudo-monotone on C if and only if the previous inequalities hold strictly, i.e., if and only if $\forall x, y \in C (x \neq y)$:

$$(x-y)^T F(y) > 0 \Rightarrow (x-y)^T F(x) > 0.$$

Karamardian and Schaible [111] also introduced the following definitions.

Definition 3.10. *F is strictly pseudo-monotone (SPM) on C if $\forall x, y \in C (x \neq y)$:*

$$(x-y)^T F(y) \geq 0 \Rightarrow (x-y)^T F(x) > 0.$$

F is quasi-monotone (QM) on C if $\forall x, y \in C$:

$$(x-y)^T F(y) > 0 \Rightarrow (x-y)^T F(x) \geq 0.$$

Every pseudo-monotone function is quasi-monotone, but the converse is not true. Consider, e.g., the function $F(x) = x^2$, $C = R$. More generally, the following relationships between the previous definitions of generalized monotonicity can be established.

$$\begin{gathered}(M) \Rightarrow (PM) \Rightarrow (QM) \\ (SM)^{\Uparrow} \Rightarrow (SPM)^{\Uparrow}.\end{gathered}$$

Karamardian and Schaible [111] show how the generalized convexity of a differentiable function f defined on the open convex set $C \subseteq R^n$ can be characterized through the generalized monotonicity of the gradient map ∇f. We have the following result.

Theorem 3.11. *Let $f : C \to R$ be differentiable on the open convex set $C \subseteq R^n$. Then f is convex (strictly convex, pseudo-convex, strictly pseudo-convex, quasi-convex) on C if and only if the gradient map ∇f is monotone (strictly monotone, pseudo-monotone, strictly pseudo-monotone, quasi-monotone) on C.*

The extension of generalized monotonicity to the invex case are more recent: see Ruiz-Garzon et al. [216], Yang et al. [248, 249], Peng [198]. The concept of invex monotone maps appeared in Parida and Sen [196], when it was called monotone. For the reader's convenience we recall here the notions of invex, pseudo-invex, strictly pseudo-invex and quasi-invex functions together with the new notion of strictly invex functions.

A function $f : X \subseteq R^n \to R$, differentiable on the open set $X \subseteq R^n$ is said to be:

(I) Invex (IX) on X if there exist $\eta : X \times X \to R^n$ such that $\forall x, y \in X$

$$f(x) - f(y) \geq \eta(x,y)^T \nabla f(y).$$

(II) Strictly invex (SIX) on X if there exist $\eta : X \times X \to R^n$ such that $\forall x, y \in X,\ x \neq y$

$$f(x) - f(y) > \eta(x,y)^T \nabla f(y).$$

(III) Pseudo-invex (PIX) on X if there exist $\eta : X \times X \to R^n$ such that $\forall x, y \in X$,

$$\eta(x,y)^T \nabla f(y) \geq 0 \Rightarrow f(x) - f(y) \geq 0.$$

(IV) Strictly pseudo-invex $(SPIX)$ on X if there exist $\eta : X \times X \to R^n$ such that $\forall x, y \in X,\ x \neq y$

$$\eta(x,y)^T \nabla f(y) \geq 0 \Rightarrow f(x) - f(y) > 0.$$

(V) Quasi-invex (QIX) on X if there exist $\eta : X \times X \to R^n$ such that $\forall x, y \in X$,

$$f(x) - f(y) \leq 0 \Rightarrow \eta(x,y)^T \nabla f(y) \leq 0.$$

We have (as already said) the following relationships

$$\begin{gathered}(IX) \Rightarrow (PIX) \Rightarrow (QIX)\\ (SIX)^{\Uparrow} \Rightarrow (SPIX)^{\Uparrow}.\end{gathered}$$

Now we shall introduce the various concepts of generalized monotonicity by means of a kernel function and shall relate these concepts to invex and generalized invex functions.

Definition 3.12. *The function $F : X \subseteq R^n \to R^n$ is said to be:*

(I) Invex monotone (IM) on X if there exist $\eta : X \times X \to R^n$ such that $\forall x, y \in X$

$$\eta(x,y)^T (F(x) - F(y)) \geq 0.$$

(II) Pseudo-invex monotone (PIM) *on* X *if there exist* $\eta : X \times X \to R^n$ *such that* $\forall x, y \in X$,

$$\eta(x,y)^T F(y) \geq 0 \Rightarrow \eta(x,y)^T F(x) \geq 0.$$

(III) Quasi-invex monotone (QIM) *on* X *if there exist* $\eta : X \times X \to R^n$ *such that* $\forall x, y \in X,\ x \neq y$

$$\eta(x,y)^T F(y) > 0 \Rightarrow \eta(x,y)^T F(x) \geq 0.$$

(IV) Strictly invex monotone (SIM) *on* X *if there exist* $\eta : X \times X \to R^n$ *such that* $\forall x, y \in X,\ x \neq y$

$$\eta(x,y)^T (F(x) - F(y)) > 0.$$

(V) Strictly pseudo-invex monotone ($SPIM$) *on* X *if there exist* $\eta : X \times X \to R^n$ *such that* $\forall x, y \in X,\ x \neq y$,

$$\eta(x,y)^T F(y) \geq 0 \Rightarrow \eta(x,y)^T F(x) > 0.$$

In accordance with the previous definitions, the following table of relationships between the various kinds of generalized invex monotonicity can be established.

$$(IM) \Rightarrow (PIM) \Rightarrow (QIM)(SIM)^{\Uparrow} \Rightarrow (SPIM)^{\Uparrow}.$$

We now connect the (following [216]) generalized invexity of f to the generalized invex monotonicity of its gradient map ∇f. We have to introduce the following definition:

Definition 3.13. *The function* $f : X \subseteq R^n \to R$ *is said to be skew symmetric on* $X \times X$ *if* $\eta(x,y) + \eta(y,x) = 0\ \forall x, y \in X \subseteq R^n$.

We describe hereafter some sufficient conditions for generalized invex monotonicity.

Theorem 3.14. *If the function* $f : X \subseteq R^n \to R$ *is invex on* X *with respect to* $\eta : X \times X \to R^n$, η *skew symmetric on* X, *then* ∇f *is invex monotone with respect to the same* η.

Proof. As f is invex on X with respect to η, we have, $\forall x, y \in X$:

$$f(x) - f(y) - \eta(x,y)^T \nabla f(y) \geq 0 \tag{3.2}$$

and

$$f(y) - f(x) - \eta(y,x)^T \nabla f(x) \geq 0.$$

As η is skew symmetric, we have

$$f(x) - f(y) + \eta(x,y)^T \nabla f(x) \geq 0. \tag{3.3}$$

By adding (3.2) and (3.3) one has

$$\eta(x,y)^T (\nabla f(x) - \nabla f(y)) \geq 0.$$

Therefore, ∇f is an invex monotone function with respect to the same η.

□

The converse of the above proposition is not true, as can be seen by considering the function $f : X \subseteq R^2 \to R$ defined by $f(x) = x_1 - \cos(x_2)$ and $X = \left\{x \in R^2 : 4x_1^2 + 4x_2^2 - 9 \leq 0,\ x \geq 0\right\}$. We have that ∇f is monotone with respect to η given by

$$\eta(x,y) = \begin{cases} \frac{\sin(x_1) - \sin(y_1)}{\cos(y_1)\cos(y_2)} \\ \sin(x_2) - \sin(y_2), \end{cases}$$

but f is not invex with respect to this η.

In a similar way one can prove the following result.

Theorem 3.15. *If the function $f : X \subseteq R^n \to R$ is strictly invex on X with respect to $\eta : X \times X \to R^n$, η skew symmetric on X, then ∇f is strictly invex monotone with respect to the same η.*

The converse of the above theorem is also not true as can be seen by considering the following example:
Let $f : (0, \pi/2) \to R$ be defined by $f(x) = -(x/2) - (\sin(2x)/4)$; $f'(x) = -\cos^2(x)$ here f' is η-strictly monotone, where $\eta(x,y) = \cos y - \cos x$ but for $x = \pi/4$ and $y = \pi/6$, we have $f(x) - f(y) \leq \eta(x,y)^T \nabla f(y)$.

As invexity and pseudo-invexity coincide, the following result is immediate:

Corollary 3.16. *If the function $f : X \subseteq R^n \to R$ is invex on X with respect to $\eta : X \times X \to R^n$, η skew symmetric on X, then ∇f is pseudo-invex monotone with respect to the same η.*

Similarly to what remarked with reference to pseudo-monotone functions, the following result holds.

Theorem 3.17. *A map $F : X \to R^n$ is η-pseudomonotone on X if and only if for every pair of distinct points $x, y \in X$ we have*

$$\eta(x,y)^T F(y) > 0 \Rightarrow \eta(x,y)^T F(x) > 0.$$

provided η is skew symmetric.

Proof. η-pseudomonotonicity is equivalent to

$$\eta(x,y)^T F(x) < 0 \Rightarrow \eta(x,y)^T F(y) < 0, \quad \forall x, y \in X.$$

As η is skew symmetric, so η-pseudomonotonicity of f is equivalent to

$$\eta(y,x)^T F(x) > 0 \Rightarrow \eta(y,x)^T F(y) > 0,$$

which is further equivalent to

$$\eta(x,y)^T F(y) > 0 \Rightarrow \eta(x,y)^T F(x) > 0, \quad \forall x, y \in X.$$

□

Now we shall establish the relationship which exists between the strict pseudo-invexity of f and the strict monotonicity of its gradient map ∇f.

Theorem 3.18. *If the function $f : X \subseteq R^n \to R$ is strictly pseudo-invex on X with respect to $\eta : X \times X \to R^n$, η skew symmetric on X, then ∇f is strictly pseudo-invex monotone with respect to the same η.*

Proof. Let f be strictly pseudo-invex, then $\forall x, y \in X,\ x \neq y$ we have

$$\eta(x,y)^T \nabla f(y) \geq 0 \Rightarrow f(x) > f(y).$$

We want to show that ∇f is strictly pseudo-invex monotone, i.e.,

$$\eta(x,y)^T \nabla f(x) > 0.$$

Suppose absurdly that

$$\eta(x,y)^T \nabla f(x) \leq 0.$$

As η is skew symmetric, then

$$\eta(y,x)^T \nabla f(x) \geq 0.$$

As f is strictly pseudo-invex, we would have

$$f(x) < f(y),$$

a contradiction with the assumption.

□

Now we shall relate quasi-invexity to quasi-invex monotonicity.

Theorem 3.19. *If the function $f : X \subseteq R^n \to R$ is quasi-invex on X with respect to $\eta : X \times X \to R^n$, η skew symmetric on X, then ∇f is quasi-invex monotone with respect to the same η.*

Proof. Suppose that f is quasi-invex, so $\forall x, y \in X$ such that

$$\eta(y,x)^T \nabla f(x) > 0$$

it holds $f(x) < f(y)$. This last inequality together with quasi-invexity of f imply

$$\eta(y,x)^T \nabla f(x) \leq 0.$$

As η is skew symmetric we have also

$$\eta(x,y)^T \nabla f(x) \geq 0.$$

Hence ∇f is quasi-invex monotone.

□

The converse of the above result is not true, as can be seen by the following example. Consider $f : R \to R$ defined by $f(x) = x^3$. Hence $\nabla f(x)$ is quasi-invex monotone with respect to $\eta(x,y) = \sin x - \sin y$. But f is not quasi-invex with respect to this η (take, e.g., $x = \pi/3,\ y = \pi/6$). Indeed f is not invex for η.

The necessary conditions for generalized monotonicity given by Ruiz-Garzon et al. [216] are not all correct. Correct conditions are given by Yang et al. [249] by means of "Condition C" (see Definition 3.3), whereas Peng [198] established correct necessary criteria for (strictly) pseudo-invex monotonicity and quasi-invex monotonicity, following the approach of Ruiz-Garzon et al. [216], i.e., without condition C. Here we follow the approach of Peng [198].

Definition 3.20. *Let $X \subseteq R^n$ be a convex set. The function $f : X \to R$ is said to be affine if*

$$f(\lambda x + (1-\lambda)y) = \lambda f(x) + (1-\lambda) f(y), \quad \forall x, y \in X \text{ and } \forall \lambda \in (0,1).$$

Lemma 3.21. *Let X be a convex subset of R^n and $\eta : X \times X \to R^n$ be a vector function. If η is affine in the first argument and skew symmetric, then η is also affine in the second argument.*

Proof. Since η is affine in the first argument and skew symmetric, $\forall x, y^1, y^2 \in X$ and $\forall \lambda \in (0,1)$,

$$\begin{aligned}\eta(x, \lambda y^1 + (1-\lambda)y^2) &= -\eta(\lambda y^1 + (1-\lambda)y^2, x)\\ &= -\left[\lambda \eta(y^1, x) + (1-\lambda)\eta(y^2, x)\right]\\ &= \left[\lambda \eta(x, y^1) + (1-\lambda)\eta(x, y^2)\right].\end{aligned}$$

□

Theorem 3.22. *Let X be an open convex subset of R^n. Suppose that:*

(I) $\nabla f : R^n \to R^n$ is strictly pseudo-invex monotone with respect to $\eta : X \times X \to R^n$;

(II) η *is affine in the first argument and skew symmetric;*
(III) For each $x \neq y$ *for some* $f(x) \geq f(y) \Rightarrow \eta(x, \bar{x})^T \nabla f(\bar{x}) \geq 0$ *for some* $\bar{x}$ *which lies on the line segment joining* x *and* y*,*

then $f : X \subseteq R^n \to R$ *is strictly pseudo-invex on* X *with respect to* η*.*

Proof. Let $x, y \in X,\ x \neq y$ be such that

$$\eta(y, x)^T \nabla f(x) \geq 0. \tag{3.4}$$

It is needed to show that

$$f(y) > f(x). \tag{3.5}$$

On the converse, we assume that

$$f(y) \leq f(x). \tag{3.6}$$

By hypothesis (III)

$$\eta(x, \bar{x})^T \nabla f(\bar{x}) \geq 0, \tag{3.7}$$

where $\bar{x} = \bar{\lambda}x + (1 - \bar{\lambda})y$ for some $0 < \bar{\lambda} < 1$. By (3.7) and the strictly pseudo-invex monotonicity of ∇f it follows that

$$\eta(x, \bar{x})^T \nabla f(\bar{x}) > 0. \tag{3.8}$$

Now, from the hypothesis (II) and Lemma 3.21, we know that η is also affine in the second argument. Hence, by (3.8), we have

$$\bar{\lambda}\eta(x, y)^T \nabla f(x) + (1 - \bar{\lambda})\eta(x, y)^T \nabla f(x) > 0. \tag{3.9}$$

The hypothesis that η is skew symmetric implies that $\eta(x, x) = 0$. Therefore, by (3.9) and the fact that $0 < \bar{\lambda} < 1$ it follows that $\eta(x, y)^T \nabla f(x) > 0$. The hypothesis that η is skew symmetric implies that

$$\eta(y, x)^T \nabla f(x) < 0.$$

But this inequality contradicts (3.4), thereby proving that the assumption (3.6) is false.

□

Theorem 3.23. *Let* X *be an open convex subset of* R^n*. Suppose that:*

(I) $\nabla f : R^n \to R^n$ *is pseudo-invex monotone with respect to* $\eta : X \times X \to R^n$*;*
(II) η *is affine in the first argument and skew symmetric;*
(III) For $x, y \in X$ $f(x) > f(y) \Rightarrow \eta(x, \bar{x})^T \nabla f(\bar{x}) > 0$ *for some* $\bar{x}$ *which lies on the line segment joining* x *and* y*,*

then $f : X \subseteq R^n \to R$ *is strictly pseudo-invex on* X *with respect to* η*.*

Proof. The proof follows the same lines of the proof of Theorem 3.22. □

Theorem 3.24. *Let X be an open convex subset of R^n. Suppose that:*

(I) $\nabla f : R^n \to R^n$ is quasi-invex monotone with respect to $\eta : X \times X \to R^n$;
(II) η is affine in the first argument and skew symmetric;
(III) For each $x \neq y$ $x, y \in X$ $f(x) \leq f(y) \Rightarrow \eta(x, \bar{x})^T \nabla f(\bar{x}) > 0$ for some $\bar{x}$ which lies on the line segment joining x and y,

then $f : X \subseteq R^n \to R$ is strictly pseudo-invex on X with respect to η.

Proof. Assume that f is not quasi-invex with respect to η. Then, there exist $x, y \in X$, such that

$$f(y) \leq f(x). \tag{3.10}$$

But

$$\eta(y, x)^T \nabla f(x) > 0. \tag{3.11}$$

By (3.10) and hypothesis (III),

$$\eta(x, \bar{x})^T \nabla f(\bar{x}) > 0, \tag{3.12}$$

where $\bar{x} = \bar{\lambda} x + (1 - \bar{\lambda}) y$, for some $0 < \bar{\lambda} < 1$. Since ∇f is quasi-invex monotone, (3.12) implies

$$\eta(x, \bar{x})^T \nabla f(\bar{x}) \geq 0. \tag{3.13}$$

From the hypothesis (II) and Lemma 3.21, we know that η is also affine in the second argument. Hence, by (3.13) we have that

$$\bar{\lambda} \eta(x, x)^T \nabla f(x) + (1 - \bar{\lambda}) \eta(x, x)^T \nabla f(x) \geq 0. \tag{3.14}$$

By the hypothesis that η is skew symmetric implies that $\eta(x, x) = 0$. Therefore, by (3.14) and the fact that $0 < \bar{\lambda} < 1$ it follows that

$$\eta(x, y)^T \nabla f(x) \geq 0. \tag{3.15}$$

Since η is skew symmetric, (3.15) becomes $\eta(y, x)^T \nabla f(x) \leq 0$, which contradicts (3.11). Hence f is a quasi-invex function with respect to η. □

4

Extensions of Invexity to Nondifferentiable Functions

4.1 Preinvex Functions

Since invexity requires the differentiabilty assumptions, in [18] and subsequently in [244], the class of preinvex functions, not necessarily differentiable, has been introduced. For the reader's convenience we recall Definition 3.2: A subset X of R^n is *η-invex* with respect to $\eta : R^n \to R$ if $x, y \in X, \lambda \in [0,1] \Rightarrow y + \lambda\eta(x,y) \in X$.

It is obvious that the above definition is a generalization of the notion of convex set. It is to be noted that any set in R^n is invex with respect to $\eta(x,y) \equiv 0, \forall x, y \in R^n$. However, the only function $f : R^n \to R$ invex with respect to $\eta(x,y) \equiv 0$ is the constant function $f(x) = c, c \in R$.

Definition 4.1. *Let f be a real-valued function defined on an η-invex set X; f is said to be preinvex with respect to η if*

$$f[y + \lambda\eta(x,y)] \leq \lambda f(x) + (1-\lambda)f(y), \quad \forall x, y \in X, \ \forall \lambda \in [0,1] \tag{4.1}$$

It is obvious that the class of convex functions is strictly contained in the class of preinvex functions: take $\eta(x,y) = x - y$. The inclusion is strict as shown by the following example, taken from [245]. Consider the function $f(x) = -|x|, x \in R$. We show that f is preinvex with respect to

$$\eta(x,y) = \begin{cases} x - y, & \text{if } xy \geq 0 \\ y - x, & \text{if } xy < 0. \end{cases}$$

Let $x \geq 0, y \geq 0$, in this case we have, for the preinvexity of f:

$$-(y + \lambda(x-y)) \leq \lambda(-x) + (1-\lambda)(-y),$$

relation which is obviously true for any $\lambda \in [0,1]$. The same result holds, if $x \leq 0, y \leq 0$. Let now be $x < 0$ and $y > 0$: we must have

$$-|y + \lambda(y-x)| = -y - \lambda(y-x) \leq \lambda x - (1-\lambda)y, \quad \forall \lambda \in [0,1].$$

We obtain $\lambda(2y) \geq 0$, so being $y > 0$, relation (4.1) holds for all $\lambda \in [0,1]$. Consider now the last case, i.e., $x > 0, y < 0$. We must have

$$-|y + \lambda(y - x)| = y + \lambda(y - x) \leq -\lambda x + (1 - \lambda)y, \quad \forall \lambda \in [0,1],$$

i.e., $\lambda(2y) \leq 0$. Being $y < 0$, relation (4.1) holds for all $\lambda \in [0,1]$. The function is therefore preinvex on R, but obviously it is not convex.

Similarly to convex functions, it is possible to characterize preinvex functions in terms of invexity of their epigraph, however not with reference to the same function η (first of all, one is an n-vector, the other is an $(n+1)$-vector).

Theorem 4.2. *Let $f : X \to R$, where $X \subset R^n$ is an η-invex set. Then f is preinvex with respect to η if and only if the set*

$$epif = \{(x, \alpha) : x \in X, \alpha \in R, f(x) \leq \alpha\}$$

is an invex set with respect to $\eta_1 : epif \times epif \to R^{(n+1)}$, where $\eta_1((y, \beta), (x, \alpha)) = (\eta(y, x), \beta - \alpha), \forall (x, \alpha), (y, \beta) \in epif$.

Proof. Necessity:

Let $(x, \alpha) \in epif$ and $(y, \beta) \in epif$, that is, $f(x) \leq \alpha$ and $f(y) \leq \beta$. From the preinvexity of f, we have

$$f[x + \lambda\eta(x, y)] \leq \lambda f(y) + (1 - \lambda)f(x) \leq (1 - \lambda)\alpha + \lambda\beta, \quad \forall \lambda \in [0,1].$$

It follows that

$$(x + \lambda\eta(x, y), (1 - \lambda)\alpha + \lambda\beta) \in epif, \quad \forall \lambda \in [0,1],$$

That is,

$$(x, \alpha) + \lambda(\eta(y, x), \beta - \alpha) \in epif, \quad \forall \lambda \in [0,1].$$

Hence $epif$ is an invex set with respect to $\eta_1((y, \beta), (x, \alpha)) = (\eta(y, x), \beta - \alpha)$.

Sufficiency:

Assume that $epif$ is an invex set with respect to $\eta_1((y, \beta), (x, \alpha)) = (\eta(y, x), \beta - \alpha)$. Let $x, y \in X$ and $\alpha, \beta \in R$ such that $f(x) \leq \alpha$ and $f(y) \leq \beta$. Then $(x, \alpha) \in epif$ and $(y, \beta) \in epif$. From the invexity of the set $epif$ with respect to $\eta_1((y, \beta), (x, \alpha)) = (\eta(y, x), \beta - \alpha)$, we have

$$(x, \alpha) + \lambda\eta_1((y, \beta), (x, \alpha)) \in epif, \quad \forall \lambda \in [0,1].$$

it follows that

$$(x + \lambda\eta(y, x), (1 - \lambda)\alpha + \lambda\beta) \in epif, \quad \forall \lambda \in [0,1].$$

That is

$$f(y + \lambda\eta(x, y)) \leq \lambda\alpha + (1 - \lambda)\beta.$$

Hence f is preinvex function with respect to $\eta_1(x, y)$ on X.

□

Other important properties of preinvex functions are given in the following results [244].

Theorem 4.3. *Let $f : X \to R$, with $X \subset R^n$ an η-invex set. If f is preinvex on X, every local minimum point is a global minimum point.*

Proof. Let $y \in X$ a local minimum point for f and suppose that y is not a global minimum point. Then there exists $x \in X$ such that $f(x) < f(y)$. Being f preinvex on X, there exists $\eta : X \times X \to R^n$ such that, for all $\lambda \in [0,1]$, we have

$$f(y + \lambda\eta(x,y)) \leq f(y) + \lambda[f(x) - f(y)] < f(y). \tag{4.2}$$

Being y a global minimum point there exists a neighbourhood I of y such that $f(z) \geq f(y)$, for each $z \in I \cap X$. As $y + \lambda\eta(x,y) \in X$ for each $\lambda \in [0,1]$, there exists $\epsilon > 0$ such that $y + \lambda\eta(x,y) \in X \cap I$ for all $\lambda \in [0,\epsilon)$, in contradiction with (4.2).

□

Similarly, it is easy to prove that if f is preinvex on the η-invex set $X \subset R^n$, then every strict local minimum point of f is a strict global minimum point.

Another result on preinvex functions, that will be used afterwards, is given here.

Theorem 4.4. *Let $f : X \to R$ be preinvex. If f has a unique global minimizer at $x^* \in X$, then f is convex at x^*.*

Proof. As f is preinvex, there exists $\eta : X \times X \to R^n$ such that for any $x, y \in X, \lambda \in [0,1]$:

$$y + \lambda\eta(x,y) \in X \quad \text{and} \quad \lambda f(x) + (1-\lambda)f(y) \geq f(y + \lambda\eta(x,y)).$$

In particular, when $x = x^*$ and $\lambda = 1$, we have

$$f(x^*) \geq f(y + \eta(x^*,y)), \quad \forall y \in X.$$

Since x^* is unique global minimizer (that is, $f(x^*) < f(x), \forall x \in X, x \neq x^*$), then $y + \eta(x^*,y) = x^*, \forall y \in X$, that is $\eta(x^*,y) = x^* - y, \forall y \in X$. Thus $\lambda f(x^*) + (1-\lambda)f(y) \geq f(\lambda x^* + (1-\lambda)y), \forall y \in X, \lambda \in [0,1]$. Hence, f is convex at x^*.

□

Now we use some algebraic properties of preinvex functions.

Theorem 4.5. *If $f : X \to R$ is preinvex on the η-invex set X, then also kf is preinvex with respect to η, for any $k > 0$.*

Proof. We have, for any $k > 0$,

$$k(f(y + \lambda\eta(x,y))) \leq (\lambda f(x) + (1-\lambda)f(y))k, \quad \lambda \in [0,1],$$

from which

$$k(f(y + \lambda\eta(x,y))) \leq k\lambda f(x) + k(1-\lambda)f(y), \quad \lambda \in [0,1].$$

□

Theorem 4.6. *Let $f, g : X \to R$ be two preinvex functions, with respect to the same kernel function η and let X be a η-invex set. The sum $f + g$ is a preinvex function, with respect to same η.*

Proof. Being f, g preinvex with respect to η, we have

$$f(y + \lambda\eta(x, y)) \leq \lambda f(x) + (1 - \lambda)f(y)$$

and

$$g(y + \lambda\eta(x, y)) \leq \lambda g(x) + (1 - \lambda)g(y).$$

By adding the above relations, we obtain

$$f(y + \lambda\eta(x, y)) + g(y + \lambda\eta(x, y)) \leq \lambda[f(x) + g(x)] + (1 - \lambda)[f(y) + g(y)].$$

□

As a direct consequence of Theorems 4.5 and 4.6, we get

Corollary 4.7. *Let $f_i : X \to R$, $i = 1, \ldots, m$, be preinvex with respect to η. Then $\sum_{i=1}^{m} k_i f_i(x)$ is preinvex with respect to same η, where $k_i > 0$, $i = 1, \ldots, m$.*

Theorem 4.8. *Let $f : X \to R$ be preinvex with respect to η on the η-invex set $X \subseteq R^n$; let $\psi : R \to R$ be an increasing and convex function. Then $\psi \circ f$ is preinvex with respect to η.*

Proof. Being f preinvex with respect to η, we have for each $x, y \in X, \lambda \in [0, 1]$,

$$f(y + \lambda\eta(x, y)) \leq \lambda f(x) + (1 - \lambda)f(y).$$

Being $\psi : R \to R$ increasing and convex, we have

$$\psi(f(y + \lambda\eta(x, y)) \leq \psi(\lambda f(x) + (1 - \lambda)f(y)) \leq \lambda(\psi(f(x)) + (1 - \lambda)\psi(f(y)).$$

□

A differentiable preinvex function is also invex (this was proved by Ben-Israel and Mond [18]) and this is the reason why functions satisfying Definition 4.1 are called preinvex by Weir and Mond in [245].

Theorem 4.9. *Let $f : X \to R$ be differentiable on the open η-invex set $X \subseteq R^n$. If f is preinvex on X with respect to η, then f is invex with respect to the same η.*

Proof. Being f preinvex, we have

$$f(y + \lambda\eta(x, y)) - f(y) \leq \lambda[f(x) - f(y)],$$

which for $\lambda \in (0, 1]$ implies

$$\frac{f(y + \lambda\eta(x, y)) - f(y)}{\lambda} \leq f(x) - f(y).$$

Taking limits for $\lambda \to 0^+$, being f differentiable, we get

$$\eta(x, y)^T \nabla f(y) \leq f(x) - f(y).$$

□

Preinvexity, for the differentiable case, is therefore a sufficient condition for invexity. Indeed, the converse is not generally true, in the sense that if f is invex with respect to a certain η, is not necessarily preinvex with respect to that function η. For example, $f(x) = e^x, x \in R$, is invex with respect to $\eta = -1$, but not preinvex with respect to the same function η. Another example is given by Pini [201]. A simple condition for the preinvexity of an invex function is given in the following result.

Theorem 4.10. *A function $f : X \to R, X \subseteq R^n$ open convex set, invex with respect to η and concave on X, is also preinvex on X with respect to the same η.*

Proof. The concavity of the differentiable function f implies, $\forall x, y \in X, \forall \lambda \in [0, 1]$:

$$f(y + \lambda\eta(x, y)) - f(y) \leq \lambda\eta(x, y)^T \nabla f(y). \tag{4.3}$$

The invexity of f implies

$$\eta(x, y)^T \nabla f(y) \leq f(x) - f(y).$$

Being $\lambda \geq 0$, we have

$$\lambda\eta(x, y)^T \nabla f(y) \leq \lambda[f(x) - f(y)],$$

And taking (4.3) into account,

$$f(y + \lambda\eta(x, y)) - f(y) \leq \lambda[f(x) - f(y)],$$

i.e.,

$$f(y + \lambda\eta(x, y)) \leq \lambda f(x) + (1 - \lambda)f(y), \quad \forall x, y \in X, \quad \forall\lambda \in [0, 1].$$

□

More general conditions, assuming that a differentiable function invex on an η-invex set X, is also preinvex on X, with respect to same η, are given by Mohan and Neogy [165], by means of the so-called "Condition C" (see Definition 3.3). For the reader's convenience we recall that definition. Let $\eta : K \times K \to R^n$ be defined on the invex set $K \subseteq R^n$ satisfies *Condition C*, if for every $x, y \in K$:

$$\eta(y, y + \eta(x, y)) = -\lambda\eta(x, y) \text{ and } \eta(x, y + \eta(x, y)) = (1 - \lambda)\eta(x, y).$$

There are many vector functions that satisfy Condition C, besides the trivial case of $\eta(x, y) = x - y$. For example, let $X = R - \{0\}$ and

$$\eta(x, y) = \begin{cases} x - y, & \text{if } x \geq 0, \ \ y \geq 0 \\ x - y, & \text{if } x \leq 0, \ \ y \leq 0 \\ -y, & \text{otherwise.} \end{cases}$$

Then X is an invex set and η satisfies Condition C.

Yang et al. [248] proved that Condition C holds if

$$\eta(x, y) = x - y + O(\|x - y\|).$$

The main result of Mohan and Neogy [165] is the following.

Theorem 4.11. *Suppose that $X \subseteq R^n$ is an η-invex set and suppose that $f : X \to R$ is differentiable on an open set containing X. Further suppose that f is invex on X, with respect to η and that η satisfies Condition C. Then f is preinvex with respect to η on X.*

Proof. Suppose that $x^1, x^2 \in X$. Let $0 < \lambda < 1$ be given and look at $\bar{x} = x^2 + \lambda\eta(x^1, x^2)$.

Note that $\bar{x} \in X$. By invexity of f, we have

$$f(x^1) - f(\bar{x}) \geq \eta(x^1, \bar{x})^T \nabla f(\bar{x}). \tag{4.4}$$

Similarly, the invexity condition applied to the pair $x^2, \bar{x}$ yields

$$f(x^2) - f(\bar{x}) \geq \eta(x^2, \bar{x})^T \nabla f(\bar{x}). \tag{4.5}$$

Now multiplying (4.4) by λ and (4.5) by $1 - \lambda$ and adding, we get

$$\lambda f(x^1) + (1-\lambda) f(x^2) - f(\bar{x}) \geq \left(\lambda\eta(x^1, \bar{x})^T + (1-\lambda)\eta(x^2, \bar{x})^T\right) \nabla f(\bar{x}).$$

However, by Condition C, $\lambda\eta(x^1, \bar{x})^T + (1-\lambda)\eta(x^2, \bar{x})^T = 0$. Hence, the conclusion of the theorem follows.

□

We have seen that an invex function, with respect to η, may not be preinvex, with respect to same η. However, Mititelu [164] and Udriste et al. [236], state that for the case of an open set $X \subseteq R^n$, the two classes coincide, in case of differentiability of f (obviously, with possible different functions η). Theorem 4.4 can be used to demonstrate that the invex function (defined and differentiable on R) in the following example is not preinvex.

Example 4.12. Consider $f : R \to R, f(x) = 1 - e^{-x^2}$, this function has a unique global minimizer at $x^* = 0$, where $f'(0) = 0$ and is therefore invex (f is also pseudo-convex and also strictly pseudo-convex). However, f is not convex at x^* and therefore not preinvex. As $x^* = 0$ and $f(x^*) = 0$, we have

$$\lambda f(x^*) + (1-\lambda) f(y) = (1-\lambda) f(y)$$

and

$$f(\lambda x^* + (1-\lambda)y) = f((1-\lambda)y).$$

Taking $y = 5, \lambda = 0.5$ for instance, we have $(1-\lambda)f(y) \approx 0.5 < f((1-\lambda)y) \approx 0.998$.

Thus, $(1-\lambda)f(y) \not\geq f((1-\lambda)y), \forall y \in R$, so the function is not convex at $x^* = 0$.

We take the opportunity to correct an example in [75], built to show that (under differentiability) preinvexity does not imply pseudo-convexity (from the previous example we have that also the converse does not hold). Example 3 in [75] must be replaced by the following.

Example 4.13. Let $f(x,y) = (x^2 - y^2)$, defined on $X = \{y \geq -x \text{ and } y \leq x\}$. Consider the two points $u = \begin{pmatrix} x_1 \\ y_1 \end{pmatrix}; v = \begin{pmatrix} x_2 \\ y_2 \end{pmatrix}$. We now show that f is preinvex with respect to $\eta(u,v) = -v$. We begin to show that for this η, X is an η-invex set, i.e., $\forall \lambda \in [0,1], \forall (u,v) \in X$, the vector

$$v + \lambda(-v) = \begin{pmatrix} x_2 - \lambda x_2 \\ y_2 - \lambda y_2 \end{pmatrix} \in X. \tag{4.6}$$

Relation (4.6) may be written as $\begin{pmatrix} 0 \\ 0 \end{pmatrix} \lambda(1-\lambda)v + \lambda(-v) = \begin{pmatrix} x_2 \\ y_2 \end{pmatrix} \in X$, being X convex and the points $(0,0), (x_2, y_2) \in X$. It follows that X is η-invex (with respect to the chosen η). Now we have

$$\begin{aligned} f(v - \lambda v) &= f\begin{pmatrix} x_2 - \lambda x_2 \\ y_2 - \lambda y_2 \end{pmatrix} \\ &= x_2^2(1-\lambda)^2 - y_2^2(1-\lambda)^2 \\ &\leq \lambda f(u) + (1-\lambda) f(v) \\ &= \lambda(x_1^2 - y_1^2) + (1-\lambda)(x_2^2 - y_2^2). \end{aligned}$$

We obtain

$$-(x_2^2 - y_2^2)(1-\lambda) \leq (x_1^2 - y_1^2),$$

which is always verified $\forall \lambda \in [0,1]$, and $\forall x, y \in X$. So, f is preinvex on X; we now show that f is not pseudo-convex. Consider, e.g., the points $(2,1)$ and $(2,0)$ for which $f(2,1) = 3 < f(2,0) = 4$, but $(2,0)^T \nabla f(2,0) = 0$, so f is not pseudo-convex.

So, we can say that there is a partial overlapping between the classes of (differentiable) preinvex functions and pseudo-convex functions. But there is also a partial overlapping between the classes of (differentiable) preinvex functions and quasi-convex functions. Indeed, as already pointed out, there are quasi-convex functions which are not invex, and therefore, not preinvex. On the grounds of Example 4.13 we can also assert the existence of preinvex functions that are not quasi-convex: recall Theorem 2.27 that a differentiable function both preinvex (and therefore invex) and quasi-convex is pseudo-convex.

In the case of vector-valued functions, $f : R^n \to R^m$ is preinvex with respect to η, if each component is preinvex with respect to that η. So, preinvex vector-valued functions are convex-like, as defined by Fan [63]. We recall that $f : R^n \to R^m$ is convex-like if there exists $z \in R^n$ such that

$$f(z) \leq \lambda f(x) + (1-\lambda) f(y), \quad \forall \lambda \in [0,1].$$

For scalar functions, this definition is not useful, as any scalar function $f : X \to R$ is convex-like, which follows by taking $z = x_1$ when $f(x_1) \leq f(x_2)$ and $z = x_2$ when $f(x_1) > f(x_2)$.

Other properties of preinvex functions are given in the paper of Yang and Li [250]. Pini [201] has introduced the following generalisations of preinvex functions, following the ideas underlying the corresponding definitions of pseudo-convex and quasi-convex functions.

Definition 4.14. *A function f is said to be pre-pseudoinvex on an η-invex set $X \subseteq R^n$ if there exists a function η and a positive function b such that*

$$f(x) < f(y) \Rightarrow f(y + \lambda\eta(x, y)) \leq f(y) + \lambda(\lambda - 1)b(x, y)$$

for every $\lambda \in (0, 1)$ and $x, y \in X$.

It is obvious that the class of pseudo-convex functions is a subset of the class of pre-pseudoinvex functions. Simply choose $\eta(x, y) = x - y$. Like in the differentiable case, the following implication holds.

Theorem 4.15. *If f is pre-invex, then f is pre-pseudoinvex, with respect to the same kernel η.*

Proof. If $f(x) < f(y)$, for every $\lambda \in (0, 1)$, we can write

$$\begin{aligned} f(y + \lambda\eta(x, y)) &\leq f(y) + \lambda[f(x) - f(y)] \\ &< f(y) + \lambda(1 - \lambda)[f(x) - f(y)] \\ &= f(y) + \lambda(\lambda - 1)[f(y) - f(x)], \end{aligned}$$

where $b(x, y) = f(y) - f(x) > 0$.

□

For pre-pseudoinvex functions, similarly to preinvex functions, local minimum points are also global and a strict local minimum point is also the strict global minimum point. We give the proof only for the first assertion.

Theorem 4.16. *Let $f : X \to R, X \subseteq R^n$ is an η-invex set. If f is pre-pseudoinvex, every local minimum point is global.*

Proof. Absurdly suppose that y is a local minimum point, but not global. Then there exists $x \in X$ such that $f(x) < f(y)$. From the definition of pre-pseudoinvexity, it follows that there exists $b(x, y) > 0$, such that, for all $\lambda \in (0, 1)$, we have

$$f(y + \lambda\eta(x, y)) \leq f(y) + \lambda(\lambda - 1)b(x, y) < f(y),$$

from which $f(y) > f(y + \lambda\eta(x, y)), \forall\lambda \in (0, 1)$, in contradiction with the assumption.

□

Finally, we give the following definition [201].

Definition 4.17. *A function f is said to be pre-quasi-invex on an η-invex set $X \subseteq R^n$ if there exists a function η such that*

$$f(y + \lambda\eta(x, y)) \leq \max\{f(x), f(y)\},$$

for every $x, y \in X$ and for every $\lambda \in [0, 1]$.

Pre-quasi-invex functions, similarly to quasi-convex functions, can be characterized in terms of lower level sets.

Theorem 4.18. *Let $X \subseteq R^n$ be η-invex and $f : X \to R$; then f is pre-quasi-invex with respect to η if and only if its lower level sets are η-invex.*

Proof. Assume that f is pre-quasi-invex on X and denoted by $L_f(\alpha)$ the subset of X given by $\{x : f(x) \leq \alpha\}$. If $L_f(\alpha)$ is empty or if it exhausts X, then the result is trivial. If $L_f(\alpha)$ is neither empty nor the whole set X, choose two points x and y in $L_f(\alpha)$. We must show that the line segment $[y, y+\eta(x,y)] \in L_f(\alpha)$. Indeed, since f is a pre-quasi-invex function, we have that

$$f(y + \lambda\eta(x,y)) \leq \max\{f(x), f(y)\} \leq \alpha,$$

for every $\lambda \in [0,1]$, thereby proving the η-invexity of $L_f(\alpha)$.

Conversely, assume that for every real number α the set $L_f(\alpha)$ is η-invex. Pick two points $x, y \in X$ and suppose that $f(x) \leq f(y)$. Consider the lower level set $L_f(f(y))$. Since $L_f(.)$ is η-invex, the segment $[x, x + \eta(x,y)] \in L_f(f(y))$. Thus

$$f(x + \lambda\eta(x,y)) \leq \max\{f(x), f(y)\} = f(y),$$

for every $\lambda \in [0,1]$.

□

Theorem 4.19. *Assume that $\eta(x,y) \neq 0$ whenever $x \neq y$ and that $X \subseteq R^n$ is an η-invex set. Let f be a pre-quasi-invex function on X with respect to η. Then every strict local minimum of f is also a strict global minimum point.*

Proof. Let y be a strict local minimum which is not global; then there exists a point $x^* \in X$ such that $f(x^*) < f(y)$. Since f is pre-quasi-invex, we have

$$f(y + \lambda\eta(x^*, y)) \leq f(y),$$

which contradicts the hypothesis that y is a strict local minimum.

□

Like in the quasi-convex case, the pre-quasi-invexity is preserved under composition with non decreasing functions $\psi : R \to R$.

Theorem 4.20. *Let f be a pre-quasi-invex function with respect to η and assume that $\psi : R \to R$ is a non decreasing function. Then $\psi \circ f$ is pre-quasi-invex with respect to η.*

Proof. Since f is pre-quasi-invex function and ψ is non decreasing we have

$$\begin{aligned}\psi \circ f(y + \lambda\eta(x,y)) &\leq \psi\left(\max\{f(x), f(y)\}\right)\\ &= \max\{\psi \circ f(x), \psi \circ f(y)\},\end{aligned}$$

which says precisely that the composite function $\psi \circ f$ is pre-quasi-invex.

□

Similar to Theorem 4.11, the following result can be proved.

Theorem 4.21. *Let $X \subseteq R^n$ be an η-invex set and let $f : X \to R$ be differentiable on an open set containing X. Suppose that f is quasi-invex with respect to η on X and that η satisfies Condition C. Then f is pre-quasi-invex on X.*

Proof. See Mohan and Neogi [165].

□

4.2 Lipschitz Invex Functions and Other Types of Nonsmooth Invex Functions

The strong growth of nonsmooth analysis, inspired above all by the work of Clarke [37,38], (see Rockafellar [211,212]) touched also the field of invex functions and its applications. Following Clarke's introduction of generalized directional derivatives and generalized subdifferentials for locally Lipschitz functions, it was natural to extend invexity to such functions. The main papers involved with nonsmooth invex functions, both in the sense of Clarke and following other treatments, are due to Craven [44], Craven and Glover [45], Giorgi and Guerraggio [71,72], Jeyakumar [102], Reiland [209,210], Tanaka [232], Ye [253]. We begin with the main concepts and definitions related to Clarke's theory (see, for a complete treatment of the theory, motivation and applications, the fundamental book of Clarke [38]).

Definition 4.22. *Let X be an open subset of R^n, the function $f : X \to R$ is said to be locally Lipschitz at $x^\circ \in x$ if there exists a positive constant K and a neighbourhood N of x° such that for any $x^1, x^2 \in N$,*

$$|f(x^1) - f(x^2)| \leq K\|x^1 - x^2\|.$$

Definition 4.23. *If $f : X \to R$ is locally Lipschitz at $x^\circ \in X$, the generalized derivative (in the sense of Clarke) of f at x° in the direction $v \in R^n$, denoted by $f^\circ(x^\circ; v)$, is given by*

$$f^\circ(x^\circ; v) = \limsup_{\substack{y \to x^\circ \\ t \downarrow 0}} \frac{f(y + tv)}{t}.$$

We shall say that a locally Lipschitz function at x° is C-differentiable at x°, with directional derivative given by $f^\circ(x^\circ; v)$. By the Lipschitz condition it follows that $|f^\circ(x^\circ; v)| \leq K\|v\|$, so $f^\circ(x^\circ; v)$ is a well defied finite quantity. Moreover, $f^\circ(x^\circ; v)$ is a sublinear function of the direction v and we have, for any $v \in R^n$

$$f^\circ(x^\circ; v) = \max\{\xi^T v : \xi \in \partial_c f(x^\circ)\},$$

where $\partial_c f(x^\circ)$ is a convex and compact set of R^n, called the Clarke subdifferential of f at $x^\circ \in X$ or Clarke generalized gradient of f at x°, given by

$$\partial_c f(x^\circ) = \{\xi \in R^n : f^\circ(x^\circ; v) \geq \xi^T v \text{ for all } v \in R^n\}.$$

We summarize hereafter some of the fundamental results concerning $\partial_c f(x^\circ)$.

(a) If f is continuously differentiable at $x^\circ \in X$, then $\partial_c f(x^\circ) = \{\nabla f(x^\circ)\}$. If f is convex, then $\partial_c f(x^\circ) = \partial f(x^\circ)$, where $\partial f(x^\circ)$, denotes the usual subdifferential of convex analysis (see [211]). For all Lipschitz functions, $\partial_c f(x^\circ) = \partial f(x^\circ, 0)$.
(b) $\partial_c(-f)(x^\circ) = -\partial_c f(x^\circ)$; if $g : X \to R$ is locally Lipschitz at $x^\circ \in X$, then $\partial_c(f+g)(x^\circ) \subseteq \partial_c f(x^\circ) + \partial_c g(x^\circ)$.
(c) Let D_f be the set of points in X at which f is not differentiable (by Rademacher's theorem D_f has Lebesgue measure zero) and let S be any other set of measure zero in R^n. Then

$$\partial_c f(x^\circ) = conv\{\lim_{k\to\infty} \nabla f(x^k) : x^k \to x^0, x^k \notin S \cup D_f\};$$

that is, $\partial_c f(x^\circ)$ is the convex hull of all points of the form $\lim \nabla f(x^k)$, where $\{x^k\}$ is any sequence which converges to x° and avoids $S \cup D_f$.
(d) For D_f and S as in (c),

$$f^\circ(x^\circ, v) = \limsup_{y \to x^\circ} \left\{(\nabla f(y))^T v : y \notin s \cup D_f\right\}$$

The following result, due to Clarke and easy consequence of property (b), gives a necessary condition for a local minimum of f.

Theorem 4.24. *Let $f : X \to R$ be C-differentiable on the open set $X \subseteq R^n$ and let $x^\circ \in X$ be a point of local minimum of f over X; then $0 \in \partial_c f(x^\circ)$.*

We justify therefore the following definition.

Definition 4.25. *The point $x^\circ \in X$ is said to be a C-min-stationary point for the C-differentiable function f if $0 \in \partial_C f(x^\circ)$.*

Definition 4.26. *Let $f : X \to R$ be locally Lipschitz on the open set $X \subseteq R^n$; then f is C-invex on X if there exists a function $\eta : X \times X \to R^n$ such that*

$$f(x) - f(x^\circ) \geq f^\circ(x^\circ, \eta(x, x^\circ)), \quad \forall x, x^\circ \in X.$$

We note that the previous definition can be equivalently given by the inequality

$$f(x) - f(x^\circ) \geq \xi^T, \eta(x, x^\circ), \quad \forall x, x^\circ \in X, \quad \forall \xi \in \partial_C f(x^\circ).$$

The above inequality was used for a more general case, by Jeyakumar [102].

Another important topic of nonsmooth analysis is the concept of *DR*-quasidifferentiabilty, introduced by Demyanov and Rubinov [55, 56] and containing, as a special case, the concept of *P*-quasidifferentiability, introduced some years before by Pshenichnyi [205], perhaps the first author to present in a systematic way a proposal for the study of nonsmooth and non-convex optimization problems.

Definition 4.27. *The function $f : X \to R$ is said to be DR-quasidifferentiable at $x^\circ \in X, X \subseteq R^n$ open set, if for any $g \in R^n$ the directional derivative*

$$f'(x^\circ; g) = \lim_{t \to o^+} \frac{f(x^\circ + tg) - f(x^\circ)}{t}$$

exists (finite) and it holds

$$f'(x^\circ, g) = \max_{v \in \underline{\partial} f(x^\circ)} \{v^T g\} + \min_{w \in \overline{\partial} f(x^\circ)} \{w^T g\},$$

where $\underline{\partial} f(x^\circ)$ and $\overline{\partial} f(x^\circ)$ are convex and compact sets of R^n.

The pair of sets $\left(\overline{\partial} f(x^\circ), \overline{\partial} f(x^\circ)\right) = Df(x^\circ)$ for which f is DR-quasidifferentiable at x° is said to be a quasidifferential of f at x°. Obviously $Df(x^\circ)$ is not unique, as the pair $\left(\overline{\partial} f(x^\circ) + C, \overline{\partial} f(x^\circ) - C\right)$ with $C \subseteq R^n$ convex and compact, is a quasidifferential too. The set $\overline{\partial} f(x^\circ)$ is called a subdifferential and the set $\overline{\partial} f(x^\circ)$ is called superdifferential of f at x°. It is easy to see that equivalently f is DR-quasidifferentiable at x° if

$$f'(x^\circ, g) = \max_{v \in \underline{\partial} f(x^\circ)} \{v^T g\} - \max_{w \in -\overline{\partial} f(x^\circ)} \{w^T g\},$$

i.e., if $f'(x^\circ, g)$ is given by the difference of two support functions of two compact and convex sets, i.e., by the difference of two sublinear (i.e., convex and positively homogeneous) functions (DSL functions) of the direction g.

Definition 4.28. *The function $f : X \to R, X \subseteq R^n$ open set, is said to be P-quasidifferentiable at $x^\circ \in X$ if for any $g \in R^n$ the directional derivative $f'(x^\circ, g)$ exists (finite) and is a positively homogeneous function of the direction g, i.e., we have*

$$f'(x^\circ, g) = \max_{v \in \partial_P f(x^\circ)} \{v^T g\},$$

with $\partial_P f(x^\circ)$ convex and compact subset of R^n.

The set $\partial_P f(x^\circ)$ is called the Pshenichnyi subdifferential of f at x°; obviously every DR-quasidifferentiable function is P-quasidifferentiable, with $\underline{\partial} f(x^\circ) = \partial_P f(x^\circ)$ and $\overline{\partial} f(x^\circ) = \{0\}$.

Theorem 4.29 (Demyanov and Vasiliev [57]; Pshemichnyi [205]).

(a) If $f : X \to R$ is DR-quasidifferentiable and $x^\circ \in X$ is a point of local minimum of f on X, then $-\overline{\partial} f(x^\circ) \subseteq \underline{\partial} f(x^\circ)$.
(b) If $f : X \to R$ is P-quasidifferentiable and $x^\circ \in X$ is a point of local minimum of f on X, then $0 \in \partial_P f(x^\circ)$.

□

The previous necessary optimality conditions justify the following definitions.

Definition 4.30. *Let $f : X \to R$ be DR-quasidifferentiable (respectively, P-quasidifferentiable) on X; then f is DR-invex (respectively, P-invex) on X if there exists a vector valued function $\eta : X \times X \to R^n$ such that*

$$f(x) - f(x^\circ) \geq f'\left(x^\circ, \eta(x, x^\circ)\right), \quad \forall x, x^\circ \in X.$$

Theorem 4.31 (Giorgi and Guerraggio [71]). *A DR-quasidifferentiable function is DR-invex if and only if there exists a function $\eta : X \times X \to R^n$ such that for each $x, x^\circ \in X$ the following inequality*

$$f(x) - f(x^\circ) \geq \left(v(x^\circ) + w^*(x, x^\circ)\right) \eta(x, x^\circ),$$

holds for any $v(x^\circ) \in \underline{\partial} f(x^\circ)$ and for some $w^*(x, x^\circ) \in \overline{\partial} f(x^\circ)$.

Proof. If f is DR-invex, obviously there exists an element $w^*(x, x^\circ) \in \overline{\partial} f(x^\circ)$ such that

$$f(x) - f(x^\circ) \geq \max_{v \in \underline{\partial} f(x^\circ)} \{v\eta(x, x^\circ)\} + w^*(x, x^\circ)\eta(x, x^\circ),$$

that is

$$f(x) - f(x^\circ) \geq v(x^\circ) + w^*(x, x^\circ)\eta(x, x^\circ), \quad \forall v(x^\circ) \in \underline{\partial} f(x^\circ).$$

Conversely if, for some $w^*(x, x^\circ) \in \overline{\partial} f(x^\circ)$, the last inequality holds, we obtain

$$f(x) - f(x^\circ) \geq \max_{v \in \underline{\partial} f(x^\circ)} \{v\eta(x, x^\circ)\} + w^*(x, x^\circ)\eta(x, x^\circ);$$

it follows

$$f(x) - f(x^\circ) \geq \max_{v \in \underline{\partial} f(x^\circ)} \{v\eta(x, x^\circ)\} + \min_{w \in \overline{\partial} f(x^\circ)} \{w^*(x, x^\circ)\eta(x, x^\circ)\}.$$

□

If f is a P-quasidifferentiable function the above theorem states that f is P-invex if and only if the inequality

$$f(x) - f(x^\circ) \geq \{v(x^\circ)\eta(x, x^\circ)\} \quad \text{holds} \quad \forall v(x^\circ) \in \partial_P f(x^\circ).$$

The next theorem extends to DR-quasidifferentiable (P-quasidifferentiable) functions the result given in Theorem 2.2 for differentiable functions.

Theorem 4.32 (Giorgi and Guerraggio [71]). *Let $f : X \to R$ be DR-quasidifferentiable (P-quasidifferentiable), then f is DR-invex (P-invex) on X if and only if every DR-min-stationary point (P-min-stationary point) is a global minimum point of f on X.*

Proof. We begin the proof for the more general case of DR-quasidifferentiability. Let f be DR-invex and $x^\circ \in X$ be a DR-min-stationary point of f, i.e., $-\overline{\partial} f(x^\circ) \subseteq \underline{\partial} f(x^\circ)$. Then it is possible to choose $v = -w^*$ in $\underline{\partial} f(x^\circ)$;

therefore from Theorem 4.31 it follows $f(x) - f(x^\circ) \geq 0, \forall x \in X$. Conversely, suppose that every DR-min-stationary point of f is a global minimum point. Let x, x° be two arbitrary points of X. If $f(x) \geq f(x^\circ)$, choose $\eta(x, x^\circ) = 0$. If $f(x) < f(x^\circ), x^\circ$ cannot be a DR-min-stationary point; then there exists $w^*(x, x^\circ) \in \overline{\partial} f(x^\circ)$ such that $0 \notin w^* + \underline{\partial} f(x^\circ)$, where $w^* + \underline{\partial} f(x^\circ)$, is a convex and compact set. This implies that $\min\{\|\xi\| : \xi \in w^* + \underline{\partial} f(x^\circ)\} = m > 0$. If ξ^* is an element of minimum norm in $w^* + \underline{\partial} f(x^\circ)$, then we have (see, e.g., Bazaraa and Shetty [12], Theorem 2.4.4) $\xi^* \cdot \xi \geq \xi^* \cdot \xi^*, \forall \xi \in w^* + \underline{\partial} f(x^\circ)$. Let us now consider the function

$$\eta(x, x^\circ) = \frac{f(x) - f(x^\circ)}{\xi^* \xi^*} \xi^*$$

for which the necessary and sufficient condition of Theorem 4.31 is satisfied. So the proof is complete.

□

Also for C-differentiable functions we have a result similar to the thesis of the previous theorem.

Theorem 4.33. *Letf* $: X \to R$ *be locally Lipschitz on* X*, then* f *is* C*-invex on* X *if and only if every* C*-min-stationary point is a global minimum point of* f *over* X.

The proof of the above result is quite similar to the one of Theorem 4.32 and hence is omitted. We remark that Theorem 4.33 was also obtained by Tanaka et al. [233] and by Reiland [209]. This author makes the superfluous condition (for finite dimensional space) that the cone

$$K(x^\circ, x) = \bigcup_{\lambda \geq 0} (\lambda \partial_C f(x^\circ) \times \{\lambda (f(x) - f(x^\circ)\})$$

is closed. As corollaries we have that if $f : X \to R$ is C-invex on X, then $x^\circ \in X$ is a global minimum of f over X if and only if $0 \in \partial_C f(x^\circ)$. Moreover if $f : X \to R$ has no stationary point, then f is C-invex on X (f locally Lipschitz on X). For $f : R^n \to R, f'(x^\circ, v)$ denotes the usual directional derivative of f at x° in the direction v:

$$f'(x^\circ, v) = \lim_{\lambda \to 0^+} \frac{f(x^\circ + \lambda v) - f(x^\circ)}{\lambda}$$

when this limit exists. If a Lipschitzian function has a directional derivative, it is not necessarily true that $f'(x^\circ, v) = f^\circ(x^\circ, v)$. This justifies the following definition.

Definition 4.34 (Clarke [38]). *A locally Lipschitz function* $f : X \to R$ *is said to be regular at* $x^\circ \in X$ *when:*

(i) For each direction $v \in R^n$ *the directional derivative* $f'(x^\circ, v)$ *exists finite.*
(ii) We have $f'(x^\circ, v) = f^\circ(x^\circ, v)$*, for every* $v \in R^n$.

For a regular function we have therefore $\partial_P f(x^\circ) = \partial_C f(x^\circ)$. To better understand the above definition it is useful to give also the following one based on two important local cone approximations.

Definition 4.35. *The set $A \subseteq R^n$ is regular at $x^\circ \in \bar{A}$ ($\bar{A}$ denotes the conjugate of A) if*

$$TC(A, x^\circ) = T(A, x^\circ),$$

where

$$TC(A, x^\circ) = \{v \in R^n : \forall\{x^k\} \to x^\circ, x^k \in A, \\ \forall\{\lambda_k\} \to 0, \lambda_k > 0, \exists\{v^k\} \to v : x^k + \lambda_k v^k \in A\}$$

is the Clarke tangent cone to A at x° and

$$T(A, x^\circ) = \{v \in R^n : \exists\{t_k\} \to 0, t_k > 0, \exists\{v_k\} \to v : x^\circ + t_k v^k \in A, \forall k\}$$

is the contingent cone to A at x° (or Bouligand tangent cone to A at x°).

The following Theorem is due to Clarke [38].

Theorem 4.36. *Let $f : X \to R$ be locally Lipschitz at $x^\circ \in X$; then f is regular at x° if and only if $epif$ is regular at $(x^\circ, f(x^\circ))$, i.e., if and only if*

$$TC\left(epif, (x^\circ, f(x^\circ))\right) = T\left(epif, (x^\circ, f(x^\circ))\right).$$

□

A sufficient condition for the regularity of a P-quasidifferentiable function on $X \subseteq R^n$ is that the multiplication $\partial_P(\cdot)$ is upper semicontinuous [38].

The previous concept is useful for the following considerations. Zang et al. [255] characterized the functions whose stationary points are global minima (i.e., the invex functions) in terms of the strict lower semicontinuity of the multifunction (lower level sets)

$$L_f(\alpha) = \{x : x \in X, f(x) \le \alpha, \alpha \in R\}.$$

Following these authors we recall Definition 2.3. We say that $L_f(\alpha)$ is strictly lower semicontinuous at a point α of its effective domain $\Xi = \{\alpha : \alpha \in R, L_f(\alpha) \neq \phi\}$ if $x \in L_f(\alpha), \{\alpha_i\} \subseteq \Xi, \alpha_i \to \alpha$ imply the existence of a natural number k, a sequence $\{x^i\}$ and a real number $\beta(x) > 0$ such that

$$x^i \in L_f\left(\alpha_i - \beta(x)\left\|x^i - x\right\|\right), \quad i = k, k+1, \dots \quad \text{and} \quad x^i \to x.$$

If $L_f(\alpha)$ is strictly lower semicontinuous at every $\alpha \in \Xi$, it is said to be strictly lower semicontinuous on Ξ.

The result of Zang et al. [255] has been generalized by Tanaka [232] to the nonsmooth case, under the regularity assumption on f.

Theorem 4.37. *Let $f : X \to R$ be regular, then f is C-invex on X (i.e., P-invex on X) if and only if $L_f(\alpha)$ is strictly lower semicontinuous.*

□

Jeyakumar [102] introduced the notion of approximately quasidifferentiable functions, according to the following definition.

Definition 4.38. *The function $f : X \to R$ is said to be approximately quasidifferentiable or J-differentiable at $x^\circ \in X$ if there exists a convex compact subset $\partial_J f(x^\circ)$ of R^n such that*

$$f^D(x^\circ, g) \leq \max_{v \in \partial_J f(x^\circ)} \{vg\}, \quad \forall g \in R^n,$$

where $f^D(x^\circ, g)$ is the Dini upper directional derivative at x° in the direction g:

$$f^D(x^\circ, g) = \limsup_{t \to 0^+} \frac{f(x^\circ + tg) - f(x^\circ)}{t}.$$

The previous definition was also considered by Ioffe [96] who defines $\psi : R^n \to R$ a first order convex approximation for f at x°, if ψ is sublinear (i.e., convex and positively homogeneous) and satisfies the condition

$$f^D(x^\circ, g) \leq \psi(g), \quad \forall g \in R^n. \tag{4.7}$$

The original definition of Ioffe contains, instead of (4.7), the equivalent condition

$$\limsup_{t \to 0^+} \frac{f(x^\circ + tg) - f(x^\circ) - t\psi(g)}{t} \leq 0.$$

We note that, according to Definition 4.38, every locally Lipschitz function defined on X is J-differentiable, since

$$f^D(x^\circ, g) \leq f^\circ(x^\circ, g) = \max_{\xi \in \partial_C f(x^\circ)} \{\xi g\}, \quad \forall g \in R^n.$$

Also every P-quasidifferentiable function is J-quasidifferentiable, since

$$f^D(x^0, g) = f'(x^0, g) = \max_{v \in \partial_P f(x^\circ)} \{vg\}, \quad \forall g \in R^n.$$

In [102] it is also given a numerical example showing that the class of J-quasidifferentiable functions strictly contains the classes of locally Lipschitz and P-quasidifferentiable functions.

Theorem 4.39. *Let $f : X \to R$ be J-quasidifferentiable; if $x^\circ \in X$ is a point of local minimum for f, then we have $0 \in \partial_J f(x^\circ)$.*

Proof. Being f J-quasidifferentiable, we have

$$f^D(x^\circ, g) = \max_{v \in \partial_J f(x^\circ)} \{vg\}, \quad \forall g \in R^n.$$

As $x^\circ \in X$ is a local minimum point for f, it holds $f^D(x^\circ, g) \geq 0$. Now let us assume, absurdly, that $0 \notin \partial_J f(x^\circ)$; so there exists a direction $\bar{g}$ such that

$$\max_{v \in \partial_J f(x^\circ)} \{v\bar{g}\} < 0.$$

But this leads to the absurd conclusion

$$0 \le f^D(x^\circ, \bar{g}) \le \max_{v \in \partial_J f(x^\circ)} \{v\bar{g}\} < 0.$$

□

The above theorem justifies the following definition.

Definition 4.40. *A point $x^\circ \in X$ is said to be a J-min-stationary point for the J-quasidifferentiable function $f : X \to R$ if $0 \in \partial_J f(x^\circ)$.*

Following Jeyakumar [102] we give the following definition.

Definition 4.41. *The J-quasidifferentiable function $f : X \to R$ is J-invex if there exists $\eta(x, x^\circ) : X \times X \to R^n$ such that*

$$f(x) - f(x^\circ) \ge v^T \eta(x, x^\circ), \quad \forall x, x^\circ \in X, \ \forall v \in \partial_J f(x^\circ).$$

It is not difficult to prove (see [71]) the following result.

Theorem 4.42. *Let $f : X \to R$ be J-quasidifferentiable; then f is J-invex if and only if every J-min-stationary point is a global minimum point of f over X.*

□

Another class of nonsmooth invex functions has been introduced by Ye [253]. See also Craven and Glover [45].

Definition 4.43. *The function $f : X \to R$ is said to be Ye-invex if there exists the directional derivative $f'(x^\circ, g), \forall g \in R^n$ and if there exists $\eta(x, x^\circ) : X \times X \to R^n$ such that*

$$f(x) - f(x^\circ) \ge f'(x^\circ, \eta(x, x^\circ)), \quad \forall x, x^\circ \in X.$$

A well known result justifies the following definition.

Definition 4.44. *Let $f : X \to R$ admit directional derivative for every direction $\forall g \in R^n$; a point $x^\circ \in X$ is said to be Ye-min-stationary point for f if $f'(x^\circ, g) \ge 0, \forall g \in R^n$.*

Theorem 4.45. *Let $f : X \to R$ admit directional derivative for any $g \in R^n$; then f is Ye-invex if and only if every Ye-min-stationary point is a global minimum point of f over X.*

Proof. If x° is a Ye-min-stationary point, the definition of Ye-invexity immediately assures that

$$f(x) - f(x^\circ) \ge 0, \quad \forall x \in X.$$

Conversely, in order to prove that f is Ye-invex, we consider two cases. If $x, x^\circ \in X$ and such that $f(x) - f(x^\circ) \ge 0$, we can take $\eta(x, x^\circ) = 0$. But if $f(x) - f(x^\circ) < 0$, x° cannot be a Ye-min-stationary point; this means that there exists a direction $\forall \bar{g} \in R^n$ such that $f'(x^\circ, \bar{g}) < 0$. In this case it is easy to verify that f is Ye-invex, with respect to

$$\eta(x, x^\circ) = \frac{f(x) - f(x^\circ)}{f'(x^\circ, \bar{g})} \bar{g}.$$

□

We recall that in Demyanov and Vasiliev [57] it is shown that if $f : X \to R$ is a locally Lipschitz DR-quasidifferentiable function, there is the following relationship between the Clarke subdifferential of f at x° and the sets $\underline{\partial} f(x^\circ)$ and $\overline{\partial} f(x^\circ)$:

$$\underline{\partial} f(x^\circ) \subseteq \partial_C f(x^\circ) - \overline{\partial} f(x^\circ).$$

We note also that the proof of the Theorem 4.32 makes use of convex analysis, namely a "projection theorem" (Theorem 2.4.4 of Bazaraa and Shetty [12]) for closed and convex sets. It turns out that it is possible to obtain results similar to the ones presented in Theorems 4.32, 4.33 and 4.42 for those directional derivatives which are support functions of a nonempty closed convex set, i.e., which are positively homogeneous sublinear functions of the direction $g \in R^n$. Besides the types already examined, this is also the case, e.g., of the Michel–Penot directional derivative and of the Rockafellar directional derivative (see, [71]).

Following these lines, Castellani [28] proposes a unifying definition of invexity for nonsmooth functions exploiting the concept of local cone approximation, introduced in an axiomatic form by Elster and Thierfelder [61, 62]. See also Ward [240, 241] and Giorgi et al. [74].

If $K(A, x^\circ) \subseteq R^n$ is a local cone approximation at $x^\circ \in \bar{A}$ ($\bar{A}$ denotes the closure of A), such as, for example, the already introduced Bouligand tangent cone and Clarke tangent cone, the set *epi f* will be locally approximated at the point $(x^\circ, f(x^\circ))$ by a local cone approximation K and a positively homogeneous function $f^K(x^\circ, \cdot)$ will be uniquely determined. More precisely, we have the following definition.

Definition 4.46. *Let $X \subseteq R^n$ be an open set. Let $f : X \to R, x^\circ \in X$ and $K(\cdot, \cdot)$ a local cone approximation. The positively homogeneous function $f^K(x^\circ, \cdot) : R^n \to [-\infty, +\infty]$ defined by*

$$f^K(x^\circ, v) = \inf\{\beta \in R : (v, \beta) \in K(epi f, (x^\circ, f(x^\circ)))\}$$

is called K-directional derivative of f at x° in the direction $v \in R^n$.

By means of Definition 4.46 we can recover most of the generalized directional derivatives used in the literature; for instance:

- The *upper Dini directional derivative* of f at x°

$$f^D(x^\circ, y) = \limsup_{t \to 0^+} \frac{f(x^\circ + ty) - f(x^\circ)}{t}$$

  is associated to the *cone of the feasible directions*

$$F(A, x^\circ) = \{v \in R^n : \forall \{t_k\} \to 0^+, x + t_k v \in A\}.$$

- The *lower Dini directional derivative* of f at x°

$$f_D(x^\circ, y) = \liminf_{t \to 0^+} \frac{f(x^\circ + ty) - f(x^\circ)}{t}$$

is associated to the *cone of the weak feasible directions*

$$WF(A, x^\circ) = \{v \in R^n : \exists \{t_k\} \to 0^+, x^\circ + t_k v \in A\}.$$

- If f is locally Lipschitz, the *Clarke directional derivative* of f at x°, $f^\circ(x^\circ, v)$, already defined, is associated to the Clarke tangent cone $TC(A, x^\circ)$, already defined.
- The Hadamard lower directional derivative

$$f_H(x^\circ, v) = \liminf_{\substack{v' \to v \\ t \to 0^+}} \frac{f(x^\circ + tv') - f(x^\circ)}{t}$$

is associated to the Bouligand tangent cone $T(A, x^\circ)$, already defined.

Definition 4.47. *Let $f : X \to R, x^\circ \in X$ and $K(\cdot, \cdot)$ be a local cone approximation.*

- *f is said to be K-sub-differentiable at x° if there exists a convex compact set $\partial^K f(x^\circ)$ such that*

$$f^K(x^\circ, v) = \max_{x \in \partial^K f(x^\circ)} \{x^*, v\}, \quad \forall v \in R^n;$$

the set $\partial^K f(x^\circ)$ is called the K-sub-differential of f at x°.
- *f is said to be K-quasidifferentiable at x° if there exist two convex compact sets $\underline{\partial}^K f(x^\circ)$ and $\overline{\partial}^K f(x^\circ)$ such that*

$$f^K(x^\circ, v) = \max_{\underline{x}^* \in \underline{\partial}^K f(x^\circ)} \{\underline{x}^* v\} - \max_{\overline{x}^* \in \overline{\partial}^K f(x^\circ)} \{\overline{x}^* v\}.$$

The sets $\underline{\partial}^K f(x^\circ)$ and $\overline{\partial}^K f(x^\circ)$ are called the K-subdifferential and K-superdifferential of f at x°, respectively.

Definition 4.48. *Let $f : X \to R$ and $K(\cdot, \cdot)$ be a local cone approximation; $x^\circ \in X$ is said to be a K-inf-stationary point for f if $f^K(x^\circ, v) \geq 0$ for each $v \in R^n$.*

Castellani [28] proves the following result.

Theorem 4.49. *Let $f : X \to R$ and $K(\cdot, \cdot)$ be a local cone approximation. If f is K-quasidifferentiable, then $x^\circ \in X$ is a K-inf-stationary point for f if and only $\overline{\partial}^K f(x^\circ) \subseteq \underline{\partial}^K f(x^\circ)$. In particular, if f is K-subdifferentiable, then $x^\circ \in X$ is a K-inf-stationary point for f if and only if $0 \in \partial^K f(x^\circ)$.*

In Castellani and Pappalardo [29] it was proved that if $K(\cdot, \cdot)$ is an isotone local cone approximation (i.e., $K(A, x^\circ) \subseteq K(B, x^\circ)$, for each $A \subseteq B$ and with $x^\circ \in A \cup B$). Then every local minimum of f over R^n is K-inf-stationary point for f. In general however, the converse does not hold; so it makes sense to introduce the following definition.

Definition 4.50. *Let $K(\cdot,\cdot)$ be a local cone approximation; the function $f : X \to R$ is said to be K-invex if there exists a function $\eta : X \times X \to R^n$ such that*

$$f(x) - f(y) \geq f^K(y, \eta(y,x)), \quad \forall y, x \in X.$$

The function η is said to be the kernel of the K-invexity.

Castellani [28] then proved the following result.

Theorem 4.51. *Let $f : X \to R$ and $K(\cdot,\cdot)$ be a local cone approximation; f is K-invex if and only if every K-inf-stationary point is a global minimum point.*

For applications of the K-invex functions to nonlinear programming problems, we refer the reader to the paper of Castellani [28]. Similarly to what done for the differentiable case, it is possible to give extensions of the invexity for the nonsmooth case. If we consider, e.g., the locally Lipschitz functions we can define the C-pseudo-invex functions, as those functions for which we have

$$f^\circ(x^\circ, \eta(x, x^\circ)) \geq 0 \Rightarrow f(x) \geq f(x^\circ), \quad \forall x, x^\circ \in X;$$

and the C-quasi-invex functions for which we have

$$f(x) \leq f(x^\circ) \Rightarrow f^\circ(x^\circ, \eta(x, x^\circ)) \leq 0, \quad \forall x, x^\circ \in X.$$

It follows easily from Theorem 4.33 that there is no distinction between C-invexity and C-pseudo-invexity. We note, moreover, that the semi-convex functions, defined by Mifflin [148] for regular Lipschitz functions, i.e., f is semi-convex at $x^\circ \in X$ if for every $d \in R^n$ such that $x^\circ + d \in X$, we have $f'(x^\circ, d) \geq 0 \Rightarrow f(x^\circ + d) \geq f(x^\circ)$, are a special case of C-pseudo-invex functions. Indeed, if f is semi-convex at x° for every $x^\circ \in X$, (i.e., f is semi-convex on X), then choosing $d = x - x^\circ$, where $x \in X$, shows that f is C-pseudo-invex (i.e., P-pseudo-invex) on X with $\eta(x, x^\circ) = x - x^\circ$.

We have previously remarked [18] that if $f : X \to R$ is differentiable, then preinvexity is a sufficient condition for invexity. We now prove a similar result for the case of DR-quasidifferentiable functions.

Theorem 4.52. *If $f : X \to R$ is a DR-differentiable preinvex function, then it is DR-invex.*

Proof. From the preinvexity of f, it follows

$$\frac{f(x^\circ + \lambda\eta(x, x^\circ)) - f(x^\circ)}{\lambda} \leq f(x) - f(x^\circ), \quad \forall \lambda \in [0,1], \ \forall x, x^\circ \in X.$$

Taking the limit as $\lambda \to 0^+$ we obtain

$$f(x) - f(x^\circ) \geq f'(x^\circ + \lambda\eta(x, x^\circ)), \quad \forall x, x^\circ \in X.$$

□

A result similar to Theorem 4.52 was given by Reiland [209] for C-invex functions, but with the following more restrictive definition of a preinvex function: there exists a neighbourhood $N(x^\circ)$ of x° and $\eta(x, x^\circ) \in R^n$ such that

$$f(y + \lambda\eta(x, x^\circ)) \leq \lambda f(x) + (1 - \lambda)f(y),$$
$$\forall y \in N(x^\circ), \quad \forall \lambda \in [0, 1], \quad \forall x, x^\circ \in X.$$

(If f is differentiable, then the above definition collapses to the usual one.)

For other definitions and concepts of nonsmooth vector-valued functions, the reader is referred to Reiland [210], Yen and Sach [254], Giorgi and Guerraggio [70, 72] and to the literature therein quoted.

5

Invexity in Nonlinear Programming

5.1 Invexity in Necessary and Sufficient Optimality Conditions

Hanson's [83] introduction of invex functions was motivated by the purpose to weaken further the class of mathematical programming problems for which the necessary optimality conditions are also sufficient. There was also the related problem of finding the widest class of functions for which weak and strong duality hold for the dual problems, such as the Wolfe dual problem or the Mond–Weir dual problem. Let us consider the following basic nonlinear programming problem:

(P)

$$\text{Minimize } f(x), \qquad x \in K$$

$$K = \{x : x \in C, g(x) \leq 0\},$$

where $f : C \to R$ and $g : C \to R^m$ are (Frechet) differentiable on the open set $C \subseteq R^n$ (if we have a problem with equality constraints, of the type $h(x) = 0, h : C \to R^p$, we could re-write these constraints as $h(x) \leq 0, -h(x) \leq 0$).

It is well known that under certain regularity assumptions on the vector function g (constraint qualifications) the Karush–Kuhn–Tucker conditions are necessary for optimality in (P), that is, if x^* is a solution of (P) or even if it is a point of local minimum of f on K, then there exists a vector $\lambda^* \in R^m$ such that

$$\nabla f(x^*) + {\lambda^*}^T \nabla g(x^*) = 0 \tag{5.1}$$

$${\lambda^*}^T g(x^*) = 0 \tag{5.2}$$

$${\lambda^*}^T \geq 0. \tag{5.3}$$

It is also well known that relations (5.1)–(5.3) become sufficient for optimality if some (generalized) convexity assumption is made on f and g.

More precisely, if (x^*, λ^*), with $x^* \in K$, satisfies (5.1)–(5.3), then x^* is optimal for (P), provided one of the following assumptions is imposed:

(a) $f(x)$ and $g_i(x)$ convex, $i = 1, \dots, m$ [133]
(b) $f(x)$ pseudo-convex and $g_i(x)$ quasi-convex, with $i \in I = \{i : g_i(x) = 0\}$ the set of the active or effective constraints at $x^* \in K$ [142, 143]
(c) $f(x)$ pseudo-convex and $\lambda^{*T} g(x)$ quasi-convex [169]
(d) $f(x) + \lambda^{*T} g(x)$ pseudo-convex [169]

Hanson [83] noted that the (generalized) convexity requirements appearing in the (a)–(d) above can be further weakened as in the related proofs of the sufficiency for problem (P) there is no explicit dependence on the linear term $(x - m)$, appearing in the definition of differentiable convex, pseudo-convex and quasi-convex functions. Thus this linear term can be substituted with an arbitrary vector-valued function.

More precisely, if $x^* \in K$ and (x^*, λ^*) satisfies (5.1)–(5.3), then x^* solves (P) if any one of the following conditions is satisfied:

(a) $f(x)$ and every $g_i(x), i \in I$, are invex with respect to the same η.
(b) $f(x)$ is pseudo-invex and every $g_i(x), i \in I$, is quasi-invex with respect to the same η.
(c) $f(x)$ is pseudo-invex and $\lambda^* g(x)$ is quasi-invex with respect to the same η.
(d) The Lagrangian function $f(x) + \lambda^{*T} g(x)$ is pseudo-invex with respect to an arbitrary η (i.e., the Lagrangian function $f(x) + \lambda^{*T} g(x)$ is invex).

The proofs are easy. We give only the proof for (a), the original result of Hanson [83]: For any $x \in C$ satisfying $g(x) \leq 0$, we have

$$\begin{aligned} f(x) - f(x^*) &\leq \eta(x, x^*)^T \nabla f(x^*) \\ &= -\eta(x, x^*)^T \nabla (\lambda^{*T} g(x)) \\ &\geq -\lambda^{*T} (g(x) - g(x^*)) \\ &= -\lambda^{*T} g(x) \\ &\geq 0. \end{aligned}$$

So x^* is minimal.

We stress the fact that f and every $g_i(x), i = 1, \dots, m$ or also $i \in I$, have to be invex with respect to a common η. By the remark to Corollary 2.13, this is equivalent to $f + \lambda^T g$ being invex for all $\lambda \in R_+^m$. So, condition (d) is more general, as it requires that $f + \lambda^{*T} g$ is invex, without having $f + \lambda^T g$ invex for all $\lambda \in R_+^m$ (the proof of (d)) is quite immediate and left it to the reader.

Jeyakumar [100] gives the following result that weakens the sufficient optimality conditions for problem (P) by means of ρ-invex functions (see, Definition 2.19).

Theorem 5.1. *Let $x^* \in K$ and let (x^*, λ^*) satisfy (5.1)–(5.3); let $f(x)$ be ρ_0-pseudo-invex at x^* and let every $g_i(x), i \in I$, be ρ_i-quasi-invex at x^*,* *with respect to the same function* η *and* θ. *Let* $\rho_0 + \sum_{i \in I} \lambda_i^* \rho_i \geq 0$. *Then* x^* *solves (P).*

Proof. As x^* satisfy the Karush–Kuhn–Tucker conditions, we have

$$\eta(x,x^*)^T \nabla f(x^*) - \sum_{i \in I} \lambda_i^* \rho_i \left(\|\theta(x,x^*)\|\right)^2 + \sum_{i \in I} \lambda_i^* \eta(x,x^*)^T \nabla g_i(x^*) + \sum_{i \in I} \lambda_i^* \rho_i \left(\|\theta(x,x^*)\|\right) = 0.$$

From the ρ_i-quasi-invexity of $g_i(x), i \in I$, we have, $\forall x \in K, \forall i \in I$:

$$g_i(x) \leq g_i(x^*) = 0 \Rightarrow \eta(x, X^*)^T \nabla g_i(x^*) + \rho_i \left(\|\theta(x,x^*)\|\right)^2 \leq 0.$$

Being $\lambda_i^* \geq 0, \forall i \in I$, we can reformulate the above relation as follows:

$$x \in K \Rightarrow \sum_{i \in I} \lambda_i^* \eta(x,x^*)^T \nabla g_i(x^*) + \sum_{i \in I} \lambda_i^* \rho_i \left(\|\theta(x,x^*)\|\right)^2 \leq 0.$$

Therefore, we have

$$\eta(x,x^*)^T \nabla f(x^*) \geq \sum_{i \in I} \lambda_i^* \rho_i \left(\|\theta(x,x^*)\|\right)^2, \quad \forall x \in K.$$

Since $\rho_0 + \sum_{i \in I} \lambda_i^* \rho_i \geq 0$, we have

$$\eta(x,x^*)^T \nabla f(x^*) + \rho_0 \left(\|\theta(x,x^*)\|\right)^2 \geq 0, \quad \forall x \in K.$$

That is, x^* solves (P), thanks to the ρ_0-pseudo-invexity assumption on $f(x)$. □

Note that the above theorem holds also with the assumption of the existence of different functions $\theta_0(x,x^*), \theta_i(x,x^*), i \in I$ with

$$\sum_{i \in IY\{0\}} \lambda_i^* \rho_i \left(\|\theta_i(x,x^*)\|\right) \geq 0, \quad \lambda_0^* = 1.$$

Invexity also allows the weakening of necessary optimality conditions, in the sense that we can weaken those constraint qualifications expressed in terms of convexity. Ben-Israel and Mond [18] have used a modified or generalized Slater constraint qualifications. This condition is described as follows:
Let $g : C \to R^m$; g is said to satisfy the modified Slater condition if $g_i(x), i = 1, \ldots, m$, is invex, with respect to a common η, and there exists $\bar{x} \in K$ such that $g(\bar{x}) < 0$.

Ben-Israel and Mond [18] proved that the Karush–Kuhn–Tucker conditions are necessary for optimality if the modified Slater condition is satisfied. We prove the corresponding result by means of a generalized Karlin constraint qualification, not equivalent, for the invex case, to the Slater constraint qualification.

Theorem 5.2. *Let x^* be optimal for (P) and let the following generalized Karlin constraint qualification be satisfied: there exists no vector $p \in R^m$, $p \geq 0$, $p \neq 0$, such that $p^T g(x) \geq 0$, $\forall x \in C$, and every $g_i(x)$, $i = 1, \dots, m$, is invex with respect to the same η. Then there exists $\lambda^* \in R^m$ such that (x^*, λ^*) satisfies condition (5.1)–(5.3).*

Proof. It is well known that x^* satisfies the Fritz John conditions; that is, there exists $r_0^* \in R$ and $r^* \in R^m$ such that

$$r_0^* \nabla f(x^*) + r^{*T} \nabla g(x^*) = 0$$

$$r^{*T} g(x^*) = 0$$

$$(r_0^*, r^*) \geq 0, (r_0^*, r^*) \neq 0.$$

We have to prove that $r_0^* > 0$, so that we can put $\lambda^* = \left(\frac{1}{r_0^*}\right) r^*$ and the thesis follows. Assume to contrary that $r_0^* = 0$; then we have $r^{*T} \nabla g(x^*) = 0$, $r^{*T} g(x^*) = 0, r^* \geq 0, r* \neq 0$ (i.e., r^* is a semipositive vector). Since every $g_i(x)$ is invex with respect to η, then

$$r^{*T} g(x) - r^{*T} g(x^*) \geq \eta(x, x^*)^T r^{*T} g(x^*), \quad \forall x \in C,$$

i.e.,

$$r^{*T} g(x^*) \geq 0, \quad \forall x \in C.$$

But this contradicts the modified Karlin constraint qualification and therefore $r_0^* > 0$.

□

Note that the modified Slater condition implies the modified Karlin condition, but the converse is not true, contrary to the convex case, where the two conditions are equivalent (see, [143]). The proof of their equivalence is made (for the convex case) by means of a generalized Gordon theorem of the alternative (theorem of Fan–Glicksberg–Hoffman [64]) and uses the fact that if g is convex, then the set

$$M = \bigcup_{x \in C} \{y : y \in R^m, g(x) < y\}$$

is convex. A separation theorem for convex sets then yields the result. However, with $g_i(x), i = 1, \dots, m$, invex functions, the set M is not in general convex, so the said separation theorem is not applicable.

Giorgi and Guerraggio [70] have considered in a more general setting the problem of constraint qualifications involving invexity. Since the original paper of Kuhn and Tucker [133] numerous constraint qualifications have been proposed for problem (P) by several authors and the relationships between them have been thoroughly investigated: see Bazaraa et al. [13], Bazaraa and Shetty [12], Giorgi et al. [74], Mangasarian [143], Peterson [199]. Some of the said constraint qualifications involve (generalized) convexity assumptions on the constraints g_i.

Let us now reformulate these last qualifications under the weaker assumptions of invexity.

(a) Arrow–Hurwicz–Uzawa first constraint qualification (CQ), denoted $(A.H.U. - I)_{in}$, where "in" just stands for "invex": the cone $L_{1,in}$ given by the solution set of the system

$$y^T \nabla g_i(x^0) \leq 0, \quad \forall i \in I_1$$

$$y^T \nabla g_i(x^0) < 0, \quad \forall i \in I - I_1,$$

is nonempty, where $I_1 = \{i : i \in I, -g_i \text{ is invex at } x^0\}$ and $x^0 \in K$. If we denote by L the linearizing cone at $x^0 \in K$:

$$L = \{y : y^T \nabla g_i(x^0) \leq 0, \forall i \in I\}.$$

It is easy to see that the $(A.H.U. - I)_{in}$ CQ can be equivalently expressed as $L \subseteq \bar{L}_{1,in}$ ($\bar{L}$ denoting the closure of the cone L).

(b) Cottle CQ, denoted (Cottle). Define the cone

$$L_0 = \{y : y^T \nabla g_i(x^0) < 0, \forall i \in I\}.$$

Then Cottle CQ is expressed as $L_0 \neq \phi$ or equivalently as: the system

$$\sum_{i \in I} \mu_i \nabla g_i(x^0) = 0, \quad \mu_i \geq 0,$$

admits the zero solution only, i.e., the vector $\nabla g_i(x^0), i \in I$, are positively linearly independent. It is easy to see that Cottle CQ can equivalently be expressed as $L \subseteq \bar{L}_0$.

(c) Slater CQ, denoted $(S)_{in}$: the functions g_i, $i \in I$, are invex at $x^0 \in K$ and there exists a vector $\bar{x}$ such that $g_i(\bar{x}) < 0, \forall i \in I$.

(d) Strict CQ, denoted $(Strict)_{in}$: the functions $g_i, i \in I$ are invex at $x^0 \in K$ and there exists at least two distinct points $x^1, x^2 \in K$ such that

$$g_i(x^2) \leq g_i(x^1) \Rightarrow g_i(x^1 + \lambda\eta(x^1, x^2)) < g_i(x^1),$$

$\forall i \in I, \forall \lambda \in (0, 1)$.

(e) Reverse CQ (weak), denoted $(R)_{in}$: the functions $-g_i, i \in I$, are invex at x^0.

(f) Zangwill CQ, denoted (Z). To introduce this CQ let us denote by D the cone of feasible directions at $x^0 \in K$:

$$D = D(K, x^0) = \{y : \exists \bar{\lambda} > 0 : x^0 + \lambda y \in K, \forall \lambda \in (0, \bar{\lambda})\};$$

note that this cone need not be either open or closed. Zangwill CQ is expressed as $L \subseteq \bar{D}$.

Obviously Cottle CQ and Zangwill CQ have been given in their original formulations, as they do not involve any generalized convexity assumptions. In the other cases the assumption of invexity substitutes the original assumption of convexity or pseudo-convexity. Moreover, in the $(Strict)_{in}$ CQ the assumption

$$g_i(x^2) \le g_i(x^1) \Rightarrow g_i(x^1 + \lambda\eta(x^1, x^2)) < g_i(x^1), \quad \forall i \in I, \ \forall \lambda \in (0,1)$$

substitutes the original assumption [13]

$$g_i(x^2) \le g_i(x^1) \Rightarrow g_i(\lambda x^1 + (1-\lambda)x^2) < g(x^1), \quad \forall i \in I, \ \forall \lambda \in (0,1).$$

Theorem 5.3. *The following implications hold:*
$(Strict)_{in} \Rightarrow (S)_{in} \Rightarrow (Cottle) \Rightarrow (A.H.U. - I)_{in} \Leftarrow (R)_{in}$.

Proof. $(Strict)_{in} \Rightarrow (S)_{in}$: By assumption there exist two distinct points $x^1, x^2 \in K$ with $g_i(x^2) \le g_i(x^1), i \in I$, such that

$$g_i(\lambda x^1 + \lambda\eta(x^1, x^2)) < g_i(x^1) \le 0, \quad \forall \lambda \in (0,1).$$

$(S)_{in} \Rightarrow (Cottle)$: For each $i \in I$, we have

$$0 > g_i(\bar{x}) = g_i(\bar{x}) - g_i(x^0) \ge \eta(\bar{x}, x^0)^T \nabla g_i(x^0).$$

Therefore, $\eta(\bar{x}, x^0) \in L_0$.
$(Cottle) \Rightarrow (A.H.U. - I)_{in}$: The implication holds trivially.
$(R)_{in} \Rightarrow (A.H.U. - I)_{in}$: As every function $-g_i, i \in I$, are invex at x^0, the system appearing in (a) reduces to $y^T \nabla g_i(x^0) \le 0, i \in I$, system that obviously always admits the solution $y = 0$.

□

The previous theorem generalizes to invex functions some implications which are well known for the original definitions of those CQ's involving generalized convexity. For this last case (i.e., the non invex case) it is also true that it holds $(A.H.U. - I) \Rightarrow (Z)$.

Now, the implication $(A.H.U. - I)_{in} \Rightarrow (Z)$ does not hold, as shown by the following counter example.

Example 5.4. Let $C = R^2, x^0 = (0,0)$; let $g = (g_1, g_2)$, with $g_1(x,y) = -x - y + y^2$; $g_2(x,y) = x + y - y^2$. It is easy to verify that $-g(x,y)$ is invex at $x^0 = (0,0)$ with respect to the function $\eta = (\eta_1, \eta_2) = (-g_1, 0)$. In this case the $(A.H.U. - I)_{in}$ CQ is satisfied, as every point (x,y), with $y = -x$, is solution of the system $y \nabla g_i(x^0) \le 0, i = 1,2$. On the contrary, Zangwill CQ is not satisfied, as the linearizing cone L is given by all the points (x,y) such that $y = -x$, but the cone of feasible directions D is empty.

We give now a sufficient condition for the validity of implication: $(A.H.U. - I)_{in} \Rightarrow (Z)$.

Theorem 5.5. *If $(A.H.U.-I)_{in}$ holds with $-g_i, i \in I$, invex at x^0 with respect to some $\eta(x, x^0)$ whose range contains $L_{1,in}$, then also Zangwill CQ holds.*

Proof. For $i \notin I$ and for $i \in I - I_1$ the proof is same as in the modified case: $(A.H.U. - I)_{in} \Rightarrow (Z)$. Now, let $i \in I_1$ and $y \in L_{1,in}$. As the last set is contained in the range of $\eta(x, x^0)$, we have $y = \eta(x, x^0)$ for some $x \in C$. Moreover, we recall that if a function f is invex at x^0, then it is also pseudo-invex at x^0, i.e., $x \in C, \eta(x, x^0)^T \nabla f(x^0) \geq 0 \Rightarrow f(x) \geq f(x^0)$. Then, thanks to the pseudo-invexity of $-g_i, i \in I$, from the inequality $\lambda\eta(x, x^0)^T \nabla g_i(x^0) \leq 0$, we obtain

$$g_i(x^0 + \lambda\eta(x, x^0)) \leq g_i(x^0) = 0, \quad \forall\lambda > 0.$$

This implies that $x^0 + \lambda\eta(x, x^0) = x^0 + \lambda y$ is such that $y \in D$.

□

If we want to preserve the whole validity of the previous implications also in the invex case, on the grounds of Theorem 5.5, we have to modify also $(R)_{in}$, by adding the same assumption made in Theorem 5.5 on the range of the function η. So, we can assure that all the new weaker definitions introduced are true constraint qualifications.

We have seen that invexity plays a non-trivial role in the sufficiency of the Kuhn–Tucker optimality conditions for (P) and also, through constraint qualifications, in the necessity of the same conditions. Martin [145] and Hanson and Mond [88] observed that by weakening slightly the requirement that all g_i be invex (with respect to a same kernel η) the modified invexity (by itself) not only remains sufficient for a Kuhn–Tucker point to be optimal for (P), but also becomes necessary as well.

Following Martin [145], let us denote the invexity of f and g with respect to the same function $\eta(x, u)$ as HC-invexity (i.e., invexity in the sense of Hanson and Craven).

HC-invexity: There exists $\eta : C \times C \to R^n$ such that, $\forall x, u \in C$

$$f(x) - f(u) - \eta(x, u)^T \nabla f(u) \geq 0$$
$$g_i(x) - g_i(u) - \eta(x, u)^T \nabla g_i(u) \geq 0, \quad i = 1, \ldots, m.$$

Kaul and Kaur [114] proved that a pointwise version of the previous condition is enough for Kuhn–Tucker sufficiency. So we can speak also of KK-invexity for (P):

KK-invexity at $u \in C$: There exists $\eta : C \times C \to R^n$ such that for any $x \in C$

$$f(x) - f(u) - \eta(x, u)^T \nabla f(u) \geq 0$$

$$g_i(x) - g_i(u) - \eta(x, u)^T \nabla g_i(u) \geq 0, \quad i = 1, \ldots, m.$$

Martin [145] pointed out that KK-invexity is not a generalized condition as it may at first sight appears to be (for problem (P)). As a matter of fact, if the constraint of (P) are linear and feasible set is bounded, then the problem is HC-invex if and only if the objective function is actually convex.

Therefore Martin [145] introduced some relaxations in the previous conditions. He denoted this weakened condition as KT-invexity (i.e., Kuhn–Tucker invexity).
KT-invexity: There exists $\eta : C \times C \to R^n$ such that, $\forall x, u \in K$ (the feasible set for (P)), we have

$$f(x) - f(u) - \eta(x,u)^T \nabla f(u) \geq 0$$
$$-\eta(x,u)^T \nabla g_i(u) \geq 0, i \in I_u, \quad \text{where,} \quad I_u = \{i : g_i(u) = 0\}.$$

The restriction to the active constraints is due to the role of complementary slackness property of a Kuhn–Tucker point. Moreover, since the sufficiency proof concerns only feasible points, another weakening has been introduced by requiring that the inequalities hold only in the feasible set of problem (P). Finally, the condition on the constraint functions is changed with respect to HC-invexity: indeed, as Mangasarian [143] proved for the convex case and Kaul and Kaur [114] for the invex case, a quasi-invexity assumption on the active constraints is sufficient. Martin [145] proved the following result.

Theorem 5.6. *Every Kuhn–Tucker point (i.e., satisfying relations (5.1)–(5.3)) of problem (P) is a global minimizer if and only if (P) is KT-invex.*

Proof. See Martin [145].

□

However, there is an open question: KT-invexity is trivially satisfied at every solution u of the problem (P), by letting $\eta(x,u) = 0$. This is a tautological condition: it is coincident with the definition itself of solution for (P). Hanson and Mond [88] were the first to study the problem of finding necessary optimality conditions of invex type that were no trivial, i.e., with $\eta(x,u)$ not identically zero for each feasible x. Hanson and Mond [88] introduced Type I invexity, a pointwise weakened notion of invexity for problem (P), which we denote here as HM-invexity (Hanson and Mond [88] introduced also Type II invexity for the study of the dual problem of (P)).
HM-invexity or Type I invexity at $u \in K$: There exists $\eta : C \to R^n$ such that, $\forall x \in K$, we have

$$f(x) - f(u) - \eta(x,u)^T \nabla f(u) \geq 0$$
$$-\eta(x,u)^T \nabla g_i(u) \geq 0, \quad i \in I_u.$$

HM-invexity at $u \in K$ is sufficient to obtain that the Kuhn–Tucker conditions imply optimality at u. This result is a simple consequence of the definition of HM-invexity and the Kuhn–Tucker conditions. Note that η depends only on x. We have therefore the following theorem.

Theorem 5.7. *Let (P) be a HM-invex problem at $u \in K$. Then u solves (P) whenever u is a Kuhn–Tucker point for problem (P).*

Proof. See Hanson and Mond [88].

□

Hanson and Mond [88] also proved that, under an additional qualification on the constraints, a vector η must exist, which is not identically zero.

Theorem 5.8. *Let u be an admissible point for (P) with card$I_u < n$. If u is a solution of problem (P), then (P) is HM-invex with respect to a function η not identically zero for each $x \in K$.*

Proof. See Hanson and Mond [88].

□

The following result is a trivial consequence of Theorems 5.7 and 5.8.

Theorem 5.9. *If $u \in K$ is a Kuhn–Tucker point for (P) and card$I_u < n$, then u solves (P) if and only if (P) is HM-invex at u with respect to a function η, not identically zero for each $x \in K$.*

The condition card$I_u < n$ cannot in general be relaxed, as it is shown by the following example, where $m = n = 1$:

Example 5.10. Minimize x, subject to $1 - x \leq 0$.
The point $u = 1$ is a Kuhn–Tucker point that solves the problem, but if HM-invexity is imposed at $u = 1$, we get $\eta(1) = 0$, in contradiction with the conclusion of the above theorem.

Now we want to weaken the modified invexity condition in the following sense: We will no more assume the existence of a common η for all the functions appearing in problem (P). We still have to impose invexity of all the functions involved but we consider a weaker relationship between the η functions.

First of all, we recall Theorem 2.22 in its pointwise version:
Let $x, u \in C, C \subseteq R^n$ open set, $\mu : C \times C \to R^n$ and let $\Lambda(x) \subseteq R^n$ be a cone with vertex at zero, whose polar is denoted by $\Lambda^*(x)$.

Theorem 5.11 (Pointwise version of Theorem 2.22). *The function $f : C \to R$ is invex at $u \in C$ with respect to $\eta : C \to R^n$ such that $\eta(x) \in \mu(x) + \Lambda(x)$ if and only if the following implication holds, for each $x \in C$:*

$$\nabla f(u) \in \Lambda^*(x) \Rightarrow f(x) - f(u) - \mu(x)^T \nabla f(u) \geq 0.$$

We will consider the case $\Lambda(x) = \Lambda$, for a fixed $u \in C$. For some fixed $u \in C$, let $\Lambda_g = \Lambda_g(u)$ be the polyhedral cone generated by the vectors $-\nabla g_i(u), i = 1, \ldots, m$ and let $\Lambda_f^i = \Lambda_f^i(u)$ be the polyhedral cone generated by the vectors $-\nabla f(u), -\nabla g_j(u), j \neq i$.

Let us introduce the following generalized version of invex problem.
Generalized KT-invexity at $u \in K$: There exists $\mu : C \to R^n$ such that, $\forall x \in C$ we have

$$f(x) - f(u) - \eta_0(x,u)^T \nabla f(u) \geq 0$$
$$-\eta_i(x,u)^T \nabla g_i(u) \geq 0, \quad i \in I_u,$$

where $\eta_0(x,u) \in \alpha_0 \mu(x) + \Lambda_g^*$ and $\eta_i(x,u) \in \alpha_i \mu(x) + \Lambda_{fi}^*, \alpha_i > 0, i \in \{0\} \bigcup I_u$.

We can prove that under generalized KT-invexity assumptions the Kuhn–Tucker conditions are sufficient for optimality in problem (P).

Theorem 5.12. *Let (P) be a generalized KT-invex problem at $u \in K$. Then, whenever u is a KT-point, u solves (P).*

Proof. If u is a KT-point, there exists $\lambda \geq 0$ such that

$$\nabla f(u) + \sum_{i \in I_u} \lambda_i \nabla g_i(u) = 0.$$

Then, we have

$$\nabla f(u) = -\sum_{i \in I_u} \lambda_i \nabla g_i(u) \in \Lambda_g;$$

it follows that, $\forall x \in K$ we have

$$f(x) - f(u) \geq \alpha_0 \mu(x,u)^T \nabla f(u).$$

Moreover, for each $i \in I_u$ such that $\lambda_i \neq 0$, we have

$$\nabla g_i(u) = -\frac{1}{\lambda_i} \nabla f(u) - \sum_{j \neq i} \frac{\lambda_j}{\lambda_i} \nabla g_i(u) \in \Lambda_f^i.$$

Then, we have

$$g_i(x) - g_i(u) \geq \alpha_i \mu(x,u)^T \nabla g_i(u),$$

and therefore,

$$\frac{f(x) - f(u)}{\alpha_0} + \sum_{i \in I_u} \frac{\lambda_i}{\alpha_i} (g_i(x) - g_i(u)) \geq \mu(x,u)^T \nabla f(u) + \sum_{i \in I_u} \lambda_i \nabla g_i(u) = 0.$$

It follows that $f(x) \geq f(u), \forall x \in K$.

□

So, taking Theorem 5.1.7 of Hanson and Mond [88] into account, we have proved the following result.

Theorem 5.13. *Let $x \in K$ and let $cardI_u < n$. Then (P) is a generalized KT-invex problem at $u \in K$, with $\eta_i(x,u) \neq 0, \forall x \in K, i \in \{0\} \bigcup I_u$, if and only if whenever u is a KT-point, u solves (P).*

Other considerations on Type I invexity (i.e., HM-invexity) and Type II invexity are made by Rueda and Hanson [214]. For example, these authors give sufficient conditions for HM-invexity, here explored in the following result (note the similarity with the result setting that a differentiable preinvex function is invex).

Theorem 5.14. *If $f : C \to R$ and $g : C \to R^m$ are differentiable at $x^0 \in C$ and there exists an n-dimensional vector function η such that*

$$f\left((x^0 + \lambda \eta(x)\right) \leq \lambda f(x) + (1-\lambda) f(x^0), \quad 0 \leq \lambda \leq 1,$$

and

$$g\left((x^0) + \beta \eta(x)\right) \leq (1-\beta) g(x^0), \quad 0 \leq \beta \leq 1,$$

for all $x \in K$, then $f(x)$ and $g(x)$ are HM-invex at x^0.

Proof. See Rueda and Hanson [214].

□

Moreover, Rueda and Hanson [214] introduced the notion of pseudo-type I (pseudo-type II) invexity and quasi-type I (quasi-type II) invexity.

Another generalization which avoids the assumption of a common η for the functions involved in (P) is presented (for a multiobjective programming problem) by Jeyakumar and Mond [105]. Following Jeyakumar and Mond [105], a vector function $f : C \to R^p$ is said to be V-invex if there exist functions $\mu : C \times C \to R^n$ and $\alpha_i : C \times C \to R^+ - \{0\}$ such that for each $x, u \in C$ and for $i = 1, \ldots, p$,

$$f_i(x) - f_i(u) - \alpha_i(x,u)\mu(x,u)^T \nabla f_i(u) \geq 0.$$

With $p = 1$ the definition of V-invexity reduces to the usual definition of invexity, by setting $\eta(x,u) = \alpha_1(x,u)\mu(x,u)$. With $p = 1$ and $\mu(x,u) = x - \mu$, the definition reduces to strong pseudo convexity (see, [169]). With $p > 1$ the above notion states that it must be possible that the kernel η is given by the product of a common part $\mu(x,u)$ and a not necessarily common part $\alpha_i(x,u) > 0$.

These authors similarly introduced the notion of V-pseudo-invexity and V-quasi-invexity and obtain optimality and duality results for a Multiobjective Optimization Problem (see also [70, 163]).

Apart from all these special structures, we stress again that in the proof of optimality theorems (and also duality theorems-see Sect. 5.3) it is required that the functions involved in (P) are to be invex with respect to a common η (or that the whole Lagrangian function is invex). Taking these considerations into account, and following the paper of Phu [200], not always invexity is a true and genuine generalization of convexity. Indeed, consider the following functions:
$\phi(x) = x_1 - x_2^2$, $\phi(x) = -x_1$, $x = (x_1, x_2) \in R^2$. These functions are of course differentiable on R^2, with $\nabla\phi(x) = (1, -2x_2)$ and $\nabla\phi(x) = (-1, 0)$. For

$$\eta(x,y)^T = x - y + \mu(x,y)^T, \quad \mu(x,y)^T = \left(-(x_2 - y_2)^2, 0\right)$$

for all $x = (x_1, x_2) \in R^2$ and $y = (y_1, y_2) \in R^2$, we have

$$\begin{aligned}\eta(x,u)^T \nabla\phi(y) &= x_1 - y_1 - (x_2 - y_2)^2 - 2y_2(x_2 - y_2) \\ &= x_1 - y_1 - x_2^2 + y_2^2 \\ &= \phi(x) - \phi(y),\end{aligned}$$

which implies that both ϕ and $-\phi$ are invex on R^2 with respect to η.

Let us consider the problems

$$(P_1) \text{ Minimize } \phi(x), \text{ subject to } \varphi(x) \leq 0$$

and

$$(P_2) \text{ Minimize } \varphi(x), \text{ subject to } \phi(x) \leq 0.$$

For $x^* = (0,0) \in R^2$, we have $\phi(x^*) = \varphi(x^*) = 0$ and $\nabla\phi(x^*) + \nabla\varphi(x^*) = (1,0) + (-1,0) = (0,0)$.

Thus, for $\lambda^* = 1$, the Kuhn–Tucker conditions are satisfied for both problems. Since $\phi(x^*) = 0$ and $\phi(x) = -x_2^2 < 0$ for all $x \in R^2$ satisfying $\varphi(x) = -x_1 = 0$ and $x \neq x^*, x^*$ is not a local minimizer of problem (P_1).

Since $\varphi(x^*) = 0$ and $\varphi(x) = -x_1 < 0$ for all $x \in R^2$ satisfying $\phi(x) = x_1 - x_2^2 = 0$ and $x \neq x^*, x^*$ is not a local minimizer of problem (P_2).

We have therefore seen that the invexity of ϕ and convexity of φ are not sufficient for a Kuhn–Tucker point of (P_1) or (P_2) to be a global minimizer. This situation may occur even if φ is a strictly convex function. For instance, if $\phi(x) = -x_1 + \frac{1}{2}(x_1^2 + x_2^2)$ then $\varphi(x^*) = 0, \phi(x^*) + \varphi(x^*) = (1,0) + (-1,0) = (0,0)$ for $x^* = (0,0)$, and for all $x \in R^2$ satisfying $x_1 = x_2^2$ and $0 < x_1 < 1$, we have $-x_1 + x_2^2 < 0$ and $\varphi(x) = -x_1 + \frac{1}{2}(x_1^2 + x_2^2) = \frac{1}{2}(-x_1 + x_2^2) < 0$, i.e., the Kuhn–Tucker point $x^* = (0,0)$ is not a local (nor a global) minimizer of the problem (P_2). It is easy to show that this vector $x^* = (0,0)$, is also a Kuhn–Tucker point of the problem (P_1), which is not a local minimizer. Note that $\phi(x) = x_1 - x_2^2$ is invex with respect to the several η which are different from $\eta(x,y)^T = x - y + \mu(x,y)^T, \mu(x,y)^T = \left(-(x_2 - y_2)^2, 0\right)$. For instance, we can consider $\eta(x,y)^T = -\left(|x_1| + x_2^2 + |y_1|\right)\nabla\varphi(y)$ and for all $x = (x_1, y_1) \in R^2, y = (y_1, y_2) \in R^2$, we have

$$
\begin{aligned}
\phi(x) - \phi(y) &= x_1 - x_2^2 - y_1 + y_2^2 \\
&\geq -\left(|x_1| + x_2^2 + |y_1|\right) \\
&\geq -\left(|x_1| + x_2^2 + |y_1|\right)\left(1 + 4y_2^2\right) \\
&= \eta(y,x)^T \nabla\phi(y),
\end{aligned}
$$

which implies that ϕ is invex on R^2 with respect to this kernel function η. But there exists no function η such that both functions ϕ and φ are invex with respect to the same η.

5.2 A Sufficient Condition for Invexity Through the Use of the Linear Programming

Here we follow the constructive approach of Hanson and Rueda [89] to check the existence of a kernel function $\eta(x,u)$ in the nonlinear programming (P). We have already mentioned that Craven [43] has given necessary and sufficient conditions for a function f to be invex, assuming that the functions f and η are twice continuously differentiable (see, Sect. 2.2). From application point of view these conditions are difficult to apply. Hanson and Rueda [89] give a sufficient condition for the existence of $\eta(x,u)$ in problem (P), through the use of linear programming, which is direct and efficient. It will be assumed throughout this section that f is the objective function and g is the constraint vector function in problem (P) of the previous section. Assume further that f and every $g_i, i = 1, \ldots, m$, are twice continuously differentiable on $C \subseteq R^n$ and that for feasible points $x, u, \frac{1}{2}(x-u)^T \cdot \nabla f(z)(x-u)$ has a lower bound K_0 and $\frac{1}{2}(x-u)^T \cdot \nabla g_i(z)(x-u)$ has a lower bound K_i for $i = 1, \ldots, m$, where $z = \alpha u + (1-\alpha)x, 0 < \alpha < 1$. Since, by Taylor's theorem, we have

$$\frac{1}{2}(x-u)^T \cdot \nabla f(z)(x-u) = f(x) - f(u) - (x-u)^T \nabla f(u)$$

for some z, then for a given u, K_0 is easily found if lower bounds for $f(x)$ and x in the feasible set K are known. Similarly $K_i, i = 1, \ldots, m$, can be found. It will also be assumed that for some fixed value of u the gradient vectors $\nabla g_i(u), i = 1, \ldots, m$, are linearly independent. Then from the theory of generalized inverses of matrices (see, e.g., [206]), it follows that each i there exists a vector $z^i(u)$ such that

$$z^i(u)^T \nabla g_i(u) = 1 \tag{5.4}$$

and

$$z^i(u)^T \nabla g_j(u) = 0, \quad \forall j \neq i. \tag{5.5}$$

A generalized inverse of the matrix $\nabla g(u)$ can be used to obtain relations (5.4) and (5.5), but linear programming is better.

Theorem 5.15. *A sufficient condition for the existence of $\eta(x, u)$ at a given feasible point u is that*

$$K_0 \geq \sum_{j=1}^{m} K_j z^j(u)^T \nabla f(u). \tag{5.6}$$

Proof. By Taylor's theorem, we have

$$\begin{aligned} f(x) - f(u) &= (x-u)^T \nabla f(u) + \frac{1}{2}(x-u)^T \cdot \nabla f(z)(x-u) \\ &\geq (x-u)^T \nabla f(u) + K_0, \end{aligned} \tag{5.7}$$

and similarly

$$g_i(x) - g_i(u) \geq (x-u)^T \nabla g_i(u) + K_i, \quad i = 1, \ldots, m. \tag{5.8}$$

We now construct an appropriate kernel function $\eta(x, u)$. Let

$$\eta(x, u) = (x-u) + \sum_{j=1}^{m} K_j z^j(u). \tag{5.9}$$

Then

$$\begin{aligned}\eta(x,u)^T\nabla g_i(u) &= (x-u)^T\nabla g_i(u) + \sum_{j=1}^{m} K_j z^j(u)^T\nabla g_i(u)(i=1,\ldots,m)\\ &= (x-u)^T\nabla g_i(u) + K_i(i=1,\ldots,m)\\ &\leq g_i(x) - g_i(u), i=1,\ldots,m.\end{aligned}$$

So every $g_i, i = 1,\ldots,m$, satisfies the definition of invexity with respect to $\eta(x,u)$. Now

$$\begin{aligned}\eta(x,u)^T\nabla f(u) &= (x-u)^T\nabla f(u) + \sum_{j=1}^{m} K_j z^j(u)^T\nabla f(u) \quad \text{by (5.9)}\\ &\leq (x-u)^T\cdot\nabla f(u) + K_0 \quad \text{(by (5.6))}\\ &\leq f(x) - f(u) \quad \text{(by (5.7))},\end{aligned}$$

which satisfies the definition of invexity with respect to η for f. So $\eta(x,u)$ exists for problem (P).

□

In general (again assuming u fixed), the vectors $z^i(u)$ are not unique. We can find the best $z^i(u)$ in the above theorem, by use of linear programming, namely, by solving the following problem:
Minimize $\sum_{j=1}^{m} K_j z^j(u)^T\nabla f(u)$
subject to $z^j(u)^T\nabla g_i(u) = 1, i = 1,\ldots,m$,
and $z^j(u)^T\nabla g_j(u) = 0, j = 1,\ldots,m, j \neq i$.

One approach to finding values for K_0 and $K_i, i = 1,\ldots,m$, is through the use of eigenvalues. If μ_0 is the smallest eigenvalue of $\nabla f(z)$, then $(x-u)^T\nabla f(z)(x-u) \geq \mu_0 \|x-u\|^2$.

Similarly one could find smallest eigenvalues μ_i for the Hessian matrices $\nabla g_i(z), i = 1,\ldots,m$.

Corollary 5.16. *A sufficient condition for the existence of $\eta(x,u)$ in the problem (P) at a given feasible point u is that*

$$\mu_0 \geq \sum_{j=1}^{m} \mu_j z^j(u)^T\nabla f(u). \tag{5.10}$$

Again linear programming can be used to find the best vector $z^i(u)$ in this expression, namely

$$\text{Minimize} \sum_{j=1}^{m} \mu_j z^j(u)^T\nabla f(u)$$

$$\text{Subject to } z^i(u)^T\nabla g_i(u) = 1, \quad i = 1,\ldots,m,$$

and

$$z^i(u)^T \nabla g_j(u) = 0, \quad i = 1, \dots, m, j \neq i.$$

If we desire to test the optimality of a feasible point u, we have, by the Kuhn–Tucker theorem, if u is optimal and a constraint qualification is satisfied, that there exists some vector $\lambda \in R^m$ such that

$$\nabla f(u) + \lambda^T \nabla g(u) = 0,$$

$$\lambda^T g(u) = 0, \lambda \geq 0.$$

That is, $\nabla f(u) = -\sum_{i=1}^m \lambda_i \nabla g_i(u)$. So we have

$$\sum_{j=1}^m K_j z^j(u)^T \nabla f(u) = \sum_{j=1}^m K_j z^j(u)^T \left[-\sum_{i=1}^m \lambda_i \nabla g_i(u) \right] = -\sum_{j=1}^m K_j \lambda_j$$

and (5.6) becomes

$$K_0 \geq -\sum_{j=1}^m K_j \lambda_j. \tag{5.11}$$

Similarly (5.10) becomes

$$\mu_0 \geq -\sum_{j=1}^m \mu_j \lambda_j. \tag{5.12}$$

So, if u is optimal, conditions (5.11) or (5.12) will imply the existence of $\eta(x, u)$ for the problem (P), without the need for calculating the vectors $z^j(u), i = 1, \dots, m$.

5.3 Characterization of Solution Sets of a Pseudolinear Problem

We consider again the contribution of Ansari et al. [3] who take into account the following problem:

$(\bar{P})$

$$\text{Minimize } f(x)$$

$$\text{Subject to } x \in K,$$

where $f : C \to R, C$ is an open subset of R^n and K is an η-convex subset of C. Moreover, we assume that the solution set of $(\bar{P})$, $\bar{S} = \text{argmin}_{x \in K} f(x)$ is non-empty.

Theorem 5.17. *If f is a preinvex function on K, then the solution set $\bar{S}$ of $(\bar{P})$ is an η-convex set.*

Proof. Let $x^1, x^2 \in \bar{S}$. Then $f(x^1) \leq f(y)$ and $f(x^2) \leq f(y)$ for all $y \in K$. Since f is preinvex, we have

$$\begin{aligned} f(x^1 + \lambda\eta(x^1, x^2)) &\leq \lambda f(x^1) + (1-\lambda)f(x^2), \quad \forall \lambda \in [0,1] \\ &\leq \lambda f(y) + (1-\lambda)f(y) \\ &= f(y). \end{aligned}$$

Hence $x^1 + \lambda\eta(x^1, x^2) \in \bar{S}$, and so, $\bar{S}$ is an η-convex set.

□

From the proof of Theorem 3.4, it is easy to show that the solution set $\bar{S}$ of the $(\bar{P})$ is η-convex if $f : C \to R$ is η-pseudolinear, where $\eta : K \times K \to R^n$ satisfies condition C of Mohan and Neogy [165].

Now we state the first order characterization of the solution set of a η-pseudolinear program, in terms of any of its solutions. This may be viewed as a generalization of results of Jeyakumar and Yang [107].

Theorem 5.18. *Let $f : C \to R$ be differentiable on an open set C and let f be η-pseudolinear on an open η-convex subset $K \subseteq D$ where η satisfies condition C (Definition 3.3) and moreover, $\eta(x,y) + \eta(y,x) = 0, \forall x,y \in K$. Let $\bar{x} \in \bar{S}$. Then $\bar{S} = \tilde{S} = \hat{S}$, where*

$$\tilde{S} = \{x \in K : \eta(\bar{x}, x)^T \nabla f(x) = 0\}$$

$$\hat{S} = \{x \in K : \eta(\bar{x}, x)^T \nabla f(\bar{x}) = 0\}.$$

Proof. The point $x \in \bar{S}$ if and only if $f(x) = f(\bar{x})$. By Theorem 3.4, we have $f(x) = f(\bar{x})$ if and only if $\eta(\bar{x}, x)^T \nabla f(x) = 0$. Also $f(x) = f(\bar{x})$ if and only if $\eta(\bar{x}, x)^T \nabla f(\bar{x}) = 0$. The latter is equivalent to $\eta(x, \bar{x})^T \nabla f(\bar{x}) = 0$, since $\eta(\bar{x}, x) = -\eta(x, \bar{x})$.

□

Corollary 5.19. *Let f and η be the same as in Theorem 5.18. Then $\bar{S} = \tilde{S}_1 = \hat{S}_1$, where*

$$\tilde{S}_1 = \{x \in K : \eta(\bar{x}, x)^T \nabla f(x) \geq 0\},$$

$$\hat{S}_1 = \{x \in K : \eta(\bar{x}, x)^T \nabla f(\bar{x}) \geq 0\}.$$

Proof. It is clear from Theorem 5.18 that $\bar{S} \subseteq \tilde{S}_1$. We prove that $\tilde{S}_1 \subseteq \bar{S}$. Assume that $x \in \tilde{S}_1$, that is, $x \in K$ such that $\eta(\bar{x}, x)^T \nabla f(x) \geq 0$. In view of Theorem 3.5, there exists a function p defined on $K \times K$ such that $p(x, \bar{x}) > 0$ and

$$f(\bar{x}) = f(x) + p(x, \bar{x})\eta(\bar{x}, x)^T \nabla f(x) \geq f(x).$$

This implies that $x \in \bar{S}$, and hence $\tilde{S}_1 \subseteq \bar{S}$. Similarly we can prove that $\hat{S}_1 = \bar{S}$, using the identity $\eta(x, \bar{x}) = -\eta(\bar{x}, x)$.

□

Theorem 5.20. *In problem $(\bar{P})$, assume that f is differentiable on C and η-pseudolinear on an η-convex set $K \subseteq C$, where η satisfies condition C (Definition 3.3) and $\eta(x,y) + \eta(y,x) = 0, \forall x, y \in K$. If $\bar{x} \in \bar{S}$, then $\bar{S} = S^* = S_1^*$, where*

$$S^* = \{x \in K : \eta(\bar{x}, x)^T \nabla f(\bar{x})\},$$
$$S_1^* = \{x \in K : \eta(\bar{x}, x)^T \nabla f(\bar{x}) \geq \eta(x, \bar{x})^T \nabla f(x)\}.$$

Proof. (1) $\bar{S} \subseteq S^*$. Let $x \in \bar{S}$. It follows from Theorem 5.18 that

$$\eta(\bar{x}, x)^T \nabla f(x) = 0 = \eta(\bar{x}, x)^T \nabla f(\bar{x}).$$

Since $\eta(\bar{x}, x) = -\eta(x, \bar{x})$, we have

$$\eta(x, \bar{x})^T \nabla f(x) = 0 = \eta(\bar{x}, x)^T \nabla f(\bar{x}).$$

Thus $x \in S^*$, and hence $\bar{S} \subseteq S^*$.
(2) $S^* \subseteq S_1^*$ is obvious.
(3) $S_1^* \subseteq \bar{S}$. Assume that $x \in S_1^*$. Then $x \in K$ satisfies

$$\eta(\bar{x}, x)^T \nabla f(\bar{x}) \geq \eta(x, \bar{x})^T \nabla f(x). \tag{5.13}$$

Suppose that $x \notin \bar{S}$. Then $f(x) > f(\bar{x})$. By pseudo-invexity of $-f$, we have

$$\eta(x, \bar{x})^T \nabla f(\bar{x}) > 0.$$

Since $\eta(\bar{x}, x) = -\eta(x, \bar{x})$, we have

$$\eta(\bar{x}, x)^T \nabla f(\bar{x}) < 0.$$

Using (5.13), we have

$$\eta(x, \bar{x})^T \nabla f(x) < 0 \text{ or } \eta(\bar{x}, x)^T \nabla f(x) > 0.$$

In view of Theorem 3.5, there exists a function p defined on $K \times K$ such that $p(x, \bar{x}) > 0$, and

$$f(\bar{x}) = f(x) + p(x, \bar{x}) \eta(\bar{x}, x)^T \nabla f(x) > f(x),$$

a contradiction. Hence $x \in \bar{S}$.

□

5.4 Duality

Hanson [83] demonstrated that invexity of f and $g_i, i = 1, \ldots, m$, with respect to a common η was also sufficient for weak and strong duality to hold between the primal problem (P) and its Wolfe dual, where the Wolfe dual is given by Wolfe [247]:

(WD)

$$\begin{aligned} &\text{Maximize}_{u,\lambda} f(u) + \lambda^T g(u) \\ &\text{Subject to } \nabla f(u) + \nabla \left(\lambda^T g(u)\right) = 0 \\ &\lambda \geq 0. \end{aligned}$$

More precisely, we have the following results.

Theorem 5.21 (Strong duality). *Under the condition of a suitable constraint qualification for (P), if x^0 is minimal in the primal problem (P), then (x^0, λ^0) is maximal in the dual problem (WD), where λ^0 is given by the Kuhn–Tucker conditions and f and $g_i, i = 1, \ldots, m$, are all invex with respect to a common η. Moreover, the extremal values are equal in the two problems.*

Proof. Let (u, λ) be any vector feasible for (WD). Then

$$\begin{aligned} &\left(f(x^0) + {\lambda^0}^T g(x^0)\right) - \left(f(u) + \lambda^T g(u)\right) \\ &\quad = f(x^0) - f(u) - \lambda^T g(u) \\ &\quad \geq \eta(x^0, u)^T \nabla f(u) - \lambda^T g(u) \\ &\quad = -\eta(x^0, u)^T \lambda^T \nabla g(u) - \lambda^T g(u) \\ &\quad \geq -\lambda^T g(x^0) \\ &\quad \geq 0. \end{aligned}$$

So (x^0, λ^0) is maximal in the dual problem, and since ${\lambda^0}^T g(x^0) = 0$, the extrema of the two problems are equal.

□

Theorem 5.22 (Weak duality). *Let x be feasible for (P) and (u, λ) be feasible for (WD) and let f and $g_i, i = 1, \ldots, m$, be all invex with respect to a common η. Then we have $f(x) \geq f(u) + \lambda^T g(u)$.*

Proof. From the invexity assumptions, we have

$$f(x) - f(u) - \eta(x, u)^T \nabla f(u) + \lambda^T \left(g(x) - g(u) - \eta(x, u)^T \nabla g(u)\right) \geq 0.$$

By regrouping terms, and taking into account the feasibility assumptions, we obtain

$$\begin{aligned} f(x) - \left(f(u) + \lambda^T g(u)\right) &\geq \left(\nabla f(u) + \lambda^T \nabla g(x)\right) \eta(x, u) - \lambda^T g(x) \\ &= \lambda^T g(u) \\ &\geq 0, \end{aligned}$$

which establishes the weak duality.

□

Jeyakumar [100] considered the Wolfe dual to problem (P) and, by means of generalized invexity, obtained various duality theorems. Martin [145] employed invexity to derive conditions necessary and sufficient for weak duality to hold between (P) and (WD).

The problem (P) is said to be weak duality invex if there exists a function $\eta : C \times C \to R^n$ such that, for all $x, u \in C$, with $g(x) \leq 0$, it holds:
either

$$f(x) - f(u) - \eta(x,u)^T \nabla f(u) \geq 0,$$

and

$$-g_i(u) - \eta(x,u)^T \nabla g_i(u) \geq 0, \quad i = 1, \dots, m;$$

or

$$-\eta(x,u)^T \nabla f(u) > 0,$$

and

$$-\eta(x,u)^T \nabla g_i(u) \geq 0, \quad i = 1, \dots, m.$$

It is shown in [145] that weak duality holds between programs (P) and (WD) if and only if (P) is weak duality invex. Again, as with Kuhn–Tucker sufficiency, the idea of invexity has allowed a characterization of weak duality for the Wolfe dual in terms of some function η. Other considerations on duality theorems by definition of Type II invexity, are made by Hanson and Mond [88]. Here, following Mond and Smart [179], we state two converse dual results. The first of these requires that f and g be twice differentiable and does not presuppose the existence of an optimal solution. It is based on the converse duality theorem of Huard [95]. The second theorem is based on Mangasarian [142], assumes the existence of an optimal solution and imposes a stronger invexity assumption.

Theorem 5.23. *Assume that f and g are twice differentiable, and that f and $g_i, i = 1, \dots, m$, are invex with respect to a common kernel function η. Let $(\bar{x}, \bar{\lambda})$ be optimal for (WD), and assume that $\nabla f(\bar{x}) + \nabla(\bar{\lambda}^T g(\bar{x}))$ is nonsingular. Then $\bar{x}$ is optimal for (P).*

Proof. From Huard [95], it follows that $\bar{x}$ is feasible for (P) and $\bar{\lambda}^T g(\bar{x}) = 0$. The invexity hypothesis implies that weak duality holds, so $\bar{x}$ is optimal for (P).

□

The next theorem requires the notion of strict invexity at a point.

Definition 5.24. *Let $f : C \to R$ be invex with respect to some function $\eta : C \times C \to R^n$; f is said to be strictly invex at $\bar{x}$ if*

$$f(x) - f(\bar{x}) > \eta(x, \bar{x})^T \nabla f(\bar{x}), \quad \forall x \in C, \quad x \neq \bar{x}.$$

Let $f : C \to R$ be pseudo-invex with respect to some function $\eta : C \times C \to R^n$; f is said to be strictly pseudo-invex at $\bar{x}$ if

$$\eta(x, \bar{x})^T \nabla f(\bar{x}) \geq 0 \Rightarrow f(x) > f(\bar{x}), \quad \forall x \in C, \quad x \neq x.$$

Theorem 5.25 (Strict converse duality). *Assume f and $g_i, i = 1, \ldots, m$, are invex with respect to a common kernel function η. Let x^* be optimal for (P) and $(\bar{x}, \bar{\lambda})$ be optimal for (WD). If a constraint qualification is satisfied for (P) and f is strictly invex for (P) at $\bar{x}$, then $x^* = \bar{x}$.*

Proof. We proceed by contradiction. Assume that $x^* \neq \bar{x}$. By Theorem 5.21 there exists λ^* such that (x^*, λ^*) is optimal for (WD). Hence

$$f(x^*) = f(x^*) + \lambda^* g(x^*) = f(\bar{x}) + \bar{\lambda}^T g(\bar{x}). \tag{5.14}$$

Now strict invexity of f at $\bar{x}$ gives

$$f(x^*) - f(\bar{x}) > \eta(x^*, \bar{x})^T \nabla f(\bar{x}), \tag{5.15}$$

and invexity of $g_i, i = 1, \ldots, m$, with $\bar{\lambda} \geq 0$, gives

$$\bar{\lambda}^T g(x^*) - \bar{\lambda}^T g(\bar{x}) \geq \eta(x^*, \bar{x})^T \nabla (\bar{\lambda}^T g(\bar{x})). \tag{5.16}$$

Adding (5.15) and (5.16), we obtain

$$f(x^*) - f(\bar{x}) + \bar{\lambda}^T g(x^*) - \bar{\lambda}^T g(\bar{x}) > \eta(x^*, \bar{x})^T \left(\nabla f(\bar{x}) + \nabla \left(\bar{\lambda}^T g(\bar{x})\right)\right) = 0,$$

as $(\bar{x}, \bar{\lambda})$ is feasible for (WD).
But, as $\bar{\lambda}^T g(x^*) \leq 0$, then

$$f(x^*) - f(\bar{x}) - \bar{\lambda}^T g(\bar{x}) > 0,$$

which contradicts (5.14). Therefore, $x^* = \bar{x}$.

□

Now, following Mond and Smart [179] we introduce the dual formulation for (P) introduced by Mond and Weir [180]. The original version of the Mond–Weir dual to (P) is defined as follows:

$$\text{Maximize } f(u)$$

$$\text{Subject to } \nabla f(u) + \lambda^T g(u) = 0$$

$$\lambda^T g(u) \geq 0, \quad \lambda \geq 0.$$

Several duality theorems with respect to this formulation are expounded by Giorgi and Molho [75]. The advantage of this formulation over the Wolfe dual is that the objective function of the dual model is the same as in the primal problem and that the duality theorems are achieved by means of further relaxations on the invexity requirements.

However, the most general form of the Mond–Weir dual is obtained by partitioning the set $M = \{1, \ldots, m\}$ into $r + 1$ subsets $I_0, I_1, \ldots, I_r, (r \leq m - 1)$, such that $I_\alpha \bigcap I_\beta = \phi, \alpha \neq \beta$ and $\bigcup_{\alpha=0}^{r} I_\alpha = M$.

The Mond–Weir general dual problem is now:
(MWD)

$$\text{Maximize } f(u) + \sum_{i \in I_0} \lambda_i g_i(u)$$

$$\text{Subject to } \nabla f(u) + \nabla \left(\lambda^T g(u)\right) = 0 \tag{5.17}$$

$$\lambda \geq 0 \tag{5.18}$$

$$\sum_{i \in I_\alpha} \lambda_i g_i(u) \geq 0, \quad \alpha = 1, \ldots, r. \tag{5.19}$$

We remark that if $I_0 = M, r = 1$ and $I_1 = \phi$, then (MWD) reduces to the Wolfe dual. If $I_0 = \phi, r = 1$ and $I_1 = M$, then (MWD) yields the previous version of the Mond–Weir dual.

A salient property of the Mond–Weir dual, and possibly one reason for its inception, is that weak duality between (P) and (WD) implies weak duality between (P) and (MWD). If we assume that weak duality for the Wolfe dual holds and that x is feasible for (P) and (u, λ) is feasible for (MWD), then as (u, λ) must also be feasible for (WD), we have

$$\begin{aligned} f(x) &\geq f(u) + \lambda^T g(u) \\ &= f(u) + \sum_{i \in I_0} \lambda_i g_i(u) + \sum_{\alpha=1}^{r} \sum_{i \in I_\alpha} \lambda_i g_i(u) \\ &\geq f(u) + \sum_{i \in I_0} \lambda_i g_i(u), \end{aligned}$$

which gives weak duality for the Mond–Weir dual.

Weak duality between (P) and (MWD) is easily established under pseudo-invexity and quasi-invexity assumptions.

Theorem 5.26 (Weak duality). *If $f + \sum_{i \in I_0} \lambda_i g_i$ is pseudo-invex with respect to some $\eta : C \times C \to R^n$ and $\sum_{i \in I_0} \lambda_i g_i$, for $\alpha = 1, \ldots, r$, is quasi-invex with respect to the same $\eta : C \times C \to R^n$, for any $\lambda \in R^m_+$, then $inf(P) \geq sup(MWD)$.*

Proof. Let x be feasible for (P), and (u, λ) be feasible for (MWD). As $g(x) \leq 0$, then by (5.18) and (5.19)

$$\sum_{i \in I_0} \lambda_i g_i(x) - \sum_{i \in I_\alpha} \lambda_i g_i(u) \leq 0, \quad \alpha = 1, \ldots, r.$$

As $\sum_{i\in I_\alpha} \lambda_i g_i$ is quasi-invex with respect to $\eta, \alpha = 1, \ldots, r$, then

$$\eta(x,u)^T \nabla \left(\sum_{i\in I_\alpha} \lambda_i g_i(u) \right) \leq 0, \quad \alpha = 1, \ldots, r.$$

Summing over α, we have

$$\eta(x,u)^T \nabla \left(\sum_{i\notin I_0} \lambda_i g_i(u) \right) \leq 0.$$

But by (5.17)

$$\nabla \left(\sum_{i\notin I_0} \lambda_i g_i(u) \right) = \nabla f(u) + \nabla \left(\sum_{i\in I_0} \lambda_i g_i(u) \right).$$

Therefore,

$$\eta(x,u)^T \left(\nabla f(u) + \nabla \left(\sum_{i\in I_0} \lambda_i g_i(u) \right) \right) \leq 0.$$

Since $f + \sum_{i\in I_0} \lambda_i g_i$ is pseudo-invex with respect to η, then

$$f(x) + \sum_{i\in I_0} \lambda_i g_i(x) \geq f(u) + \sum_{i\in I_0} \lambda_i g_i(u).$$

Now, $\lambda_i \geq 0$ and $g_i(x) \leq 0, \forall i \in I_0$, so $\sum_{i\in I_0} \lambda_i g_i(x) \leq 0$, and hence

$$f(x) \geq f(u) + \sum_{i\in I_0} \lambda_i g_i(u).$$

As x and (u, λ) were arbitrary feasible solutions of (P) and (MWD) respectively, then inf(P) $\geq$ sup(MWD).

□

Strong duality holds with the assumption of a suitable constraint qualification.

Theorem 5.27 (Strong duality). *Let x^* be optimal for (P), and assume the invexity assumptions of Theorem 5.26 are satisfied. Assume also that a suitable constraint qualification is satisfied for (P). Then there exists $\lambda^* \in R^m$ such that (x^*, λ^*) is optimal for (MWD), and the objective values are equal.*

Proof. As x^* is optimal for (P) and a suitable constraint qualification is satisfied, then there exists $\lambda^* \in R^m$ such that the Kuhn–Tucker conditions for (P) are satisfied. Since $\lambda^* \geq 0$ and $g(x^*) \leq 0$, then the complementarity relations $\lambda_i^* g_i(x^*), i = 1, \ldots, m$, implies that

$$\sum_{i\in I_\alpha} \lambda_i^* g_i(x^*) = 0, \quad \alpha = 0, 1, \ldots, r.$$

Hence, (x^*, λ^*) is feasible for (MWD), with

$$f(x^*) + \sum_{i \in I_0} \lambda_i^* g_i(x^*) = f(x^*),$$

which implies that (x^*, λ^*) is optimal for (MWD), by Theorem 5.26.

□

Similar to Wolfe duality, two types of converse dual theorems will be given here. The first generalizes that of Mond and Weir [180] to pseudo-invexity and quasi-invexity, whereas the second is a strict converse duality theorem in the spirit of Theorem 5.25.

Theorem 5.28. *Assume that f and g are twice differentiable and that the invexity assumptions of Theorem 5.26 hold. Let $(\bar{x}, \bar{\lambda})$ be optimal for (MWD). Assume also that $\nabla^2 f(\bar{x}) + \nabla^2 \left(\bar{\lambda}^T g(\bar{x})\right)$ is non-singular, and that the vectors $\sum_{i \in I_\alpha} \nabla \bar{\lambda} g_i(\bar{x}) = 0, \alpha = 1, \ldots, r$, are linearly independent. Then, $\bar{x}$ is optimal for (P).*

Proof. Mond and Weir [180] demonstrated that $\bar{x}$ is feasible for (P). Optimality of $\bar{x}$ follows by weak duality (Theorem 5.26).

□

Theorem 5.29 (Strict converse duality). *Assume the invexity assumptions of Theorem 5.26 hold. Let x^* be optimal for (P) and $(\bar{x}, \bar{\lambda})$ be optimal for (MWD). If a constraint qualification for (P) is satisfied and $f + \sum_{i \in I_0} \bar{\lambda}_i g_i$ is strictly pseudo-invex at $\bar{x}$, then $x^* = \bar{x}$.*

Proof. By contradiction. Assume that $x^* \neq \bar{x}$. From Theorem 5.27, there exists $\lambda^* \in R^m$ such that (x^*, λ^*) is optimal for (MWD). Thus

$$f(x^*) = f(x^*) + \sum_{i \in I_0} \lambda_i^* g_i(x^*) = f(\bar{x}) + \sum_{i \in I_0} \bar{\lambda}_i g_i(\bar{x}). \tag{5.20}$$

As $g(x^*) \leq 0$ and $\bar{\lambda} \geq 0$, then

$$\sum_{i \in I_\alpha} \bar{\lambda}_i g_i(x^*) - \sum_{i \in I_\alpha} \bar{\lambda}_i g_i(\bar{x}) \leq 0, \quad \alpha = 1, \ldots, r.$$

By quasi-invexity of $\sum_{i \in I_\alpha} \bar{\lambda}_i g_i, \alpha = 1, \ldots, r$, we have

$$\eta(x^*, \bar{x})^T \nabla \left(\sum_{i \notin I_0} \bar{\lambda}_i g_i(\bar{x}) \right) \leq 0.$$

By (5.17), this implies that

$$\eta(x^*, \bar{x})^T \left(\nabla f(\bar{x}) + \nabla \left(\sum_{i \in I_0} \bar{\lambda}_i g_i(\bar{x}) \right) \right) \leq 0.$$

Now, strict pseudo-invexity of $f + \sum_{i \in I_0} \bar{\lambda}_i g_i$ at $\bar{x}$ gives

$$f(x^*) + \sum_{i \in I_0} \bar{\lambda}_i g_i(x^*) > f(\bar{x}) + \sum_{i \in I_0} \bar{\lambda}_i g_i(\bar{x}).$$

But $\bar{\lambda} \geq 0$ and $g(x^*) \leq 0$ imply that $\sum_{i \in I_0} \bar{\lambda}_i g_i(x^*) \leq 0$, so

$$f(x^*) > f(\bar{x}) + \sum_{i \in I_0} \bar{\lambda}_i g_i(\bar{x}).$$

However, this contradicts (5.20), so $x^* = \bar{x}$.

□

Invexity can be used also to weaken the usual convexity assumptions for symmetric duality. A pair of program is said to be symmetric if, when the dual is cast in the form of the primal, its dual is the primal. The simplest and the classical example of this occurs for the case of linear programming. Dantzig et al. [53] presented a more general nonlinear symmetric dual pair in terms of bifunctions defined on $R^n \times R^m$.

Consider a differentiable bifunction $K : R^n \times R^m \to R$. Denote by $\nabla_1 K$ the gradient with respect to the first vector and $\nabla_2 K$ the gradient with respect to the second vector, so that $\nabla_1 K(x, y) \in R^n$ and $\nabla_2 K(x, y) \in R^m$. When K is twice differentiable, $\nabla_{11} K$ and $\nabla_{22} K$ denote the Hessians with respect to the first and second vectors, respectively.

The pair of programs given by Dantzig et al. [53] are:

(SP)

$$\text{Minimize } K(x, y) - y^T \nabla_2 K(x, y)$$

$$\text{Subject to } \nabla_2 K(x, y) \leq 0$$

$$x \geq 0$$

$$y \geq 0$$

(SD)

$$\text{Maximize } K(u, v) - u^T \nabla_1 K(u, v)$$

$$\text{Subject to } \nabla_1 K(u, v) \geq 0$$

$$u \geq 0$$

$$v \geq 0.$$

Weak duality between (SP) and (SD) was established under the assumption that K is convex-concave; that is, $K(\cdot, y)$ convex for each $y \in R^m$ and $K(x, \cdot)$ concave for each $x \in R^n$. A more general convexity requirement for weak duality, namely that $K(\cdot, y)$ is invex with respect to some function $\eta_1 : R^n \times R^n \to R^n$ (η_1 need not be independent of $y \in R^m_+$,) $-K(x, \cdot)$ is invex with respect to some function $\eta_2 : R^m \times R^m \to R^m$ (η_2 need not be independent of $x \in R^n_+$) and whenever (x, y) is feasible for (SP) and (u, v) is feasible for (SD),

$$\eta_1(x, u) + u \geq 0$$

and

$$\eta_2(v, y) + y \geq 0.$$

The invexity requirements may be weakened further by employing the approach of Mond and Weir [180], using the idea of Mond–Weir dual with $r = 1, I_0 = \phi$ and $I_1 = M$. This gives the pair of problems:

(MWSP)

$$\begin{aligned} &\text{Minimize } K(x, y) \\ &\text{Subject to } \nabla_2 K(x, y) \leq 0 \\ &y^T \nabla_2 K(x, y) \geq 0 \\ &x \geq 0 \end{aligned}$$

(MWSD)

$$\begin{aligned} &\text{Maximize } K(u, v) \\ &\text{Subject to } \nabla_1 K(u, v) \geq 0 \\ &u^T \nabla_1 K(u, v) \leq 0 \\ &v \geq 0. \end{aligned}$$

We have the following results, whose proofs are left to the reader.

Theorem 5.30 (Weak duality). *Assume that for each (u, v) is feasible for (MWSD) for some $u \in R^n, K(\cdot, v)$ is pseudo-invex with respect to some function $\eta_v : R^n \times R^n \to R^n$ satisfying $\eta_v(x, u) + u \geq 0$ whenever (x, y) is feasible for (MWSP). Assume also that for each $x \in R^n$ such that (x, y) is feasible for (MWSP) for some $y \in R^m, -K(x, \cdot)$ is pseudo-invex with respect to some function $\xi_x : R^m \times R^m \to R^m$ satisfying $\xi_x(v, y) + y \geq 0$ whenever (u, v) is feasible for (MWSD). Then inf(MWSP) $\geq$ sup(MWSD).*

Theorem 5.31 (Strong duality). *Assume the pseudo-invexity conditions of Theorem 5.30 hold. Let (x^*, y^*) be an optimal solution of (MWSP) and assume that $\nabla_{22} K(x^*, y^*)$ is non-singular and $\nabla_2 K(x^*, y^*) \neq 0$. Then (x^*, y^*) is an optimal solution for (MWSD), and the respective objective values are equal.*

For other treatments of symmetric duality with invexity, see, e.g., Nanda [189] and Nanda and Das [190]. Balas [11] examined the symmetric duality results of Dantzig et al. [53] when some primal and dual variables are constrained to belong to some arbitrary set for example, the set of integers.

Let R^n denote the n-dimensional Euclidean space and let R^n_+ be its non-negative orthant. Let $f(x, y)$ be a real valued thrice continuously differentiable function defined on an open set in $R^n \times R^n$. Let U and V be two arbitrary sets of integers in R^n and R^m, respectively. As in Balas [11] we constrain some of the components of x and y to belong to arbitrary sets of integers. Suppose the first $n_1(0 \leq n_1 \leq n)$ components of x belong to U and the first $m_1(0 \leq m_1 \leq m)$ components of y belong to V. Then we write $(x, y) = (x^1, x^2, y^1, y^2)$, where $x^1 = (x_1, x_2, \ldots, x_{n1}) \in U$ and $y^1 = (y_1, y_2, \ldots, y_{m1}) \in V$, x^2 and y^2 being the vectors of the remaining components of x and y, respectively. Let $\nabla_{x^2} K(\bar{x}, \bar{y})$ denote the gradient vector of $K(x, y)$ with respect to x^2 at $(\bar{x}, \bar{y})$. Also let $\nabla_{x^2 x^2} K(\bar{x}, \bar{y})$ denote the Hessian matrix with respect to x^2 evaluated at $(\bar{x}, \bar{y})$. $\nabla_{y^2} K(\bar{x}, \bar{y})$ and $\nabla_{y^2 y^2} K(\bar{x}, \bar{y})$ are defined similarly.

Let $s^1, s^2, \ldots, s^r$ be elements of an arbitrary vector space. A vector function $G(s^1, \ldots, s^r)$ will be called additively separable with respect to s^1 if there exist vector functions $H(s^1)$ (independent of $s^1, \ldots, s^r$) and $K(s^2, \ldots, s^r)$ (independent of s^1), such that

$$G(s^1, \ldots, s^p) \equiv H(s^1) + K(s^2, \ldots, s^p).$$

With $C_1 = R_+^{n-n_1}, C_2 = R_+^{m-m_1}$ and $\lambda = \mu = 1$, the symmetric dual formulation of Mishra et al. [152]:

$$\mathrm{Max}_{x^1} \mathrm{Min}_{x^2, y} K(x, y) - (y^2)^T \nabla_2 K(x, y)$$

$$\text{Subject to } \nabla_2 K(x, y) \leq 0$$

$$x^1 \in U, \quad y^1 \in V,$$

$$x^2 \geq 0, \quad y^2 \geq 0,$$

$$\mathrm{Min}_{v^1} \mathrm{Max}_{u, v^2} K(x, y) - (x^2)^T \nabla_1 K(x, y)$$

$$\text{Subject to } \nabla_1 K(x, y) \geq 0$$

$$x^1 \in U, \quad y^1 \in V,$$

$$x^2 \geq 0, \quad y^2 \geq 0.$$

The above formulation of Mishra et al. [152] is incorrect as pointed out by Kumar et al. [135] as can be seen from the following example due to Kumar et al. [135]

Example 5.32. Let $K(x, y) = e^{x^1} + e^{x^2 - y}$ be defined on some open set of $R^2 \times R$ containing $R_+^2 \times R_+ . K(x, y)$ is pseudo-convex in x^2 for fixed (x^1, y) and pseudo-concave in y for fixed x and satisfies all assumptions on the kernel function of Mishra et al. [152]. With this $K(x, y)$, the above pair of problem take the following form:

$$\mathrm{Max}_{x^1} \mathrm{Min}_{x^2, y} e^{x^1} + e^{x^2 - y} + y e^{x^2 - y}$$

$$\text{Subject to } \; - e^{x^2 - y} \leq 0,$$

$$0 \le x^1 \le L, \;\; x^1 \text{integer},$$
$$x^2 \ge 0, \;\; y \ge 0,$$

$$\text{Min}_{v^1} \text{Max}_{u,v^2} e^{x^1} + e^{x^2-y} - x^2 e^{x^2-y}$$
$$\text{Subject to } e^{x^2-y} \ge 0$$
$$0 \le x^1 \le L, \;\; x^1 \text{integer}$$
$$x^2 \ge 0, \;\; y \ge 0.$$

Here the arbitrary set U is taken to be the set of integers greater than or equal to zero and less than or equal to L. Taking $f(x,y) = e^{x^1} + e^{x^2-y} + ye^{x^2-y}$ and $g(x,y) = e^{x^1} + e^{x^2-y} - x^2 e^{x^2-y}$, one may easily observe that the supremum infimum of $f(x,y)$ for any (x,y) feasible to the primal problem is e^L and infimum supremum of $g(x,y)$ over all (x,y) feasible to the dual problem is $e^L + 1$. Obviously $e^L < e^L + 1$ and hence, the symmetric duality Theorem of Mishra et al. [152] failed.

Kumar et al. [135] modified the model of Mishra et al. [152]:

(MMSP)

$$\text{Max}_{x^1} \text{Min}_{x^2,y} K(x,y)$$
$$\text{Subject to } \nabla_2 K(x,y) \le 0$$
$$y^T \nabla_2 K(x,y) \ge 0$$
$$x^1 \in U, \;\; y^1 \in V,$$
$$x^2 \ge 0, \;\; y^2 \ge 0,$$

(MMSD)

$$\text{Min}_{v^1} \text{Max}_{u,v^2} K(u,v)$$
$$\text{Subject to } \nabla_1 K(u,v) \ge 0$$
$$u^T \nabla_1 K(u,v) \le 0$$
$$u^1 \in U, \;\; v^1 \in V,$$
$$u^2 \ge 0, \;\; v^2 \ge 0.$$

Kumar et al. [135] established symmetric duality theorem for (MMSP) and (MMSD) under pseudo-convexity/pseudo-concavity assumptions. We state the theorem without proof.

Theorem 5.33 (Symmetric duality). *Let $K(x,y)$ be separable with respect to x^1 or y^1 and twice differentiable in x^2 or y^2. Let $K(x,y)$ be pseudo-convex in x^2 for each (x^1,y) and pseudo-concave in y^2 for each (x,y^1). Let $(\bar{x},\bar{y})$ solves (MMSP) and $\nabla_1^2 K(x,y)$ and $\nabla_2^2 K(x,y)$ be positive or negative definite together with $\nabla_1 K(\bar{x},\bar{y}) \neq 0$ and $\nabla_2 K(\bar{x},\bar{y}) \neq 0$. Then $(\bar{x},\bar{y})$ also solves (MMSD) and the two optimal values are equal.*

5.5 Second and Higher Order Duality

Consider the nonlinear programming problem introduced in Sect. 5.1:

(P)

$$\text{Minimize } f(x)$$

$$\text{Subject to } g(x) \leq 0,$$

where $x \in C, C$ open set $C \subseteq R^n$ and f and g are twice differentiable functions from $C \to R$ and R^m, respectively. The Wolfe dual of (P) is (see Sect. 5.4)

(WD)

$$\text{Maximize}_{u,y} f(u) + y^T g(u)$$

$$\text{Subject to } \nabla f(u) + y^T \nabla g(u) = 0,$$

$$y \geq 0,$$

where $y \in R^m$.

By introducing an additional vector $p \in R^n$, Mangasarian [144] formulated the following second order dual:

(WD2)

$$\text{Maximize}_{(u,y,p)} f(u) + y^T g(u) - \frac{1}{2} p^T \nabla^2 \left(f(u) + y^T g(u) \right) p$$

$$\text{Subject to } \nabla f(u) + y^T \nabla g(u) + \nabla^2 \left(f(u) + y^t g(u) \right) p = 0,$$

$$y \geq 0.$$

Under appropriate conditions on f and g involving convexity and rather complicated restrictions on p, Mangasarian established duality theorems for (P) and (WD2). At the same time Mond [168] gave rather simple conditions than Mangasarian [144], using a generalized form of convexity. This type of generalization was also studied by Mahajan and Vartak [141]. Here we follow the approach of Hanson [84]; for another more general approach see Mishra [154]. One significant practical use of duality is that it provides bounds for the value of the objective function when approximations are used. Second order duality may provide tighter bounds than first order duality because there are more parameters involved.

The dual considered is of the form

(WD3)

$$\text{Maximize } f(u) + y^T g(u) - \frac{1}{2} q^T \nabla^2 \left(f(u) + y^T g(u) \right) r$$

$$\text{Subject to } \nabla f(u) + y^T \nabla g(u) + \nabla^2 \left(f(u) + y^T g(u) \right) p = 0, \tag{5.21}$$

$$y \geq 0, \tag{5.22}$$

where $p, q, r \in R^n$. In general, p, q, r can be regarded as functions, although the operators ∇ and ∇^2 in the above operate only on f and g. Note that the constraints of (WD3) are the same as in (WD2).

Hanson and Mond [88] introduced a slight generalization of the class of invex functions, when applied to problem (P), called by them Type I functions. Here we generalize further to second order Type I functions.

Definition 5.34. *Let Y be the constraint set of (WD3), and let $\eta(x,u)$, $p(x,u), q(x,u)$ and $r(x,u)$ be vector functions: $Y \times Y \to R^n$. The objective function $f(x)$ is said to be a second order Type I objective function and $g_i(x), i = 1,\ldots,m$, is said to be second order Type I constraint function at $u \in Y$ with respect to the functions $\eta(x,u), p(x,u), q(x,u)$ and $r(x,u)$ if for all $x \in Y$,*

$$f(x) - f(u) \geq \eta(x,u)^T \nabla f(u) + \eta(x,u)^T \nabla^2 f(u) p(x,u) - \frac{1}{2} q(x,u)^T \nabla^2 f(u) r(x,u) \tag{5.23}$$

and

$$-g_i(u) \geq \eta(x,u)^T \nabla g_i(u) + \eta(x,u)^T \nabla^2 g_i(u) p(x,u) - \frac{1}{2} q(x,u)^T \nabla^2 g_i(u) r(x,u), \tag{5.24}$$

where $i = 1,\ldots,m$.

Note that if p, q, r are identically zero functions (i.e., zero vectors), then (5.23) and (5.24) are the definitions of Type I functions (or Type I invexity or HM-invexity) given by Hanson and Mond [88]. Now we obtain for (P) and (WD3) weak and strong second order duality theorems.

Theorem 5.35 (Weak duality). *Let x satisfy the constraints of (P) and u, y, p, q, r satisfy the constraints of (WD3). Let f and $g_i, i = 1,\ldots,m$, be second order Type I functions defined over the constraint sets of (P) and (WD3). Then inf(P) $\geq$ sup(WD3).*

Proof.

$$\begin{aligned}
&f(x) - f(u) - y^T g(u) + \frac{1}{2} q(x,u)^T \nabla^2 \left[f(u) + y^T g(u) \right] r(x,u) \\
&\geq \eta(x,u)^T \nabla f(u) + \eta(x,u)^T \nabla^2 f(u) p(x,u) - \frac{1}{2} q(x,u)^T \nabla^2 f(u) r(x,u) \\
&\quad - y^T g(u) + \frac{1}{2} q(x,u)^T \nabla^2 \left[f(u) + y^T g(u) \right] r(x,u) \\
&= \eta(x,u)^T \nabla f(u) + \eta(x,u)^T \nabla^2 f(u) p(x,u) \\
&\quad - y^T g(u) + \frac{1}{2} q(x,u)^T \nabla^2 \left(y^T g(u) \right) r(x,u)
\end{aligned}$$

$$= -\eta(x,u)^T\nabla\left(y^Tg(u)\right) - \eta(x,u)^T\nabla^2\left(y^Tg(u)\right)p(x,u)$$
$$- y^Tg(u) + \frac{1}{2}q(x,u)^T\nabla^2\left(y^Tg(u)\right)r(x,u) \quad \text{by (5.21)}$$
$$\geq 0, \quad \text{by (5.24) and (5.22).}$$

□

Theorem 5.36 (Strong duality). *Suppose x^* is optimal in (P) and x^* satisfies one of the usual constraint qualifications for (P). Then there exists $y \in R^m$ such that $(x^*, y, p = q = r = 0)$ is feasible for (WD3) and the corresponding values of (P) and (WD3) are equal. If in addition (5.23) and (5.24) are satisfied for all feasible solutions of (WD3), then x^* and $(x^*, y, p = q = r = 0)$ are optimal for (P) and (WD3), respectively.*

Proof. The Kuhn–Tucker conditions for a minimum at x^* are that there exists $y \in R^m$ such that

$$\nabla f(x^*) + \nabla y^Tg(x^*) = 0,$$
$$y^Tg(x^*) = 0,$$
$$y \geq 0.$$

Therefore, the points $(x^*, y, p = q = r = 0)$ is feasible for (WD3) and the values of (P) and (WD3) are equal, and it follows from Theorem 5.35 that x^* and $(x^*, y, p = q = r = 0)$ are optimal for (P) and (WD3).

□

Since $p = q = r = 0$ at the optimum, in which case the second order dual reduces to the first order dual, there may seem to be no point in having the additional complication of introducing the extra functions p, q and r. However, if an appropriate value for x^* is used then the optimal values for p, q and r in the approximating dual are not necessarily zero, and the second order dual may be used to give a tighter bound than the first order dual for the value of the primal objective function. Note that in the proof of Theorem 5.36 it would be sufficient for either q or r to be zero and not necessarily both.

Higher order duality problems and invexity have been examined by Mond and Zhang [182] and by Mishra and Rueda [160]. Mishra and Rueda [161] have also treated he case of higher order (generalized) invexity and duality for a nondifferentiable mathematical programming problem.

For higher order duality the starting paper is Mangasarian [144]. If (P)

$$\text{Minimize}_x \left\{f(x) : g(x) \leq 0\right\}$$

is the usual primal problem, the Mangasarian second order dual is (WD2). By introducing two differentiable functions $h : R^n \times R^n \to R$ and $k : R^n \times R^n \to R^m$ Mangasarian [144] formulated the following higher order dual:
(HD)

$$\text{Maximize}_{u,y,p} f(u) + h(u,p) + y^Tg(u) + y^Tk(u,p)$$
$$\text{Subject to } \nabla_p h(u,p) + \nabla_p y^Tk(u,p) = 0$$
$$y \geq 0,$$

where $\nabla_p h(u,p)$ denotes the gradient of h with respect to p and $\nabla_p \left(y^T k(u,p)\right)$ denotes the gradient of $y^T k$ with respect to p.

Note that if $h(u,p) = p^T \nabla f(u), k(u,p) = p^T \nabla g(u)$, then (HD) becomes (WD), and if $h(u,p) = p^T \nabla f(u) - \frac{1}{2} p^T \nabla^2 f(u) p$, and $k(u,p) = p^T \nabla g(u) - \frac{1}{2} p^T \nabla^2 g(u) p$, then (HD) becomes (WD2).

Mangasarian [144], however, did not prove a weak duality theorem for (P) and (HD) and only gave a limited version of strong duality. The following theorems and proofs are taken from Mond and Zhang [182].

Theorem 5.37 (Weak duality). *Let x be feasible for (P) and (u,y,p) feasible for (HD). If for all feasible (x,u,y,p) there exists a function $\eta : R^n \times R^n \to R^n$ such that*

$$f(x) - f(u) \geq \eta(x,u)^T \nabla_p h(u,p) + h(u,p) + p^T \nabla_p h(u,p) \tag{5.25}$$

and

$$g_i(x) - g_i(u) \geq \eta(x,u)^T \nabla_p k_i(u,p) + k_i(u,p) + p^T \nabla_p k_i(u,p), \tag{5.26}$$

for $i = 1, \ldots, m$, then, inf(P) $\geq$ sup(HD).

Proof.

$$\begin{aligned}
&f(x) - f(u) - h(u,p) - y^T g(u) - y^T k(u,p) \\
&\geq \eta(x,u)^T \nabla_p h(u,p) + p^T \nabla_p h(x,u) - y^T g(u) - y^T k(u,p) \quad \text{by (5.25)} \\
&= -\eta(x,u)^T \nabla_p \left(y^T k(u,p)\right) - p^T \nabla_p \left(y^T k(u,p)\right) - y^T g(u) - y^T k(u,p) \\
&\geq -y^T g(x) \quad \text{by (5.26)} \\
&\geq 0.
\end{aligned}$$

□

The following strong duality follows on the lines of Mond and Weir [181].

Theorem 5.38 (Strong duality). *Let x^0 be a local or global optimal solution of (P) at which a constraint qualification is satisfied and let $h(x^0,0) = 0$, $k(x^0,0) = 0, \nabla_p h(x^0,0) = \nabla_x f(x^0), \nabla_p k(x^0,0) = \nabla_x g(x^0)$. Then there exists $y \in R^m$ such that $(x^0, y, p = 0)$ is feasible for (HD) and the corresponding values of (P) and (HD) are equal. If (5.25) and (5.26) are satisfied for all feasible (x,u,y,p), then x^0 and $(x^0, y, p = 0)$ are global optimal solutions for (P) and (HD).*

5.6 Saddle Points, Optimality and Duality with Nonsmooth Invex Functions

In Sect. 5.4, we have described the Wolfe and the Mond–Weir duals for problem (P). Another approach to duality, which does not necessarily require that the functions involved have to be differentiable, is through Lagrangian duality and saddle points of the Lagrangian function. The general theory of Lagrangian duality is concerned with functions $f : C \to R$, $g : C \to R^m$, where $C \subseteq R^n$. The general Lagrangian function (suitable for problem (P)) may be written as

$$L(x,y) = \begin{cases} f(x) + y^T g(x), & x \in C, \ y \geq 0 \\ -\infty, & x \in C, \ y < 0 \\ \infty, & x \notin C. \end{cases}$$

Two problems associated with this Lagrangian function can be considered.

(P1)

$$\inf_{x \in R^p} \sup_{y \in R^m} L(x,y)$$

(D1)

$$\sup_{y \in R^m} \inf_{x \in R^n} L(x,y).$$

Weak duality holds between (P1) and (D1) without any convexity assumptions; it is of more interest to see how the properties of this pair of problems related to Wolfe duality. Necessary and sufficient conditions for optimality of (P) may be expressed using (P1) and (D1). First we require conditions for optimality of (P1) and (D1), which calls for the notion of a saddle point.

Definition 5.39. *The point* $(x^0, y^0), y^0 \geq 0$, *is said to be a saddle point of* $L(x,y)$ *if* $L(x^0, y) \leq L(x^0, y^0) \leq L(x^0, y), \forall x \in R^n, \forall y \in R^m$.

Stoxer and Witzgall [228] have proved the following two properties with no further assumption on $L(x,y)$:

(a) If (x^0, y^0) is a saddle point of $L(x,y)$, then (x^0, y^0) is optimal in programs (P1) and (D1).
(b) If (x^1, y^1) is optimal in (P1), and (x^2, y^2) is optimal in (D1), with $L(x^1, y^1) = L(x^2, y^2)$, then (x^1, y^2) is a saddle point of $L(x,y)$ (and consequently, (x^1, y^2) is optimal in both (P1) and (D1)).

Now, program (P1) is actually equivalent to (P) since

$$L(x,y) = \begin{cases} f(x), & x \in C, \ g(x) \leq 0 \\ \infty, & x \in C, \ g(x) \nleq 0, \ \text{ or } x \in C. \end{cases}$$

Now, program (P1) corresponds to minimize $f(x)$ subject to $x \in C$ and $g(x) \leq 0$. This leads to the conclusion that if (x^0, y^0) is a saddle point of L, then x^0 is optimal in (P). A converse result regarding conditions for an optimal solution x^0 of (P) to yield a y^0 such that the pair (x^0, y^0) is a saddle point of L has been given in Ben-Israel and Mond [18], using invexity of the Lagrangian function L on C along with the modified Slater condition. That Theorem may be extended to admit the modified Karlin condition.

Theorem 5.40. *Assume that x^0 is an optimal solution of (P) and that a suitable constraint qualification holds. If f and $g_i, i = 1, \dots, m$, are invex with respect to a common η, then there exists $y^0 \geq 0$ such that*

$$f(x^0) + y^T g(x^0) \leq f(x^0) + {y^0}^T g(x^0) \leq f(x) + {y^0}^T g(x), \tag{5.27}$$

$\forall x \in C, \forall y \in R^m, y \geq 0$ (i.e., (x^0, y^0) is a solution of the saddle point problem, (here given in a less general formulation than before).

Proof. Since x^0 is optimal for (P) and a constraint qualification is satisfied, then by the Kuhn–Tucker necessary conditions, there exists $y^0 \in R^m$ such that $y^0 \geq 0$ and $\nabla f(x^0) + \nabla({y^0}^T g(x^0)) = 0$, as well as ${y^0}^T g(x^0) = 0$. Since $f(x)$ and every $g_i(x), i = 1, \dots, m$, are invex with respect to a common η, the Kuhn–Tucker conditions imply

$$f(x^0) + {y^0}^T g(x^0) \leq f(x) + {y^0}^T g(x)$$

$\forall x \in C$, which is the right-hand side of (5.27). The left side holds since $y \geq 0$, $g(x^0) \leq 0$, ${y^0}^T g(x^0) = 0$.

□

We note that the assumption that $f(x)$ and $g_i(x), i = 1, \dots, m$, are invex with respect to a common η, is equivalent to the assumption in Ben-Israel and Mond [18] that $f + y^T g$ be invex for all $y \in R^m_+$ (recall what observed after Corollary 2.13). Moreover, the invexity assumptions of Theorem 5.40 may be weakened. It is sufficient to suppose that $f + y^T g$ is invex only for those $y \in R^m_+$ for which (x^0, y^0) is a Kuhn–Tucker point.

We now turn to the dual problem (D1) in order to relate it to the Wolfe dual. Note that

$$\inf_{x \in R^n} L(x, y) = \begin{cases} \inf_{x \in C} \left(f(x) + y^T g(x)\right), & y \geq 0 \\ -\infty, & y \ngeq 0 \end{cases}$$

Therefore, (D1) is equivalent to

$$\sup_{y \in R^m} \inf_{x \in R^n} \left(f(x) + y^T g(x)\right)$$

$$\text{Subject to } y \geq 0.$$

Now, if $\inf_{x \in C} \left(f(x) + y^T g(x)\right)$ is attained, being C open and f and g differentiable, then the infimum is attained at a stationary point x^0; that is, for some $x^0 \in C$ satisfying

$$\nabla f(x^0) + \nabla y^T g(x^0) = 0.$$

If f and $g_i, i = 1, \ldots, m$, are invex with respect to a common η, then this condition is also sufficient for the infimum to be attained at x^0. Thus, with the invexity assumptions, (D1) is equivalent to

$$\sup_{y \in R^m} f(x) + y^T g(x)$$

$$\text{Subject to } \nabla f(x) + \nabla \left(y^T g(x)\right) = 0, \quad y \geq 0,$$

which corresponds to the Wolfe dual (WD).

Another approach to the saddle point problem (5.27) may be given through the assumptions of preinvexity of the functions involved. This has been done by Weir and Jeyakumar [244] and by Weir and Mond [245]. We follow the last two authors. Obviously in this approach no differentiability assumption is required, so this approach is more alike to the original saddle point theorems for problem (P), than Theorem 5.40.

We have already mentioned in Sect. 4.1 that every preinvex vector-valued function is convex-like, i.e., $f : S \to R^m, S \subseteq R^n$ (S an η-invex set) satisfies the property:

$$\exists z \in S : f(x) \leq \lambda f(x) + (1 - \lambda) f(y), \quad \forall x, y \in S, \ \forall \lambda \in [0, 1].$$

Hayashi and Komiya [90] and also Jeyakumar [101] developed theorems of the alternative involving convexlike functions and in addition considered Lagrangian duality for convexlike programs. In particular, Hayasi and Komiya [90] obtained, under convexlike assumptions, the Fan–Glickeberg–Hoffman [64] theorem of the alternative (see also [143]) which is the nonlinear version of the classical Gordan theorem of the alternative for linear inequalities. Obviously the theorem of Hayashi and Komiya [90] holds also for the preinvex case, but for the reader's convenience we prove this theorem under preinvexity assumptions. The saddle point theorem then follows easily, in the same way it is obtained for the classical convex case (see, [143]).

Theorem 5.41. *Let S be a non-empty η-invex set in R^n and let $f : S \to R^m$ be a preinvex function on S (with respect to η). Then either*

$$f(x) < 0 \quad \textit{has a solution} \quad x \in S$$

or

$$p^T f(x) \geq 0, \quad \forall x \in S, \quad \textit{for some} \quad p \in R^m, \ p \geq 0, \ p \neq 0.$$

But both alternatives are never true.

Proof. Following Mangasarian [143], the proof depends on establishing the convexity of the set $A = \bigcup\{A(x) : x \in S\}$, where $A(x) = \{u \in R^m : u > f(x)\}, x \in S$. Under our assumptions this is immediate, for if u^1 and u^2 are in A, then for $0 \leq \lambda \leq 1$,

$$\lambda u^1 + (1 - \lambda) u^2 > \lambda f(x^1) + (10\lambda) f(x^2) \geq f(x^2 + \lambda \eta(x^1, x^2)).$$

□

We recall the problem (P):
(P)

$$\text{Minimize} f(x)$$
$$\text{Subject to } g(x) \leq 0,$$

where $f : S \to R$ and $g : S \to R^m$, $S \subseteq R^n$ will be said to satisfy the generalized Slater constraint qualification if g is preinvex (with respect to η) on the η-invex set S and there exists $x^1 \in S$ such that $g(x^1) < 0$.

Theorem 5.42. *For the problem (P), assume that f is preinvex (with respect to η) and g is preinvex (with respect to same η) on the η-invex set S. Assume that the generalized Slater constraint qualification holds. If (P)attains a minimum at $x = x^0 \in S$, then there exists $v^0 \in R^m, v^0 \geq 0$, such that (x^0, y^0) is a saddle point for the Lagrangian function $L(x, v) = f(x) + v^T g(x)$, i.e., it holds*

$$L(x^0, v) \leq L(x^0, v^0) \leq L(x, v^0), \quad \forall x \in S, \ \forall v \in R^m, \ \ v \geq 0. \tag{5.28}$$

We recall also that if condition (5.28) is satisfied for some (x^0, v^0), then x^0 is a (global) minimum for (P); in other words, the saddle point condition (5.28) is sufficient for global minimality without any (generalized) invexity assumption.

Another fruitful approach to investigate optimality, duality and saddle point conditions for problem (P), under some invexity assumption, but in absence of differentiability, is through nonsmooth analysis. For the Lipschitz case (i.e., making use of Clarke generalized derivatives and generalized gradients) this has been done by Reiland [209] and by Kaul et al. [116], even if it must be said that also Jeyakumar [103] treated the case of a nonsmooth nonconvex problem in which the functions are locally Lipschitz and satisfy an invexity assumption. The reader is referred to Sect. 4.2 for the basic definitions and properties. Consider the usual problem (P)
(P)

$$\text{Minimize } f(x)$$
$$\text{Subject to } g(x) \leq 0,$$

where $f : C \to R$, $g : C \to R^m, C \subseteq R^n$ is an open set in R^n and f and every $g_i, i = 1, \ldots, m$, are Lipschitz on C.

We have the following necessary optimality conditions; in terms of Clarke generalized gradients (see, [37,38,92]).

Theorem 5.43. *Let x^0 be a local minimum for (P) and a suitable constraint qualification be satisfied. Then the following nonsmooth form of the Kuhn–Tucker conditions hold:*

$$0 \in \partial_C f(x^0) + \sum_{i=1}^{m} \lambda_i \partial_C g_i(x^0) \tag{5.29}$$

$$\lambda_i g_i(x^0) = 0, \quad i = 1, \ldots, m, \tag{5.30}$$

$$\lambda_i \geq 0, \quad i = 1, \ldots, m. \tag{5.31}$$

Following Hiriart-Urruty [92], a constraint qualification assuring the thesis of the previous theorem is the following one:

$$\exists v \in R^n : g_i^0(x^0, v) < 0, \quad \forall i \in I(x^0),$$

where $I(x^0) = \{i : g_i(x^0) = 0\}$. In the presence of inequality constraints, from Fritz John type optimality condition (always in terms of generalized gradients), Clarke [37, 38] established Kuhn–Tucker type conditions (5.29)–(5.31) under the assumptions of "calmness" of the optimization problem. This regularity condition has the advantage to be present in most problems, even if it seems difficult to verify it in general. For a more recent and general analysis of constraint qualifications for the Lipschitz case see Jourani [108] and the references there quoted.

The following theorem shows that if the functions involved in (P) and C-invex (Definition 4.26) with respect to a common η, then (5.29)–(5.31) are also sufficient for optimality.

Theorem 5.44. *Suppose x^0 is feasible for (P) and that the generalized Kuhn–Tucker conditions (5.29)–(5.31) are satisfied at x^0. If f and $g_i, i = 1, \dots, m$, are C-invex on C for the same kernel η, then x^0 is a global minimum for (P).*

Proof. Let x be any other feasible point for (P); then

$$\begin{aligned} f(x) - f(x^0) &\geq f(x) + \sum_{i=1}^{m} \lambda_i g_i(x) - f(x^0) - \sum_{i=1}^{m} \lambda_i g_i(x^0), \\ &\quad \text{by feasibility of } x \text{ and (5.30)} \\ &\geq f^0\left(x^0, \eta(x, x^0)\right) + \sum_{i=1}^{m} \lambda_i g_i^0\left(x^0, \eta(x, x^0)\right), \end{aligned}$$

by (5.29) there exists $\xi \in \partial_C f(x^0)$ and $\varsigma_i \in \partial_C g_i(x^0), i = 1, \dots, m$, such that

$$\xi + \sum_{i=1}^{m} \lambda_i \varsigma_i = 0.$$

Therefore,

$$\begin{aligned} f(x) - f(x^0) &\geq \xi^T \eta(x, x^0) + \sum_{i=1}^{m} \lambda_i \varsigma_i^T \eta(x, x^0) \\ &= \left(\xi + \sum_{i=1}^{m} \lambda_i \varsigma_i^T\right) \eta(x, x^0) \\ &= 0. \end{aligned}$$

□

Hiriart-Urruty [93] obtained improved necessary optimality conditions for x^0 to be a local minimum of (P) by substituting

$$0 \in \partial_C \left(f + \sum_{i=1}^{m} \lambda_i g_i \right) (x^0) \tag{5.32}$$

for (5.29) with (5.30) and (5.31). We recall (Sect. 4.2) that it holds

$$\partial_C (f+g)(x^0) \subseteq \partial_C f(x^0) + \partial_C g(x^0).$$

The equality being in general not satisfied (unless, e.g., f and g are regular). So (5.32) is a stronger requirement for x^0 than (5.29). In fact, when (5.32) is substituted for (5.29), we can weaken the C-invexity assumptions in Theorem 5.44.

Theorem 5.45. *Suppose x^0 is feasible for problem (P) and the generalized Kuhn–Tucker conditions (5.32), (5.30) and (5.31) are satisfied at x^0. If $f + \sum_{i=1}^{m} \lambda_i g_i$ is C-invex on C, then x^0 is a global minimum for (P).*

Proof. Let x be any other feasible point for (P); then by (5.30) and (5.31)

$$\begin{aligned} f(x) - f(x^0) &\geq f(x) + \sum_{i=1}^{m} \lambda_i g_i(x) - f(x^0) - \sum_{i=1}^{m} \lambda_i g_i(x^0) \\ &\geq \left(f(x) + \sum_{i=1}^{m} \lambda_i g_i \right)^0 (x^0, \eta(x, x^0)) \\ &\geq 0, \quad \text{by (5.32).} \end{aligned}$$

□

We note, following Reiland [209], that also for the Lipschitz case, the C-invexity assumption of Theorem 5.45 is weaker than the C-invexity assumption in Theorem 5.44. Indeed, if f and $g_i : C \to R$ are C-invex on C for the same η, then since the limsup of sum is bounded above by the sum of the limsups, for $y \in R^m, y \geq 0$, we have

$$\begin{aligned} &f(x) + \sum_{i=1}^{m} y_i g_i(x) - f(u) - \sum_{i=1}^{m} y_i g_i(u) \\ &\qquad \geq f^0 (u, \eta(x, u)) + \sum_{i=1}^{m} y_i g_i^0 (u, \eta(x, u)) \\ &\qquad \geq \left(f + \sum_{i=1}^{m} y_i g_i \right)^0 (u, \eta(x, u)); \end{aligned}$$

hence $f + \sum_{i=1}^{m} y_i g_i$ is C-invex on C. Kaul et al. [116] have generalized the results of Reiland [209] by introducing the following definitions for a Lipschitz function $f : C \to R$ (C open set of R^n). See also Sect. 4.2.

Definition 5.46. *The Lipschitz function $f : C \to R$ is:*

(a) C-pseudo-invex on C if there exists a function $\eta : C \times C \to R^n$ such that, with $x, u \in C$,

$$f^0\,(u, \eta(x,u)) \geq 0 \Rightarrow f(x) \geq f(u);$$

(b) C-quasi-invex on C if there exists a function $\eta : C \times C \to R^n$ such that, with $x, u \in C$,

$$f(x) \leq f(u) \to f^0\,(u, \eta(x,u)) \leq 0;$$

(c) C-strictly pseudo-invex on C if there exists a function $\eta : C \times C \to R^n$ such that, with $x, u \in C, x \neq u$,

$$f^0(u, \eta(x,u)) \geq 0 \Rightarrow f(x) > f(u).$$

The said authors proved various sufficient optimality conditions under various assumptions. We report only the following one.

Theorem 5.47. *Let x^0 be feasible for (P) and the Kuhn–Tucker conditions (5.29)–(5.31) be satisfied at x^0; let I denote the set of the active constraints at x^0 ($I = \{i : g_i(x^0) = 0\}$). If any one of the following conditions hold:*

(a) f is C-pseudo-invex on C and every $g_i, i \in I$, is C-quasi-invex on C,
(b) f and every $g_i, i \in I, i \neq s$, are C-quasi-invex on C, and g_i is strictly C-pseudo-invex on C, with $\lambda_s > 0$, for some $s \in I$,

with respect to the same η, then x^0 is a global minimum of (P).

Further generalizations of sufficient optimality conditions for (P), obtained by means of the K-directional derivatives (Sect. 4.2), are given by Castellani [28].

Consider now the following dual problems related to problem (P):

(D1)

$$\text{Maximize } f(x) + \sum_{i=1}^{m} y_i g_i(x)$$

$$\text{Subject to } 0 \in \partial_C f(x) + \sum_{i=1}^{m} y_i \partial_C g_i(x),$$

$$y_i \geq 0, \quad i = 1, \ldots, m.$$

(D2)

$$\text{Maximize } f(x) + \sum_{i=1}^{m} y_i g_i(x)$$

$$\text{Subject to } 0 \in \partial_C \left(f + \sum_{i=1}^{m} y_i g_i \right)(x),$$

$$y \geq 0, \quad i = 1, \ldots, m.$$

We note that the feasible region for (D1) is larger than the feasible region for (D2), hence the optimal value of the objective function in (D1) will be larger than the optimal value in (D2). Weak duality between (P) and (D1) thus requires a stronger assumption than weak duality between (P) and (D2).

Theorem 5.48 (Weak duality).

(a) If for any fixed $y \in R_+^m$, $(f + \sum_{i=1}^m y_i g_i)$ is C-invex on the open set $C \subseteq R^n$, then weak duality holds between (P) and (D2).
(b) If f and $g_i, i = 1, \dots, m$, are C-invex on the open set $C \subseteq R^n$ for a common η, then weak duality holds for (P) and (D1).

Proof. (1) Let $\bar{x}$ and (x^0, y^0) be feasible for (P) and (D2), respectively. Then

$$\begin{aligned} f(\bar{x}) - f(x^0) - \sum_{i=1}^m y_i^0 g_i(x^0) \\ &\geq f(\bar{x}) + \sum_{i=1}^m y_i^0 g_i(\bar{x}) - f(x^0) - \sum_{i=1}^m y_i^0 g_i(x^0), \\ &\quad \text{since } y_i^0 \geq 0, \text{ for each } i \text{ and } \bar{x} \text{ is feasible for (P)} \\ &\geq \left(f + \sum_{i=1}^m y_i^0 g_i\right)^0 (x^0, \eta(\bar{x}, x^0)) \\ &\geq 0 \quad \left(\text{since } 0 \in \partial_C \left(f + \sum_{i=1}^m y_i^0 g_i\right)(x^0)\right). \end{aligned}$$

(2) Let $\bar{x}$ and (x^*, y^*) be feasible for (P) and (D1), respectively. Then

$$\begin{aligned} f(\bar{x}) - f(x^*) - \sum_{i=1}^m y_i^* g_i(x^*) \\ &\geq f(\bar{x}) + \sum_{i=1}^m y_i^* g_i(\bar{x}) - f(x^*) - \sum_{i=1}^m y_i^* g_i(x^*) \\ &\quad (\text{since } y_i^* \geq 0, \text{ for each } i \text{ and } \bar{x} \text{ is feasible for (P)}) \\ &\geq f^0(x^*, \eta(\bar{x}, x^*)) + \sum_{i=1}^m y_i^* g_i^0(x^*, \eta(\bar{x}, x^*)). \end{aligned}$$

Since (x^*, y^*) is feasible for (D1), there exist $\xi \in \partial_C f(x^*)$ and $\varsigma_i \in \partial_C g_i(x^*)$ such that $0 = \xi + \sum_{i=1}^m y_i^* \varsigma_i^*$, hence

$$\begin{aligned} f^0(x^*, \eta(\bar{x}, x^*)) + \sum_{i=1}^m y_i^0 g_i^0(x^8, \eta(\bar{x}, x^*)) \\ &\geq \xi^T \eta(\bar{x}, x^*) + \sum_{i=1}^m y_i^* \varsigma_i^* \eta(\bar{x}, x^*) \end{aligned}$$

$$\geq \left(\xi^T + \sum_{i=1}^{m} y_i^* \varsigma_i^*\right)^T \eta(\bar{x}, x^*)$$
$$= 0.$$

□

Theorem 5.49.

(a) *Assume that* $f + \sum_{i=1}^m y_i g_i$ *is* C*-invex on the open set* $C \subseteq R^n$ *for* $y \in R_+^m$*. If* $\bar{x}$ *is optimal for (P) and the Kuhn–Tucker conditions (5.30)–(5.32) are satisfied at* $(\bar{x}, \bar{y})$*, then (D2) is maximized at* $(\bar{x}, \bar{y})$ *and the optimal values of (P) and (D2) are equal.*

(b) *Assume* f *and* $g_i, i = 1, \ldots, m$*, are* C*-invex on the open set* $C \subseteq R^n$ *for the same* η*. If* $\bar{x}$ *is optimal for (P) and the Kuhn–Tucker conditions (5.29)–(5.31) are satisfied at* $(\bar{x}, \bar{y})$*, then (D1) is maximized at* $(\bar{x}, \bar{y})$ *and the optimal values of (P) and (D1) are equal.*

Proof. (1) Since $(\bar{x}, \bar{y})$ satisfies (5.32) and (5.31), $(\bar{x}, \bar{y})$ is feasible for (D2). By (5.30) and weak duality

$$f(\bar{x}) + \sum_{i=1}^{m} \bar{y}_i g_i(\bar{x}) = f(\bar{x})$$
$$\geq f(x) + \sum_{i=1}^{m} y_i g_i(x), \quad \forall (x, y) \text{ feasible for (D2).}$$

Thus $(\bar{x}, \bar{y})$ is optimal for (D2) and the optimal values of (P) and (D2) are equal.
(2) The proof is similar to part (1).

□

(D1) and (D2) are Wolfe type dual problems for a Lipschitz primal problem (P). Kaul et al. [116] consider a Mond–Weir type dual problem for (P), i.e., the problem:
(D3)

$$\text{Maximize} f(u)$$
$$\text{Subject to } 0 \in \partial_C f(u) + \sum_{i=1}^{m} y_i \partial_C g_i(u)$$
$$y_i g_i(u) \geq 0, \quad i = 1, \ldots, m,$$
$$y_i \geq 0, \quad i = 1, \ldots, m.$$

These authors, under various C-invexity assumptions, obtained weak and strong duality results for (P) and (D3).

The saddle point problem, when the functions involved are nonsmooth and C-invex, has been treated by Jeyakumar [103] and by Brandao et al. [24]. In particular, the last authors obtain also a generalization of the Gordan theorem of the alternative (i.e., a Fan–Glicksberg–Hoffman theorem for C-invex functions). The problem considered is the dual one:

(P)

$$\text{Minimize} f(x)$$

$$\text{Subject to } g_i(x) \leq 0, \quad i = 1, \ldots, m, \quad x \in C,$$

where C is a non-empty open subset of R^n, and f and $g_i, i = 1, \ldots, m$, are Lipschitz (or even locally Lipschitz) functions on C. Brandao et al. [24] obtained necessary Kuhn–Tucker conditions for (P) under a Slater-type constraint qualification: every $g_i, i = 1, \ldots, m$, is C-invex on C for the same η and there exists $\hat{x} \in C$ such that $g_i(\hat{x}) < 0, i = 1, \ldots, m$.

Brandao et al. [24] proved the following necessary optimality conditions.

Theorem 5.50. *Let $\bar{x}$ be an optimal solution for (P) and suppose that f and $g_i, i = 1, \ldots, m$, are C-invex on C, for the same η. If the constraints of (P) satisfy the Slater-type constraint qualification, then there exists $\bar{\lambda} = (\bar{\lambda}_1, \ldots, \bar{\lambda}_m)$ such that*

$$0 \in \partial_C \left(f + \sum_{i=1}^{m} \bar{\lambda}_i g_i \right) (\bar{x}) \tag{5.33}$$

$$\bar{\lambda}_i g_i(\bar{x}) = 0, \quad i = 1, \ldots, m, \tag{5.34}$$

$$\bar{\lambda}_i \geq 0, \quad i = 1, \ldots, m. \tag{5.35}$$

Then they proved that if $\bar{x}$ is feasible for (P) and f and $g_i, i = 1, \ldots, m$, are C-invex on C for the same η, the previous Kuhn–Tucker type conditions are sufficient for the global optimality of $\bar{x}$.

Recall that a point $(\bar{x}, \bar{y}) \in C, \bar{\lambda} \in R_+^m$, is said to be a saddle point for (P) if

$$L(\bar{x}, \lambda) \leq L(\bar{x}, \bar{\lambda}) \leq L(x, \bar{\lambda}), \tag{5.36}$$

for all $x \in C$ and $\lambda \in R_+^m$, being $L(x, \lambda) = f(x) + \sum_{i=1}^{m} \lambda_i g_i(x)$. Following Brandao et al. [24] we can prove the following saddle point theorem.

Theorem 5.51. *Suppose $\bar{x}$ is an optimal solution for (P), f and $g_i, i = 1, \ldots, m$, are C-invex on C with respect to the same η, the constraints of (P) satisfy the Slater-type constraint qualification. Then there exists $\bar{\lambda} \in R_+^m$, such that $(\bar{x}, \bar{\lambda})$ is a saddle point for (P).*

Proof. If $\bar{x}$ is an optimal solution for (P), being the Slater-type condition verified, by Theorem 5.50 there exists $\bar{\lambda}_i \in R$ such that the Kuhn–Tucker conditions (5.33)–(5.35) hold. Therefore, invoking the invexity assumptions these conditions are also sufficient for $\bar{x}$ to be a global minimum for $f + \sum_{i=1}^{m} \bar{\lambda}_i g_i(x)$ over C, that is the right side inequality of (5.36) holds. The left side inequality of (5.36) follows from (5.34) since $g_i(\bar{x}) \leq 0$ and $\lambda \in R_+^m$. □

6

Invex Functions in Multiobjective Programming

6.1 Introduction

In general, the unconstrained vector optimization problem (VOP) is presented as follows:

(VOP)

$$\text{Minimize} f(x) = (f_1(x), \ldots, f_p(x))$$

$$\text{Subject to } x \in S \subseteq R^n.$$

Unlike problems with a single objective in which there may exist an optimal solution to the effect that it minimizes the objective function, in the multiobjective programming problem there does not necessarily exist a point which may be optimal for all objectives. To this effect the solution concept introduced by Pareto [194], where the concept of efficient points is given as follows:

Definition 6.1. *A feasible point $\bar{x}$, is said to be an efficient solution if and only if there does not exist another $x \in S$ such that $f_i(x) \leq f_i(\bar{x})$ for $i = 1, \ldots, p$ with strict inequality holding for at least one i.*

At times locating the efficient points is quite costly. As a result there appears a more general concept such as that of weakly efficient solution, which under certain conditions, presents topological properties that are not given in the set of efficient points see, e.g., Naccache [186].

Definition 6.2. *A feasible point $\bar{x}$ is said to be a weakly efficient solution if and only if there does not exist another $x \in S$ such that $f_i(x) < f_i(\bar{x}), \forall i = 1, \ldots, p$.*

It is easy to verify that every efficient point is a weakly efficient point.

The following convention for equalities and inequalities will be used. If $x, y \in R^n$, then

$x = y$ iff $x_i = y_i, i = 1, \ldots, n$;

$x \leqq y$ iff $x_i = y_i, i = 1, \ldots, n$;

$x \leq y$ iff $x_i \leqq y_i, i = 1, \ldots, n$; with strict inequality holding for at least one i.
$x < y$ iff $x_i < y_i, i = 1, \ldots, n$.

The study of the solutions of a multiobjective problem may be approached from two aspects:

(a) Trying to relate them with the solutions to he scalar problems, whose resolution has been studied extensively (see, e.g., [238]). Vartak and Gupta [238] have given three types of scalar optimization problems to be associated with vector optimization problem (VOP) for this purpose, namely the weighting problem, the k-th objective Lagrangian problem and the k-th objective ϵ-constraint problem.
(b) Trying to locate conditions which are easier to deal with computationally and which guarantee efficiency.

As much in one case as in the other, the convexity plays an important role as a fundamental condition in obtaining the desired results. For the reader's convenience, let us provide the concept of invex function for p-dimensional case.

Definition 6.3. *Let $f : S \subseteq R^n \rightarrow R^p$ be a differentiable function on the open set S. Then f is a vector invex function on S with respect to η if $\forall x, y \in S$ there exists $\eta(x, y) \in R^n$ such that*

$$f(x) - f(y) \geq \nabla f(y)\eta(x, y),$$

where $\nabla f(y) \in M^{p\times n}$ whose rows are gradient vectors of each component function valued at the point y.

Constrained multiobjective programming problem (CVOP) is more useful than the unconstrained vector optimization problem:

(CVOP)

$$\text{Minimize} f(x) = (f_1(x), \ldots, f_p(x))$$

$$\text{Subject to } g(x) \leqq 0$$

$$x \in S \subseteq R^n,$$

where $f : S \rightarrow R^p$ and $g : S \rightarrow R^m$ are differentiable functions on the open set $S \subseteq R^n$.

In Chap. 5, we studied the optimality conditions of invex functions for scalar programming problems. In this chapter, we study Kuhn–Tucker necessary and sufficient optimality conditions for differentiable and non-differentiable vector optimization problems. Second and higher order duality results are also discussed for multiobjective case. Moreover, we discuss multi-objective symmetric duality involving invex and generalized invex functions.

In multiobjective (vector) optimization problems, multiobjectives are usually noncommensurable and cannot be combined into a single objective. Moreover, often the objectives conflict with each other. Consequently, the concept of optimality for single-objective (scalar) optimization problems cannot be applied directly to vector optimization problems. The concept of Pareto optimality, characterizing an efficient solution, has been introduced for vector optimization problems. Osuna-Gomez et al. [192] have characterized weakly efficient solutions of a multiobjective programming problem with differentiable functions via the generalization of the optimality conditions established by Kuhn–Tucker for scalar programming problems. These authors have given a necessary and sufficient optimality conditions under generalized convexity characterizing weakly efficient solutions which is similar to that established for optimal solutions in scalar programming problems.

6.2 Kuhn–Tucker Type Optimality Conditions

Osuna-Gomez et al. [193] established conditions for multiobjective problems, similar to those given by Kuhn–Tucker for the scalar problems, for which Osuna-Gomez et al. [193] provided a concept analogous to the stationary point or critical point for the scalar function.

Definition 6.4. *A feasible point $\bar{x} \in S$ is said to be a vector critical point to (VOP) if there exists a vector $\lambda \in R^p$ with $\lambda \geq 0$ such that $\lambda^T \nabla f(\bar{x}) = 0$.*

Note that stationary points for a scalar optimization problem are those whose vector gradients are zero. However, vector critical points are those such that there exists a non-negative linear combination of the gradient vectors of each component objective function, valued at that point, equal to zero. Moreover, every weakly efficient solution is a vector critical point (see, e.g., [193]).

Theorem 6.5. *Let $\bar{x}$ be a weakly efficient solution for problem (VOP). Then there exists $\bar{\lambda} \geq 0$ such that $\lambda^T \nabla f(\bar{x}) = 0$.*

For the converse of the above theorem, we need some convexity hypothesis.

Theorem 6.6. *Let $\bar{x}$ be a vector critical point to (VOP) and f is invex at $\bar{x}$ with respect to η, then $\bar{x}$ is a weakly efficient solution for (VOP).*

Proof. See Osuna-Gomez et al. [193].

As mentioned before, one way to solve multiobjective programming problems is to relate its weakly efficient solutions to the optimal solutions for scalar problems whose resolution has already been studied. One of the most known scalar problems associated with multiobjective programming problems is the weighting problem (see [238]) whose formulation has the following form:

(P_λ)

$$\text{Minimize } \lambda^T f(x)$$

$$\text{Subject to } x \in S \subseteq R^n,$$

where $\lambda \in R^p$. It has been proved that every solution of weighting scalar problem with $\lambda \geq 0$ is a weakly efficient solution, but the converse is not always true see, e.g., Geoffrion [67].

Theorem 6.7. *If f is invex on an open set S, then all weakly efficient solutions solve a weighting scalar problem with $\lambda \geq 0$.*

Proof. Let $\bar{x}$ be a weakly efficient point, then there exists $\lambda \in R^p$ with $\lambda \geq 0$ such that $\lambda^T \nabla f(\bar{x}) = 0$. As f is invex at $\bar{x}$ with respect to η, so is $\lambda^T f$, that is,

$$\lambda^T f(x) - \lambda^T f(\bar{x}) \geq \lambda^T \nabla f(\bar{x}) \eta(x, \bar{x}) = 0.$$

Then $\lambda^T f(x) \geq \lambda^T f(\bar{x})$. Therefore, $\bar{x}$ is optimal solution for (P_λ) with $\lambda \geq 0$. □

It is interesting to note that under the invexity hypothesis, the vector critical point, the weakly efficient solutions and the optimal solutions for weighting scalar problems coincide. Osuna-Gomez et al. [193] have shown that the invexity hypothesis is not only a sufficient but also a necessary condition for all these classes of points to be equivalent.

Theorem 6.8. *Each vector critical point is a weakly efficient solution and solve a weighting scalar problem if and only if the objective function is invex.*

Proof. The sufficient part has already been proved in Theorems 6.6 and 6.7. For the necessary part see Osuna-Gomez et al. [193]. □

Osuna-Gomez et al. [193] characterized weakly efficient solutions for the constrained vector optimization problem (CVOP) using concepts similar to Fritz-John and Kuhn–Tucker optimality condition concepts.

Definition 6.9. *A feasible point $\bar{x} \in S$, is said to be a vector Fritz-John point to the problem (CVOP) if there exists a vector $(\bar{\lambda}, \bar{\mu}) \in R^{p+m}$ with $(\bar{\lambda}, \bar{\mu}) \geq 0$ such that*

$$\bar{\lambda}^T \nabla f(\bar{x}) + \bar{\mu}^T \nabla g(\bar{x}) = 0 \tag{6.1}$$

$$\bar{\mu}^T g(\bar{x}) = 0. \tag{6.2}$$

Definition 6.10. *A feasible point, $\bar{x} \in S$, is said to be a vector Kuhn–Tucker point to the problem (CVOP) if there exists a vector $(\bar{\lambda}, \bar{\mu}) \in R^{p+m}$ with $(\bar{\lambda}, \bar{\mu}) \geq 0$ and $\bar{\lambda} \neq 0$ such that*

$$\bar{\lambda}^T \nabla f(\bar{x}) + \bar{\mu}^T \nabla g(\bar{x}) = 0 \tag{6.3}$$

$$\bar{\mu}^T g(\bar{x}) = 0. \tag{6.4}$$

The following results due to Osuna-Gomez et al. [193] extend the scalar case in a natural way. In fact, the above definitions coincides with the Fritz-John and Kuhn–Tucker conditions when f is a real-valued function.

Theorem 6.11. *Let $\bar{x}$ be a weakly efficient solution for problem (CVOP); then there exist $\bar{\lambda}$ and $\bar{\mu}$ such that $\bar{x}$ is a vector Fritz-John point for (CVOP).*

Theorem 6.12. *Let $\bar{x}$ be a weakly efficient solution for problem (CVOP) and the Kuhn–Tucker constraint qualification is satisfied at $\bar{x}$. Then there exist $\bar{\lambda}$ and $\bar{\mu}$ such that $\bar{x}$ is a vector Kuhn–Tucker point for problem (CVOP).*

Many of the results given above in this section are even true under weaker invexity assumption, namely pseudo-invex functions:

Definition 6.13. *Let $f : S \subseteq R^n \to R^p$ be a differentiable function on the open set S. Then f is pseudo-invex on S with respect to η if for all $x, y \in S$, there exists $\eta(x, y) \in R^n$ such that*

$$f(x) - f(y) < 0 \Rightarrow \nabla f(y)\eta(x, y) < 0. \tag{6.5}$$

Osuna-Gomez et al. [192,193] obtained the Theorem 6.6 under pseudo-invexity hypothesis.

Theorem 6.14. *Let $\bar{x}$ be a vector critical point for (VOP), and let f be a pseudo-invex function at $\bar{x}$ with respect to η. Then, $\bar{x}$ is a weakly efficient solution.*

Proof. Let $\bar{x}$ be a vector critical point for (VOP), i.e., there exists $\lambda \geq 0$ such that $\lambda^T \nabla f(\bar{x}) = 0$. If there exists another $x \in S$ such that $f(x) < f(\bar{x})$, then there exists $\eta(x, \bar{x}) \in R^n$ such that $\nabla f(\bar{x})\eta(x, \bar{x}) < 0$, i.e., the system

$$\lambda^T \nabla f(\bar{x}) = 0,$$

$$\lambda \geq 0, \quad \lambda \in R^p,$$

has no solution for λ.

□

Osuna-Gomez et al. [192, 193] established a stronger result than that of Theorem 6.14.

Theorem 6.15. *A vector function f is pseudo-invex of and only if every vector critical point of f is a weakly efficient solution on S.*

Proof. See the proof of Theorem 2.2 of Osuna-Gomez et al. [192] or the proof of Theorem 2.5 of Osuna-Gomez et al. [193].

□

Gulati and Islam [77] considered a constrained vector optimization problem with inequality and equality constraints. Even though as it is pointed out in previous chapters that the equality constraints are already taken care by inequality constraints as $h(x) = 0$ can be written as $h(x) \leq 0$ and $-h(x) \leq 0$.

Martin [145] established that it is not possible to infer (f and g)-invexity from the equivalence of minimum points and Kuhn–Tucker points for scalar programming problems. So Martin [145] defined KT-invexity, which is a condition weaker than (f and g)-invexity. Similar question was investigated by Osuna-Gomez et al. [192, 193]. KT-invex functions are extended to multiobjective case by Osuna-Gomez et al. [192, 193].

Definition 6.16. *Problem (CVOP) is said to be KT-invex on the feasible set if there exists a vector function* $\eta : S \times S \to R^n$ *such that* $\forall x, y \in S$, *with* $g(x) \leq 0$ *and* $g(y) \leq 0$,

$$f(x) - f(y) \geq \nabla f(y)\eta(x, y) \tag{6.6}$$

$$-\nabla g(y)\eta(x, y) \geq 0, \quad \forall i \in I(y). \tag{6.7}$$

Osuna-Gomez et al. [193] gave the following characterization for the weakly efficient points.

Theorem 6.17. *Every vector Kuhn–Tucker point is a weakly efficient solution if problem (CVOP) is KT-invex.*

Osuna-Gomez et al. [193] used Theorem 6.17 to prove that if the problem (CVOP) is KT-invex, then all weakly efficient solutions can be found as solutions for a scalar problem. Thus under KT-invexity condition and if constraint qualification is satisfied, vector-Kuhn–Tucker point, weakly efficient points and optimal solutions for weighting problem coincides.

Theorem 6.18. *If (CVOP) is KT-invex and the Kuhn–Tucker constraint qualification is satisfied at all weakly efficient solutions, then every weakly efficient solution solves a weighting scalar problem.*

An analogous theorem for vector Kuhn–Tucker points is as follows (see, [193]):

Theorem 6.19. *Every vector Kuhn–Tucker point solves a weighting scalar problem if (CVOP) is KT-invex.*

Even a stronger result is given by Osuna-Gomez et al. [193].

Theorem 6.20. *Every vector Kuhn–Tucker point is a weakly efficient point and solves a weighting scalar problem if and only if the problem (CVOP) is KT-invex.*

These authors have given a weaker KT-invexity namely, KT-pseudo-invexity in [193], however the same definition is called KT-invex in [192].

Definition 6.21. *The problem (CVOP) is said to be a KT-pseudo-invex with respect to* η *if for any* $x, y \in S$ *with* $g(x) \leq 0$ *and* $g(y) \leq 0$ *there exists* $\eta(x, y) \in R^n$ *such that:*

$$f(x) < f(y) \Rightarrow \nabla f(y)\eta(x, y) < 0 \tag{6.8}$$

$$-\nabla g_i(y) \geq 0, \quad \forall i \in I(y), \tag{6.9}$$

where $I(y) = \{i : i = 1, \ldots, m \text{ such that } g_i(y) = 0\}$.

It is interesting to see (see [193]) that this definition is necessary and sufficient for the set of vector Kuhn–Tucker points and the set of weakly efficient point to be the same.

Theorem 6.22. *Every vector Kuhn–Tucker point is weakly efficient for the problem (CVOP) if and only if (CVOP) is a KT-pseudo-invex problem.*

For the proof, see [193] or [192].

It is worth noting that invexity allows us to give the necessary and sufficient conditions for locating the solutions to the general problem starting from the solutions of a scalar problem or verifying certain conditions of optimality of the type defined by Kuhn–Tucker and Fritz-John. Several results from scalar optimization problems can be extended to vector optimization problem.

The properness of the efficient solution of the optimal problem with multicriteria has been introduced at the early stage of the study of this problem [133]. Geoffrion [67] defined the properness for the purpose of eliminating an undesirable possibility in the concept of efficiency, namely the possibility of the criterion functions being such that efficient solutions could be found for which the marginal gain for one function could be made arbitrarily large relative to the marginal losses for the others. Geoffrion [67] gave a theorem describing the relation of the Kuhn–Tucker proper efficient solutions and his proper efficient solution. In this section, we summarize briefly the known results of proper (improper) efficient solutions for (VP), and apply them to five examples.

The problems discussed in the papers of Kuhn and Tucker [133], Geoffrion [67], Tamura and Arai [231] and Singh and Hanson [223] are of the following nature:

$$V - Maximize f(x)$$

$$\text{Subject to } g(x) \leq 0,$$

where $f(x)$ is p-dimensional vector function and $g(x)$ is m-dimensional vector function.

Let K denote the set of feasible solutions of the above vector maximum problem.

Definition 6.23. *An efficient solution x^0 is called a proper efficient solution if there exists no $x \in K$ such that*

$$\nabla f(x^0)x \geq 0, \quad \nabla g_I(x^0)x \geq 0,$$

where $\nabla g_I(x^0)$ is a matrix whose row vector is a gradient function of an active constraint.

This solution is also known as KT-proper efficient solution.

Definition 6.24 ([67]). *An efficient solution x^0 is called a proper efficient solution if there exists a scalar $M > 0$ such that, for each i,*

$$\frac{f_i(x) - f_i(x^0)}{f_j(x^0) - f_j(x)} \leq M$$

for some j such that $f_j(x) < f_j(x^0)$, whenever, $x \in K$ and $f_i(x) > f_i(x^0)$.

In case the problem is a minimization problem, we have the following inequalities:

$$\frac{f_i(x^0) - f_i(x)}{f_j(x) - f_j(x^0)} \leq M$$

for some j such that $f_j(x) > f_j(x^0)$, whenever, $x \in K$ and $f_i(x) < f_i(x^0)$. Geoffrion's [67] definition of proper efficiency M is independent of x, and it may happen that if f is unbounded such an M may not exist. Also an optimizer might be willing to trade different levels of losses for different levels of gains by different values of the decision variable x. Singh and Hanson [223] extended the concept to situation where M depends on x.

Definition 6.25. *The point x^0 is said to be conditionally properly efficient if x^0 is efficient and there exists a positive function $M(x)$ such that, for each i, we have*

$$\frac{f_i(x) - f_i(x^0)}{f_j(x^0) - f_j(x)} \leq M(x),$$

for some j such that $f_j(x) < f_j(x^0)$ whenever $x \in X$ and $f_i(x) > f_i(x^0)$.

If the problem is of vector minimization, we have:
The point x^0 is said to be conditionally properly efficient if x^0 is efficient and there exists a positive function $M(x)$ such that, for each i, we have

$$\frac{f_i(x^0) - f_i(x)}{f_j(x) - f_j(x^0)} \leq M(x),$$

for some j such that $f_j(x) > f_j(x^0)$ whenever $x \in X$ and $f_i(x) < f_i(x^0)$. For more discussion on different concepts of efficiency (properly efficient, conditionally properly efficient, etc.) see Mishra [153].

In Chap. 4, Definition 4.1, we have discussed pre-invex functions for real-valued functions. A p-dimensional vector-valued function $f : S \to R^p$ is pre-invex on S with respect to η if each of its components is pre-invex on S with respect to the same η. For other algebraic properties of pre-invex functions, see Sect. 4.1 of Chap. 4. Geoffrion [67] showed that the properly efficient solutions of (CVOP) may be characterized in terms of the solutions to a scalar valued parametric programming problem under convexity assumptions. Weir and Mond [245] showed that Geoffrion's assumption of convexity may be replaced by pre-invexity to characterize properly efficient solutions of

(CVOP). In relation to (CVOP), Geoffrion [67] considered the following scalar minimization problem:

$$(CVOP_{\lambda})\text{Minimize}\lambda^T f(x) \text{ Subject to } x \in S,$$

where $\lambda \in \Lambda^+ = \{\lambda \in R^p : \lambda > 0, \sum_{i=1}^p \lambda_i = 1\}$. Geoffrion established the following fundamental result:

Theorem 6.26. *Let $\lambda > 0, i = 1, \ldots, p$, be fixed. If x_0 is optimal in $(CVOP_{\lambda})$, then x_0 is properly efficient for (CVOP).*

Weir and Mond [245] established an extension of Gordna's alternative theorem involving pre-invex functions:

Theorem 6.27. *Let S be a non-empty set in R^n and let $f : S \to R^m$ be a pre-invex function on S with respect to η. Then either*

$$f(x) < 0 \text{ has a solution } x \in S$$

or

$$p^T f(x) \geqq 0 \text{ for all } x \in S, \text{ for some } p \in R^m, p \geqq 0,$$

but both alternatives are never true.

Using above Theorem in place of Gordan's alternative, Weir and Mond [245] gave the converse of Theorem 6.26 under pre-invexity assumption thus extending "Comprehensive Theorem" of Geoffrion [67].

Theorem 6.28. *Let f be pre-invex on S with respect to η. Then x_0 is properly efficient for (CVOP) if and only if x_0 is optimal for $(CVOP_{\lambda})$.*

Later Weir [242] established the following result:

Theorem 6.29. *Suppose there exists a feasible point x^* and scalars $\lambda_i > 0$, $i = 1, \ldots, p, \mu_j \geq 0, j \in I(x^*)$ such that*

$$\sum_{i=1}^{p} \lambda_i \nabla f_i(x^*) + \sum_{j \in I(x^*)} \mu_j \nabla g_j(x^*) = 0. \tag{6.10}$$

Then, if each $f_i, i = 1, \ldots, p$ and $g_j, j \in I(x^*)$ are invex with respect to the same η, x^* is a properly efficient solution for (CVOP).

Proof. Since the gradient of the invex function $\psi(x) = \sum_{i=1}^p \lambda_i f_i(x) + \sum_{j \in I(x^*)} \mu_j g_j(x)$ vanishes at x^*, then x^* is an unconstrained global minimizer of ψ. Since x^* is feasible for (CVOP), it follows that x^* minimizes $\sum_{i=1}^p \lambda_i f_i(x)$ subject to $g_j(x) \leqq 0, j = 1, \ldots, m$. The proper efficiency of x^* for (CVOP) then follows as in Theorem 1 of [67].

□

Weir [242] also obtained the following theorem:

Theorem 6.30. *Assume that x^* is a properly efficient solution for (CVOP). Assume also, that there exists a point $\bar{x}$ such that $g_j(\bar{x}) < 0, j = 1, \ldots, m$ and that each $g_j, j \in I(x^*)$ are invex with respect to the same η. Then there exists scalars $\lambda_i, i = 1, \ldots, p$ and $\mu_j \geqq 0, j \in I(x^*)$ such that $(x^*, \lambda_i, \mu_j), i = 1, \ldots, p, j \in I(x^*)$ satisfies (6.10).*

Remark 6.31. Geoffrion [67] established that (6.10) holds under the Kuhn–Tucker constraint qualification without any convexity assumptions. Since every convex function is invex, Geoffrion [67] established that (6.10) is a necessary condition for x^* to be properly efficient under the Slater constraint qualification.

Kaul et al. [116] considered a multi-objective optimization problem involving type I, quasi-type I, pseudo-type I, quasi-pseudo-type I, pseudo-quasi-type I objective and constrained functions and discussed Kuhn–Tucker type necessary and sufficient optimality conditions for a feasible point to be an efficient or properly efficient solution. They have also obtained duality results for Wolfe type and Mond–Weir type duals under the aforesaid weaker invexity assumptions. For the reader's convenience let us recall the following concepts from Kaul et al. [116].

Definition 6.32. *For $i = 1, \ldots, p$, (f_i, g) is said to be type I with respect to η at $x^0 \in S$ if there exists a vector function $\eta(x, x^0)$ defined on $K \times S$ such that, $\forall x \in K$,*

$$f_i(x) - f_i(x^0) \geqq \nabla f_i(x^0)\eta(x, x^0), \tag{6.11}$$

$$-g(x^0) \geqq \nabla g(x^0)\eta(x, x^0). \tag{6.12}$$

If in the above definition, (6.11) is a strict inequality, then we say that (f_i, g) is semi-strictly-type I at x^0.

Definition 6.33. *For $i = 1, \ldots, p$, (f_i, g) is said to be quasi-type I with respect to η at $x^0 \in S$ if there exists a vector function $\eta(x, x^0)$ defined on $K \times S$ such that, $\forall x \in K$,*

$$f_i(x) \leqq f_i(x^0) \Rightarrow \nabla f_i(x^0)\eta(x, x^0) \leqq 0, \tag{6.13}$$

$$-g(x^0) \leqq 0 \Rightarrow \nabla g(x^0)\eta(x, x^0) \leqq 0. \tag{6.14}$$

Definition 6.34. *For $i = 1, \ldots, p$, (f_i, g) is said to be pseudo-type I with respect to η at $x^0 \in S$ if there exists a vector function $\eta(x, x^0)$ defined on $K \times S$ such that, $\forall x \in K$,*

$$\nabla f_i(x^0)\eta(x, x^0) \geqq 0 \Rightarrow f_i(x) \geqq f_i(x^0), \tag{6.15}$$

$$\nabla g(x^0)\eta(x, x^0) \geqq 0 \Rightarrow -g(x^0) \geqq 0. \tag{6.16}$$

Definition 6.35. *For $i = 1, \ldots, p$, (f_i, g) is said to be quasi-pseudo-type I with respect to η at $x^0 \in S$ if there exists a vector function $\eta(x, x^0)$ defined on $K \times S$ such that, $\forall x \in K$,*

$$f_i(x) \leqq f_i(x^0) \Rightarrow \nabla f_i(x^0)\eta(x, x^0) \leqq 0, \tag{6.17}$$

$$\nabla g(x^0)\eta(x, x^0) \geqq 0 \Rightarrow -g(x^0) \geqq 0. \tag{6.18}$$

Definition 6.36. *For $i = 1, \ldots, p$, (f_i, g) is said to be pseudo-quasi-type I with respect to η at $x^0 \in S$ if there exists a vector function $\eta(x, x^0)$ defined on $K \times S$ such that, $\forall x \in K$,*

$$\nabla f_i(x^0)\eta(x, x^0) \geqq 0 \Rightarrow f_i(x) \geqq f_i(x^0), \tag{6.19}$$

$$-g(x^0) \leqq 0 \Rightarrow \nabla g(x^0)\eta(x, x^0) \leqq 0. \tag{6.20}$$

Kaul et al. [116] obtained necessary and sufficient conditions for a feasible solution to be efficient or properly efficient for the following vector optimization problem:

(VP)

$$\text{Minimize} f(x) = (f_1(x), \ldots, f_p(x))$$

$$\text{Subject to } g(x) \leqq 0,$$

$$x \in S \subseteq R^n,$$

where $f : S \to R^p$ and $g : S \to R^m$ are differentiable functions on a set $S \subseteq R^n$ and minimization means obtaining efficient solutions of (VP). For a feasible point $x^* \in K$, we denote by $I(x^*)$ the set

$$I(x^*) = \{j : g_j(x^*) = 0\}.$$

We give some results from Kaul et al. [116].

Theorem 6.37. *Suppose that there exists a feasible solution x^* for (VP) and scalar $\lambda_i^* > 0, i = 1, \ldots, p$, $\mu_j^* \geqq 0, j \in I(x^*)$ such that*

$$\sum_{i=1}^{p} \lambda_i^* \nabla f_i(x^*) + \sum_{j \in I(x^*)} \mu_j^* \nabla g_j(x^*) = 0. \tag{6.21}$$

If for $i = 1, \ldots, p$, (f_i, g_I) are type at x^* with respect to same η, then x^* is a properly efficient solution for (VP).

Proof. See Kaul et al. [116].

□

These authors have obtained several sufficient optimality conditions under different types of generalized invexity assumptions (such as, semi-strictly-type

I, pseudo-quasi-type I, etc.). They have also given the following necessary optimality conditions.

Theorem 6.38. *Assume that x^* is a properly efficient solution for (VP). Assume also that there exists a point $\bar{x}$ such that $g_(\bar{x}) < 0, i = 1, \ldots, m$, such that $g_j, j \in I(x^*)$, satisfies*

$$-g_j(x^*) > \nabla g_j(x^*)\eta(x, x^*), \quad \forall x \in S. \tag{6.22}$$

Then, there exist scalars $\lambda_i^* > 0, i = 1, \ldots, p$. $\mu_j \geqq 0, \in I(x^*)$, such that the triplet $(x^*, \lambda_i^*, \mu_j^*)$ satisfies

$$\sum_{i=1}^{p} \lambda_i^* \nabla f_i(x^*) + \sum_{j \in I(x^*)} \mu_j^* \nabla g_j(x^*) = 0. \tag{6.23}$$

Theorem 6.39. *Assume that x^* is an efficient solution for (VP) at which the Kuhn–Tucker constraint qualification is satisfied. Then, there exist scalar $\lambda_i^* \geqq 0, i = 1, \ldots, p, \sum_{i=1}^{p} \lambda_i^* = 1, \mu_j^* \geqq 0, j = 1, \ldots, m$, such that*

$$\sum_{i=1}^{p} \lambda_i^* \nabla f_i(x^*) + \sum_{j=1}^{m} \mu_j^* \nabla g_j(x^*) = 0, \tag{6.24}$$

$$\sum_{j=1}^{m} \mu_j^* g_j(x^*) = 0. \tag{6.25}$$

Invex, Type I invex and their natural generalizations require the same kernel function for the objective and the constraints. This requirement turns out to be a severe restriction in applications. Because of this restriction, pseudo-linear multiobjective problems [36] and certain nonlinear multiobjective fractional programming problems require separate treatment as far as optimality and duality properties are concerned. In order to avoid this restriction, Jeyakumar and Mond [105] introduced a new class of functions, namely V-invex functions. See also Sect. 5.1 of the present monograph. For detailed study on V-invex functions and applications see, Mishra [153]. The V-invex functions can be formed from certain nonconvex functions (in particular from convex-concave or linear fractional functions) by coordinate transformations,(see, e.g., [105, 153]). Following Kaul et al. [116] and Jeyakumar and Mond [105], Hanson et al. [85], introduced V-type I functions, quasi V-type I functions, pseudo V-type I functions, etc.

Definition 6.40. *(f_i, g_j) is said to be V-type I objective and constraint functions at $y \in K$ if there exist positive real-valued functions α_i and β_j defined on $S \times S$ and n-dimensional vector-valued function $\eta : S \times S \to R^n$ such that*

$$f_i(x) - f_i(y) \geqq \alpha_i(x, y)\nabla f_i(x, y)\eta(x, y) \tag{6.26}$$

and

$$-g_j(x, y) \geqq \beta_j(x, y) \nabla g_j(x, y) \eta(x, y), \tag{6.27}$$

for every $x \in K$ *and* $\forall i = 1, \ldots, p$, *and* $j = 1, \ldots, m$.

If strict inequality holds in (6.26) (whenever $x \neq y$) we say (f_i, g_j) is of semi-strictly V-type I at y.

Hanson et al. [85] have given other generalization namely, quasi V-type I, pseudo V-type I, quasi-pseudo V-type I and pseudo-quasi V-type I on the lines of Kaul et al. [116]. Hanson et al. [85] have obtained several sufficient optimality conditions and a necessary optimality conditions similar to Kaul et al. [116].

Later Aghezzaf and Hachimi [1] introduced new classes of functions called weak strictly pseudo-quasi-type I and strong pseudo-quasi-type I be relaxing the definitions of type I, weak strictly pseudoconvex (see, [147]), the classes of weak quasi-strictly-pseudo-type I and weak strictly pseudo-type I functions are introduced as a generalization of quasi-pseudo-type I and strictly pseudo-type I functions [116]. However, these authors have only discussed duality results under these new classes of functions. Aghezzaf and Hachimi [2] introduced some new classes of nonconvex functions by relaxing the definitions of invex and preinvex functions. Let us recall the definition of weak pre-quasi-invex functions from [2].

Definition 6.41. *A vector-valued function* f *is said to be weak pre-quasi-invex at* $\bar{x} \in X$ *with respect to* η *if* X *is invex at* $\bar{x}$ *with respect to* η *and for each* $x \in X$,

$$f(x) \leq f(\bar{x}) \Rightarrow f(\bar{x} + \lambda\eta(x, \bar{x})) \leqq f(\bar{x}), 0 < \lambda \leqq 0.s$$

The following example shows that the converse of every pre-quasi-invex function is weak pre-quasi-invex with respect to the same η is not true in general.

Example 6.42. The function $f : R \to R^2$ defined by $f_1(x) = x(x-2)^2$ and $f_2(x) = x(x-2)$ is weak pre-quasi-invex at $\bar{x} = 0$ with respect to $\eta(x, \bar{x}) = x - \bar{x}$, but it is not pre-quasi-invex at $\bar{x}$ with respect to the same η, because f_1 is not pre-quasi-invex at $\bar{x}$ with respect to the same η.

Aghezzaf and Hachimi [2] further introduced weak strictly-pseudo-invex, strong pseudo-invex, weak quasi-invex and weak strictly-pseudo-invex functions and obtained various first order and second order sufficient optimality conditions involving the newly introduced classes of functions. Let us give the following second order optimality conditions for multiobjective optimization problem (vector minimization problem):

Theorem 6.43. *Suppose that* f *is weak pre-quasi-invex and* g *is pre-quasi-invex with respect to the same* η *at* $\bar{x} \in A = \{x \in S : g(x) \leqq 0\}$ *and are twice* *continuously differentiable at $\bar{x}$. Further suppose that $\eta(x,y) \neq 0$ for all $x \neq y$. If for each critical direction $y \neq 0$, there exist $\bar{u} \in R^p$ and $\bar{v} \in R^l$ satisfying*

$$\bar{\lambda}^T \nabla f(\bar{x}) + \bar{\mu}^T \nabla g_I(\bar{x}) = 0,$$

$$\left(\sum_{i=1}^{p} \bar{\lambda}_i \nabla^2 f_i(\bar{x}) + \sum_{j \in I} \bar{\mu}_j \nabla^2 g_j(\bar{x}) \right) \eta(x,y) > 0,$$

$$\bar{\lambda} \geq 0, \quad \bar{\mu} \geqq 0,$$

then $\bar{x}$ is (weak) Pareto minimum for the vector minimization problem.

Mishra [155] considered the usual vector minimization problem:
(VP)

$$\text{Minimize } f(x) = (f_1(x), \ldots, f_p(x))$$

$$\text{Subject to } g(x) \leqq 0,$$

$$x \in S \subseteq R^n,$$

where $f : X \to R^p$ and $g : S \to R^m$ are differentiable functions and minimization means obtaining efficient solution of (VP).

Mishra [155] obtained several sufficient optimality conditions for (VP) under some new type of generalized convexity, by combining the concepts of type I, pseudo-type I, quasi-type I, quasi-pseudo-type I, pseudo-quasi-type I, strictly pseudo-quasi-type I (see above in this section and also [116]) and the class of univex functions defined by Bector et al. [15]:

Definition 6.44. *Let S be a non-empty open set in R^n, $f : S \to R$, $\eta : S \times S \to R^n$, $\phi : R \to R$, and $b : S \times [0,1] \to R_+$, the function f is said to be univex at $u \in S$ with respect to η, ϕ and b if $\forall x \in X$, we have*

$$b(x,u)\phi\left[f(x) - f(u)\right] \geqq \eta(x,u)\nabla f(u).$$

For further discussion on univex functions see also Rueda et al. [215].

6.3 Duality in Vector Optimization

In this section we consider three different types of dual models for the vector minimization problem (VP) discussed in Sect. 6.2. The following Wolfe type dual model is an extension of the Wolfe dual problem (see Sect. 5.4) to multiobjective case. (WD)

$$\text{Maximize} f(u) + \sum_{j=1}^{m} \mu_j g_j(u)$$

$$\text{Subject to } \sum_{i=1}^{p} \lambda_i \nabla f_i(u) + \sum_{j=1}^{m} \mu_j \nabla g_j(u) = 0, \tag{6.28}$$

$$\mu_j \geqq 0, \quad j = 1, \ldots, m, \tag{6.29}$$
$$\lambda_i \geqq 0, \quad i = 1, \ldots, p, \tag{6.30}$$
$$\sum_{i=1}^{p} \lambda_i = 1. \tag{6.31}$$

Assuming each $f_i, i = 1, \ldots, p$ and $g_j, j = 1, \ldots, m$ to be invex for the same η. Weir [242] proved that weak duality and strong duality hold for (VP) and (WD).

Theorem 6.45 (Weak duality). *Let x be feasible for (CVOP) and (u, λ, μ) feasible for (WD). If each $f_i, i = 1, \ldots, p$ and $g_j, j = 1, \ldots, m$ are invex for the same η, then*

$$f(x) \nleq f(u) + \sum_{i=1}^{m} \mu_j g_j(u).$$

Theorem 6.46 (Strong duality). *Let each $f_i, i = 1, \ldots, p$ and each $g_j, j = 1, \ldots, m$ be invex with respect to the same η, and let x^* be a properly efficient solution for (CVOP). Assume that there exists an $\bar{x}$ such that $g_j(\bar{x}) < 0, j = 1, \ldots, m$. Then there exist (λ, μ) such that (x^*, λ, μ) is properly efficient for (WD) and the objective values of (CVOP) and (WD) are equal.*

Weir and Mond [245] obtained the above two duality relation under pre-invexity assumptions as well. Following Weir [242] Kaul et al. [116] proved various duality results for (VP) and (WD) under weaker invexity conditions.

Theorem 6.47. *Let x be feasible solution for (VP) and let the triplet (u, λ, μ) be feasible for (WD). Let either (a) or (b) of below holds:*

(a) For $i = 1, \ldots, p, \lambda_i > 0$ and (f_i, g) are type I objective and constraint functions at u with respect to the same η;
(b) For $i = 1, \ldots, p$, (f_i, g) are semi-strictly-type I objective and constraint functions at u with respect to the same η.

Then

$$f(x) \nleq f(u) + \sum_{j=1}^{m} \mu_j g_j(u).$$

Proof. (a) Note that

$$\begin{aligned} &\sum_{i=1}^{p} \lambda_i \left(f_i(x) - \left(f_i(u) + \sum_{j=1}^{m} \mu_j g_j(u) \right) \right) \\ &\geqq \sum_{i=1}^{p} \lambda_i \nabla f_i(u) \eta(x, u) + \sum_{j=1}^{m} \mu_j \nabla g_j(u) \eta(x, u) \\ &= 0, \end{aligned}$$

by the type I invexity hypothesis of (a) and (6.28). Thus,

$$f(x) \nleq f(u) + \sum_{j=1}^{m} \mu_j g_j(u).$$

The proof of part (b) is also based on the definition, see [116].

□

Kaul et al. [116] gave the following strong duality theorem under type I invexity assumptions.

Theorem 6.48. *Let $(f_i, g), i = 1, \ldots, p$, be type I with respect to the same η at u and let x^* be a properly efficient solution for (VP). Assume also that there exists a point $\bar{x}$ such that $g_j(\bar{x}) < 0, j = 1, \ldots, m$, and each $g_i, i \in I(x^*)$ satisfying*

$$-g_j(x^*) > \nabla g_j(x^*)\eta(x, x^*), \quad \forall x \in S.$$

Then, there exists $(\lambda^*, \mu^*) \in R^p \times R^m$ such that (x^*, λ^*, μ^*) is properly efficient for (WD) and the objective function values of (VP) and (WD) are equal.

The Mond–Weir type dual for (VP) has the advantage over the Wolfe dual is that it allows further weakening of the invexity conditions in order to obtain usual duality results. Note that the following Mond–Weir type dual for (VP) is the vector case of the Mond–Weir dual discussed in Chap. 5. (MWD)

$$Maximize f(u) = (f_1(u), \ldots, f_p(u))$$

Subject to

$$\sum_{i=1}^{p} \lambda_i \nabla f_i(u) + \sum_{j=1}^{m} \mu_j \nabla g_j(u) = 0, \tag{6.32}$$

$$\sum_{j=1}^{m} \mu_j g_j(u) \geqq 0, \tag{6.33}$$

$$\mu_j \geqq 0, \quad j = 1, \ldots, m, \tag{6.34}$$

$$\lambda_i \geqq 0, \quad i = 1, \ldots, p, \quad \sum_{i=1}^{p} \lambda_i = 1. \tag{6.35}$$

Kaul et al. [116] established the following weak duality for (VP) and (MWD):

Theorem 6.49. *Let x be feasible for (VP) and let (u, λ, μ) be feasible for (MWD). Let either (a) or (b) of below hold:*

(a) For $i = 1, \ldots, p, \lambda_i > 0$ and $(\sum_{i=1}^{p} \lambda_i f_i, \sum_{j=1}^{m} \mu_j g_j)$ is pseudo-quasi-type I at u with respect to the same η;
(b) For $i = 1, \ldots, p, \lambda_i > 0$ and $(\sum_{i=1}^{p} \lambda_i f_i, \sum_{j=1}^{m} \mu_j g_j)$ is quasi-strictly-pseudo-type I at u with respect to the same η.

Then, $f(x) \nleq f(u)$.

Theorem 6.50. *Let* $(\sum_{i=1}^{p} \lambda_i f_i, \sum_{j=1}^{m} \mu_j g_j)$ *be pseudo-quasi-type I for all feasible solutions* (u, λ, μ) *for (MWD) with respect to* η *at* u. *Let* x^* *be properly efficient solution for (VP). Assume that there exists a point* $\bar{x}$ *such that* $g_j(\bar{x}) < 0, \forall j = 1, \ldots, m$, *and each* $g_j, j \in I(x^*)$ *satisfying*

$$-g_j(x^*) \leqq 0 \Rightarrow \eta(x, x^*)\nabla g_j(x^*), \quad \forall x \in S.$$

Then, there exists $(\lambda^*, \mu^*) \in R^p \times R^m$ *such that* (x^*, λ^*, μ^*) *is properly efficient for (MWD) and the objective function values of (VP) and (MWD) are equal.*

Kaul et al. [116] obtained strong duality under further weakened type I invexity assumptions for (VP) and (MWD). Aghezzaf and Hachimi [2] considered the Mond–Weir type dual for (VP) and obtained weak, strong and converse duality results under strong pseudo-invexity/quasi-invexity, weak strictly pseudo-invexity/quasi-invexity. For the reader's convenience, we recall the new definitions from Aghezzaf and Hachimi [2]:

Definition 6.51. f *is said to be weak strictly pseudo-invex with respect to* η *at* $\bar{x} \in S$ *if there exists a vector function* $\eta(x, \bar{x})$ *defined on* $S \times S$ *such that,* $\forall x \in S$

$$f(x) \leq f(\bar{x}) \Rightarrow \nabla f(\bar{x})\eta(x, \bar{x}) < 0.$$

Aghezzaf and Hachimi [2] have observed that every strictly pseudo-invex function is weak strictly pseudo-invex with respect tot he same η. However, the converse is not necessarily true, as can be seen from the following example from [2].

Example 6.52. The function $f = (f_1, f_2)$ defined on $S = R$, by $f_1(x) = x(x+2)$ and $f_2(x) = x(x+2)^2$ is weak strictly pseudo-invex with respect to $\eta(x, \bar{x}) = x + 2$ at $\bar{x} = 0$ but it is not strictly pseudo-invex with respect to the same η at $\bar{x} = -2$

$$f(x) \leqq f(\bar{x}) \quad \text{but} \quad \nabla f(\bar{x})\eta(x, \bar{x}) = 0.$$

Now we are in position to state the weak duality given by Aghezzaf and Hachimi [2].

Theorem 6.53. *Let* x *be feasible for (VP) and let* (u, λ, μ) *be feasible for (MWD). If* $\lambda \geqq 0, f$ *is weak strictly pseudo-invex and* $\mu^T g$ *is quasi-invex at* u *with respect to the same* η. *Then,* $f(x) \nleq f(u)$.

Proof. Suppose contrary to the result, i.e., $f(x) \leq f(u)$. By feasibility and duality constraint (6.33), we have

$$\nabla f(u)\eta(x, u) < 0, \tag{6.36}$$

and

$$\nabla \mu^T \nabla g(u)\eta(x, u) \leqq 0. \tag{6.37}$$

Since $\lambda \geqq 0$, from (6.36) and (6.37), we get

$$\left(\lambda^T \nabla f(u) + \mu^T \nabla g(u)\right) \eta(x,u) < 0, \tag{6.38}$$

which contradicts the duality constraint (6.32).

□

These authors have given several other weaker invexity conditions and obtained weak duality under those weaker invexity assumptions. We refer the reader to Aghezzaf and Hachimi [2]. We give the following strong and converse duality theorems from Aghezzaf and Hachimi [2] under the invexity assumptions used in the above weak duality theorem.

Theorem 6.54. *Let $\bar{x}$ be Pareto minimal for (VP) and assume that $\bar{x}$ satisfies a suitable constraint qualification for (VP). Then there exist $\bar{\lambda} \in R^p$ and $\bar{\mu} \in R^m$ such that $(\bar{x}, \bar{\lambda}, \bar{\mu})$ is feasible for (MWD). If also the invexity hypothesis used in the weak duality Theorem 6.71 holds, then $(\bar{x}, \bar{\lambda}, \bar{\mu})$ is Pareto minimal (efficient) for (MWD).*

Theorem 6.55. *Let $(\bar{x}, \bar{\lambda}, \bar{\mu})$ be Pareto minimal (efficient solution) for (MWD), and let the invexity hypothesis of weak duality Theorem 6.71 holds. If the $n \times n$ Hessian matrix $\nabla^2 \left[\bar{\lambda}^T f(\bar{x}) + \bar{\mu}^T g(\bar{x})\right]$ is negative definite and if $\nabla \bar{\mu}^T g(\bar{x}) \neq 0$, then $\bar{x}$ is Pareto minimal for (VP).*

These authors have given several strong and converse duality theorems under various generalized invexity assumptions. Aghezzaf and Hachimi [1] have extended the results of Aghezzaf and Hachimi [2] to generalized type I invexity assumptions. Let us recall the following concept of weaker type I invexity from [2].

Definition 6.56. *(f, g) is said to be weak strictly-pseudo-quasi-type I with respect to η at u if there exists a vector function $\eta(x, u)$ such that $\forall x \in S$*

$$f(x) \leq f(u) \Rightarrow \nabla f(u)\eta(x,u) < 0,$$

$$-g(u) \leqq 0 \Rightarrow \nabla g(u)\eta(x,u) \leqq 0.$$

Aghezzaf and Hachimi [2] have shown that every strictly-pseudo-quasi-type I functions are weak strictly-pseudo-quasi-type I but the converse is not necessarily true. See, e.g., Example 2.1 of [2]. These authors have obtained several weak, strong and converse duality for (VP) and (MWD) under various weaker type I invexity assumptions. We give here one weak duality for (VP) and (MWD) from [2].

Theorem 6.57. *Assume that for all feasible x for (VP) and all feasible (u, λ, μ) for (MWD) and $(f, \mu g)$ is weak strictly pseudo-quasi-type I at u with respect to the η. Then, $f(x) \nleq f(u)$.*

The proof is very similar to the proof of Theorem 6.53. Aghezzaf and Hachimi [2] have obtained several other weak duality results under other weaker invexity assumptions. Based on the Theorem 6.49, we give the following strong and converse duality theorems from [2].

Theorem 6.58. *Let x^0 be an efficient solution for (VP) and assume that x^0 satisfies a constraint qualification ([147]) for (VP). Then there exist $\lambda^0 \in R^p$ and $\mu^0 \in R^m$ such that (x^0, λ^0, μ^0) is feasible for (MWD). If the generalized invexity hypothesis of weak duality Theorem 6.49 holds, then (x^0, λ^0, μ^0) is efficient solution for (MWD).*

Theorem 6.59. *Let (x^0, λ^0, μ^0) be efficient solution for (MWD), and let the invexity hypothesis of Theorem 6.49 holds. If the $n \times n$ Hessian matrix $\nabla^2 \left[\lambda^0 f(x^0) + \mu^0 g(x^0)\right]$ is negative-definite and if $\nabla \mu^0 g(x^0) \neq 0$, then x^0 is an efficient solution for (VP).*

Hanson et al. [85] obtained various duality results for (VP) and (MWD) under different types of generalized invexity. They have extended the concept of pseudo-quasi-type I functions to pseudo-quasi-V-type I by combining the concept of pseudo-quasi-type I functions and V-invex functions. For the reader's convenience, let us recall the same over here from [85].

Definition 6.60. *The problem (VP) is said to be pseudo-quasi-V-type I at u if there exist positive real-valued functions α_i and β_j defined on $S \times S$ and an n-dimensional vector-valued function $\eta : S \times S \to R^n$ such that for some $\lambda \in R^p, \lambda \geqq 0$ and $\mu \in R^m, \mu \geqq 0$, we have*

$$\sum_{i=1}^{p} \lambda_i \nabla f_i(u) \geqq 0$$

$$\Rightarrow \sum_{i=1}^{p} \lambda_i \alpha_i(x,u) \left(f_i(x) - f_i(u)\right) \geqq 0, \quad \forall x \in S \tag{6.39}$$

and

$$\sum_{j=1}^{m} \mu_j \beta_j(x,u) g_j(u) \geqq 0 \Rightarrow \sum_{j=1}^{m} \lambda_j \eta(x,u) \nabla g_j(u) \leqq 0, \quad \forall x \in S. \tag{6.40}$$

Hanson et al. [85] obtained weak duality for (VP) and (MWD) under pseudo-quasi-V-type I invexity.

Theorem 6.61. *Suppose that x is feasible for (VP) and (u, λ, μ) is feasible for (MWD) and the problem (VP) is pseudo-quasi-type I at u with respect to $\alpha_i, i = 1, \ldots, p$, $\beta_j, j = 1, \ldots, m$ and $\lambda > 0$. Then $f(x) \nleq f(u)$.*

For proof, see Hanson et al. [85]. These authors have obtained weak duality under several other weaker invexity assumptions. They also obtained usual strong and converse duality theorems as well. Mishra [155] obtained usual duality results for (VP) and (MWD) under type I univex (by combining type I functions and univex functions) functions.

Theorem 6.62. *Let x be feasible for (VP) and let the triplet (u, λ, μ) be feasible for (VD). Let for $i = 1, \ldots, p, \lambda > 0, \mu_j \geqq 0, j = 1, \ldots, m$ $\eta, b_0, b_1, \phi_0, \phi_1$ such that*

$$b_0(x,u)\phi_0\left[\sum_{i=1}^{p}\lambda_i f_i(x) - \sum_{i=1}^{p}\lambda_i f_i(u)\right] \geqq \sum_{i=1}^{p}\lambda_i \nabla f_i(u)\eta(x,u), \tag{6.41}$$

and

$$-b_1(x,u)\phi_1\left[\sum_{j=1}^{m}\mu_j g_j(u)\right] \geqq \sum_{j=1}^{m}\mu_j \nabla_i(u)\eta(x,u), \tag{6.42}$$

at u. Further suppose

$$a \leqq 0 \Rightarrow \phi_0(a) \leqq 0, \tag{6.43}$$

$$\phi_1(u) \leqq 0 \Rightarrow a > 0, \tag{6.44}$$

$$b_0(x,u) > 0, \quad b_1(x,u) \geqq 0. \tag{6.45}$$

Then, $f(x) \nleq f(u)$.

Mishra [155] obtained two more weak duality theorems and usual strong and converse duality theorems under further weakened invexity assumptions on the objective and constraint functions.

We continue our discussion on duality by considering the general Mond–Weir type dual (see [180]):

(GMWD)

$$\text{Maximize} f(u) + \mu_{J_0} g_{J_0}$$

$$\text{Subject to } \lambda\nabla f(u) + \mu\nabla g(u) = 0, \tag{6.46}$$

$$\mu_{J_t} g_{J_t}(u) \geqq 0, 1 \leqq t \leqq r, \tag{6.47}$$

$$\mu \geqq 0, \tag{6.48}$$

$$\lambda \geqq 0, \lambda^T e = 1, e = (1, \ldots, 1)^T \in R^p, \tag{6.49}$$

$J_t, 0 \leqq t \leqq r$ are partitions of the set $P = \{1, \ldots, p\}$.

6.4 Invexity in Nonsmooth Vector Optimization

Recently there has been an increasing interest in developing optimality conditions and duality relations for nonsmooth multiobjective programming problems involving locally Lipschitz functions; see, e.g., Egudo and Hanson [60], Giorgi and Guerraggio [72, 73], Kim and Schaible [125], Lee [136], Mishra and Mukherjee [158].

In this section we consider the following nonsmooth multiobjective programming problem involving locally Lipschitz functions: (NVOP)

$$\text{Minimize } f(x) = (f_1(x), \dots, f_p(x))$$

$$\text{Subject to } g_j(x) \leqq 0, j = 1, \dots, m,$$

$$x \in R^n,$$

where $f_i : R^n \to R, i = 1, \dots, p$, $g_j : R^n \to R, j = 1, \dots, m$ are locally Lipschitz functions.

Definition 6.63. *Let $f_i : R^n \to R$ be locally Lipschitz function for $i = 1, \dots, p$ is invex with respect to $\eta : R^n \times R^n \to R^n$ if for any $x, u \in R^n$ and $\xi_i \in \partial f_i(u)$,*

$$f_i(x) - f_i(u) \geqq \xi_i \eta(x, u), \quad \textit{for} \quad i = 1, \dots, p.$$

Theorem 6.64 (Sufficient optimality conditions). *Let $(u, \lambda, \mu) \in R^n \times R^p \times R^m$ satisfy the following generalized Karush–Kuhn–Tucker conditions:*

$$0 \in \sum_{i=1}^{p} \lambda_i \partial f_i(u) + \sum_{j=1}^{m} \mu_j \partial g_j(u),$$

$$g_j(u) \leqq 0, \quad j = 1, \dots, m,$$

$$\lambda_i \geqq 0, \quad i = 1, \dots, p, \quad \lambda^T e > 0,$$

$$\mu_j \geqq 0, \quad j = 1, \dots, m.$$

If f and g are invex with respect to the same η, then u is a weak minimum for (NVOP).

Proof. Since $0 \in \sum_{i=1}^{p} \lambda_i \partial f_i(u) + \sum_{j=1}^{m} \mu_j \partial g_j(u)$, there exist $\xi_i \in \partial f_i(u)$ and $\zeta_j \in \partial g_j(u)$ such that

$$\sum_{i=1}^{p} \lambda_i \xi_i + \sum_{j=1}^{m} \mu_j \zeta_j = 0. \tag{6.50}$$

Suppose that u is not a weak minimum for (NVOP). Then there exists $x \in X$ such that $f_i(x) < f_i(u), i = 1, \dots, p$. So we have

$$\xi_i \eta(x, u) < 0, \quad \text{for each} \quad \xi_i \in \partial f_i(u).$$

Therefore, we have

$$\sum_{i=1}^{p} \lambda_i \xi_i \eta(x, u) < 0.$$

Hence, by (6.50), we have

$$\sum_{j=1}^{m} \mu_j \zeta_j \eta(x,u) > 0.$$

Then by the invexity of g, we have

$$\sum_{j=1}^{m} \mu_j \left[g_j(x) - g_j(u) \right] > 0.$$

Since $\mu_j g_j(u) = 0, j = 1, \ldots, m$, we have

$$\sum_{j=1}^{m} \mu_j g_j(u) > 0,$$

which contracts the condition $g_j(x) \leqq 0$. Thus, u is a weak minimum for (NVOP).

□

Definition 6.65. *Let $f_i : R^n \to R$ be locally Lipschitz function for $i = 1, \ldots, p$ is pseudo-invex with respect to $\eta : R^n \times R^n \to R^n$ if for any $x, u \in R^n$ and $\xi_i \in \partial f_i(u)$,*

$$\xi_i \eta(x,u) \geqq 0 \Rightarrow f_i(x) \geqq f_i(u), \quad for \quad i = 1, \ldots, p.$$

Definition 6.66. *Let $f_i : R^n \to R$ be locally Lipschitz function for $i = 1, \ldots, p$ is quasi-invex with respect to $\eta : R^n \times R^n \to R^n$ if for any $x, u \in R^n$ and $\xi_i \in \partial f_i(u)$,*

$$f_i(x) \leqq f_i(u) \Rightarrow \xi_i \eta(x,u) \leqq 0, \quad for \quad i = 1, \ldots, p.$$

The subgradient sufficient optimality conditions given in above theorem can be obtained under weaker invexity assumptions as well.

Theorem 6.67. *Let $(u, \lambda, \mu) \in R^n \times R^p \times R^m$ satisfy the following generalized Karush–Kuhn–Tucker conditions:*

$$0 \in \sum_{i=1}^{p} \lambda_i \partial f_i(u) + \sum_{j=1}^{m} \mu_j \partial g_j(u),$$

$$g_j(u) \leqq 0, \quad j = 1, \ldots, m,$$

$$\lambda_i \geqq 0, \quad i = 1, \ldots, p, \quad \lambda^T e > 0,$$

$$\mu_j \geqq 0, \quad j = 1, \ldots, m.$$

If f is pseudo-invex and g is quasi-invex with respect to the same η, then u is a weak minimum for (NVOP).

Proof. As in above theorem, we have (6.50). Suppose that u is not a weak minimum for (NVOP). Then there exists $x \in X$ such that $f_i(x) < f_i(u), i = 1, \ldots, p$. Then by pseudo-invexity of f, we get

$$\xi_i \eta(x, u) < 0, \quad \text{for each} \quad \xi_i \in \partial f_i(u), \ i = 1, \ldots, p.$$

Hence, we have

$$\sum_{i=1}^{p} \lambda_i \xi_i \eta(x, u) < 0. \tag{6.51}$$

By feasibility and $\mu_j g_j(u) = 0, j = 1, \ldots, m$, we have

$$\mu_j g_j(x) \leqq \mu_j g_j(u), \quad j = 1, \ldots, m.$$

By quasi-invexity of g, we get

$$\sum_{j=1}^{m} \mu_j \zeta_j \eta(x, u) \leqq 0, \quad \text{for each} \quad \zeta_j \in \partial g_j(u), \ j = 1, \ldots, m. \tag{6.52}$$

From (6.51) and (6.52), we get

$$\sum_{i=1}^{p} \lambda_i \xi_i \eta(x, u) + \sum_{j=1}^{m} \mu_j \zeta_j \eta(x, u) \leq 0, \tag{6.53}$$

which contradicts (6.50). Hence u is a weak minimum for (NVOP).

□

Egudo and Hanson [60] extended the concept of V-invex functions to nonsmooth case. Following Egudo and Hanson [60], Mishra and Mukherjee [158] extended the concepts of V-pseudo-invexity and V-quasi-invexity to nonsmooth case and obtained optimality and duality results for (NVOP) under V-pseudo-invexity and V-quasi-invexity assumptions. Readers are suggested to see Mishra et al. [163] for detailed study on V-invexity and related concepts and results. Following Mishra and Mukherjee [158] Kuk et al. [134] obtained optimality and duality results under $V - \rho$-invexity assumptions.

Definition 6.68. *Let $f_i : R^n \to R$ be locally Lipschitz function for $i = 1, \ldots, p$ is $V - \rho$-invex with respect to $\eta : R^n \times R^n \to R^n$ and $\theta : R^n \times R^n \to R^n$ if there exists $\alpha_i : R^n \times R^n \to R_+ - 0$ and $\rho_i \in R, i = 1, \ldots, p$ such that for any $x, u \in R^n$ and $\xi_i \in \partial f_i(u)$,*

$$\alpha_i(x, u) \left[f_i(x) - f_i(u) \right] \geqq \xi_i \eta(x, u) + \rho_i \left\| \theta(x, u) \right\|^2, \quad \textit{for} \quad i = 1, \ldots, p.$$

Kuk et al. [134], obtained the following sufficient optimality conditions:

Theorem 6.69. *Let $(u, \lambda, \mu) \in R^n \times R^p \times R^m$ satisfy the following generalized Karush–Kuhn–Tucker conditions:*

$$0 \in \sum_{i=1}^{p} \lambda_i \partial f_i(u) + \sum_{j=1}^{m} \mu_j \partial g_j(u),$$

$$g_j(u) \leqq 0, \quad j = 1, \ldots, m,$$
$$\lambda_i \geqq 0, \quad i = 1, \ldots, p, \quad \lambda^T e > 0,$$
$$\mu_j \geqq 0, \quad j = 1, \ldots, m.$$

If f and g are $V - \rho$-invex with respect to the same η, then u is a weak minimum for (NVOP).

Proof. See the proof of Theorem 2.1 in Kuk et al. [134].

□

Recall from Mishra and Mukherjee [158] the concept of V-pseudo-invexity and V-quasi-invexity:
Let $f_i : R^n \to R$ be locally Lipschitz function for $i = 1, \ldots, p$ is V- pseudo-invex with respect to $\eta : R^n \times R^n \to R^n$ if there exists $\alpha_i : R^n \times R^n \to R_+ - 0$ such that for any $x, u \in R^n$ and $\xi_i \in \partial f_i(u)$,

$$\sum_{i=1}^{p} \xi_i \eta(x, u) \geqq 0 \Rightarrow \sum_{i=1}^{p} \alpha_i(x, u) \left[f_i(x) - f_i(u)\right] \geqq 0, \quad \forall \xi_i \in \partial f_i(u).$$

Let $f_i : R^n \to R$ be locally Lipschitz function for $i = 1, \ldots, p$ is V- quasi-invex with respect to $\eta : R^n \times R^n \to R^n$ if there exists $\alpha_i : R^n \times R^n \to R_+ - 0$ such that for any $x, u \in R^n$ and $\xi_i \in \partial f_i(u)$,

$$\sum_{i=1}^{p} \alpha_i(x, u) \left[f_i(x) - f_i(u)\right] \leqq 0 \Rightarrow \sum_{i=1}^{p} \xi_i \eta(x, u) \leqq 0, \quad \forall \xi_i \in \partial f_i(u).$$

Mishra [154] obtained the following more general result on sufficiency:

Theorem 6.70. *Let $(u, \lambda, \mu) \in R^n \times R^p \times R^m$ satisfy the following generalized Karush–Kuhn–Tucker conditions:*

$$0 \in \sum_{i=1}^{p} \lambda_i \partial f_i(u) + \sum_{j=1}^{m} \mu_j \partial g_j(u),$$

$$g_j(u) \leqq 0, \quad j = 1, \ldots, m,$$
$$\lambda_i \geqq 0, \quad i = 1, \ldots, p, \quad \lambda^T e > 0,$$
$$\mu_j \geqq 0, \quad j = 1, \ldots, m.$$

If $(\lambda_1 f_1, \ldots, \lambda_p f_p)$ is V-pseudo-invex and $(\mu_1 g_1, \ldots, \mu_m g_m)$ is V-quasi-invex with respect to the same η, then u is a properly efficient solution for (NVOP).

Proof. See the proof of Theorem 3.1 in Mishra [154].

□

One can easily extend the above result of Mishra [154] to the class of functions given by Kuk et al. [134] as well as the result of Kuk et al. [134] to weaker class of functions, say $V-\rho$-pseudo-invexity and $V-\rho$-quasi-invexity. For more results on optimality in nonsmooth multiobjective programming problems involving invex and generalized invex functions see Kim and Schaible [125].

Mond–Weir type nonsmooth dual to (NVOP) is:

(NVD)

$$\text{Maximize}\,(f_1(u),\ldots,f_p(u))$$

$$\text{Subject to } 0\in\sum_{i=1}^{p}\lambda_i\partial f_i(u)+\sum_{j=1}^{m}\mu_j\partial g_j(u),$$

$$\mu_j g_j(u)\geqq 0,\quad j=1,\ldots,m,$$

$$\lambda_i\geqq 0,\quad i=1,\ldots,p,\quad \lambda^T e=1,$$

$$\mu_j\geqq 0,\quad j=1,\ldots,m,$$

where $e=(1,\ldots,1)^T\in R^p$. Weak duality for (NVOP) and (NVD) is:

Theorem 6.71. *Let x be feasible for (NVOP) and (u,λ,μ) feasible for (NVD). If f and g are invex with respect to the same η, then the following cannot hold:*

$$f_i(x)\leqq f_i(u),\quad \forall i=1,\ldots,p$$

and

$$f_{i_0}(x)<f_{i_0}(u),\quad \text{for some}\quad i_0\in\{1,\ldots,p\}.$$

Proof. From the first duality constraint, we get

$$0=\sum_{i=1}^{p}\lambda_i\xi_i+\sum_{j=1}^{m}\mu_j\zeta_j,$$
$$\forall\xi_i\in\partial f_i(u),\ \zeta_j\in\partial g_j(u),\ i=1,\ldots,p,\ j=1,\ldots,m.\tag{6.54}$$

From feasibility, we get

$$\mu_j g_j(x)\leqq\mu_j g_j(u),\quad j=1,\ldots,m.$$

Then by invexity hypothesis on g, yields

$$\sum_{j=1}^{m}\mu_j\zeta_j\leqq 0,\quad \forall\zeta_j\in\partial g_j(u),\ j=1,\ldots,m.\tag{6.55}$$

Therefore, from (6.54), we get

$$\sum_{i=1}^{p}\lambda_i\xi_i\geqq 0,\quad \forall\xi_i\in\partial f_i(u),\ i=1,\ldots,p.\tag{6.56}$$

By invexity hypothesis on f, we get

$$\sum_{i=1}^{p} \lambda_i f_i(x) \geqq \sum_{i=1}^{p} \lambda_i f_i(u).$$

Since $\lambda_i \geqq 0$ and $\lambda^T e = 1$, we get the result.

□

Theorem 6.72. *Let $\bar{x}$ be a weak minimum for (NVOP) at which a constraint qualification is satisfied. Then there exist $\bar{\lambda} \in R^p$ and $\bar{\mu} \in R^m$ such that $(\bar{x}, \bar{\lambda}, \bar{\mu})$ is feasible for (NVD). If also f and g are invex with respect to the same η, then $(\bar{x}, \bar{\lambda}, \bar{\mu})$ is a weak minimum for (NVD).*

Proof. Since $\bar{x}$ is a weak minimum for (NVOP) and a constraint qualification is satisfied at $\bar{x}$, from the generalized Karush–Kuhn–Tucker theorem (see, for example, Theorem 6.1.3 of [38]), there exist $\lambda \in R^p$ and $\mu \in R^m$ such that

$$0 \in \sum_{i=1}^{p} \lambda_i \partial f_i(\bar{x}) + \sum_{j=1}^{m} \mu_j \partial g_j(\bar{x}),$$

$$\mu_j g_j(\bar{x}) = 0, \quad j = 1, \ldots, m,$$

$$\lambda_i \geqq 0, \quad i = 1, \ldots, p, \quad \lambda^T e > 0,$$

$$\mu_j \geqq 0, \quad j = 1, \ldots, m.$$

Since $\lambda_i \geqq 0, i = 1, \ldots, p$ and $\lambda^T e > 0$, we can write

$$\bar{\lambda}_i = \frac{\lambda_i}{\sum_{i=1}^{p} \lambda_i}$$

and

$$\bar{\mu}_j = \frac{\mu_j}{\sum_{j=1}^{m} \mu_j}.$$

Then $(\bar{x}, \bar{\lambda}, \bar{\mu})$ is feasible for (NVD). Since $\bar{x}$ is feasible for (NVOP), it follows from weak duality that $(\bar{x}, \bar{\lambda}, \bar{\mu})$ is a weak maximum for (NVD).

□

The above two theorems can be extended to pseudo-invexity/quasi-invexity of f and g as well. Kuk et al. [134] extended the above results to the class of $V - \rho$-invexity conditions.

Theorem 6.73. *Let x be a feasible solution for (NVOP) and (u, λ, μ) feasible for (NVD). Assume that*

$$\sum_{i=1}^{p} \lambda_i \rho_i + \sum_{j=1}^{m} \mu_j \sigma_j \geqq 0.$$

If f is $V-\rho$-invex and g is $V-\sigma$-invex with respect to the same η and θ, then the following cannot hold:

$$f_i(x) \leqq f_i(u), \quad \forall i = 1, \ldots, p$$

and

$$f_{i_0}(x) < f_{i_0}(u), \quad \text{for some} \quad i_0 \in \{1, \ldots, p\}.$$

For proof, see the proof of Theorem 2.2 of Kuk et al. [134].

Theorem 6.74. *Let $\bar{x}$ be a weak minimum for (NVOP) at which a constraint qualification is satisfied. Then there exist $\bar{\lambda} \in R^p$ and $\bar{\mu} \in R^m$ such that $(\bar{x}, \bar{\lambda}, \bar{\mu})$ is feasible for (NVD). If also f is $V-\rho$-invex and g is $V-\sigma$-invex with respect to the same η and θ, then $(\bar{x}, \bar{\lambda}, \bar{\mu})$ is a weak minimum for (NVD).*

For proof see the proof of Theorem 2.3 of Kuk et al. [134]. Mishra and Mukherjee [158] have obtained weak and strong duality under V-pseudo-invexity and V-quasi-invexity assumptions. Kim and Schaible [125] obtained weak and strong duality under invexity conditions for a vector optimization problem with inequality and equality constraints and its Wolfe and Mond–Weir type dual models.

6.5 Nonsmooth Vector Optimization in Abstract Spaces

Vector optimization problems discussed in above sections deal with the finite dimensional spaces and assume that the functions involved are differentiable. But in many situations, we have to solve a non-differentiable and multiobjective optimization problem in abstract spaces. Several authors have discussed optimality and duality in Banach spaces, see, e.g., Coladas and Wang [39], Luc [140], Li [138], Minami [149]. Throughout this section let X be a real Banach space with norm $\|\cdot\|$ and Y a real locally convex separable topological vector space. We denote the topological dual space of X and Y by X^* and Y^* respectively, which are assumed to be equipped with the weak* topology. Let ω be a non-empty subset of X and K a closed convex cone of Y. For set $A \subset Y$, its interior and its closure are denoted by $intA$ and clA respectively, its dual cone A^+ and its strict dual cone A^{s+} are respectively defined as

$$A^+ = \{y^* \in Y* : \langle y, y^* \rangle \geq 0, \forall y \in A\} \tag{6.57}$$

and

$$A^{s+} = \{y^* \in Y^* : \langle y, y^* \rangle > 0, \forall y \in A - \{0\}\}. \tag{6.58}$$

The cone generalised by A is

$$P(A) = \{\alpha y : y \in A, \alpha \geq 0\} \tag{6.59}$$

and the tangent cone to A at $y \in clA$ is

$$T(A,y) = \{d \in Y : \exists t_k > 0 \quad \text{and} \quad y_k \in A \quad \text{such that} \\ y_k \to y \quad \text{and} \quad t_k(y_k - y) \to d\}. \tag{6.60}$$

Let h be a locally Lipschitz function defined on X. We denote by $h^0(x;v)$ and $\partial^0 h(x)$ the Clarke's generalized directional derivative of h at x in the direction v and the Clarke's generalized subdifferential of h at x, respectively. For the reader's convenience, we quote the following properties of $h^0(x;v)$ and $\partial^0 h(x)$ from [38]:

(a) $\partial^0 h(x)$ is a non-empty convex weak* compact subset of X.
(b) $h^0(x;v) = \max\{\langle v, x^* \rangle : x^* \in \partial^0 h(x)\}$.
(c) $h^0(x;\cdot)$ is a positively homogeneous, convex and continuous function on X.
(d) $\partial^0 h(x) = \partial h(x), \forall x \in X$ when h is convex on X, where $\partial h(x)$ denotes the subdifferential of h at x.

The vector optimization problem considered in this section are like the one given by Coladas and Wang [39]:
(AVP)

$$V - Minimize f(x)$$

$$\text{Subject to } x \in \Omega,$$

where $f : X \to Y$ satisfies that $k * f = k * \circ f$ is locally Lipschitz and regular on ω for each $k^* \in K^+, \omega$ is a non-empty subset of X and K is a closed convex cone in Y with $intK \neq \phi$.

For further discussion we need he following definitions:

Definition 6.75 ([244]). *$f : X \to Y$ is said to be K-pre-invex on ω if there exists a function $\eta : \omega \times \omega \to X$ such that for any $x, y \in \omega$ and $\lambda \in [0,1], y + \lambda\eta(x,y) \in \omega$ and*

$$\lambda f(x) + (1-\lambda) f(y) - f(y + \lambda\eta(x,y)) \in K. \tag{6.61}$$

Definition 6.76. *$f : X \to Y$ is said to be K-convexlike on ω if for any $x, y \in \omega$ and $\lambda \in [0,1]$ there exists a $z \in \omega$ such that*

$$\lambda f(x) + (1-\lambda) f(y) - f(z) \in K. \tag{6.62}$$

Definition 6.77. *$f : X \to Y$ is said to be K-subconvexlike on ω if there exists $\theta \in intK$ such that for any $x, y \in \omega, \lambda \in [0,1]$ and $\epsilon > 0$ we can find a $z \in \omega$ satisfying*

$$\epsilon\theta + \lambda f(x) + (1-\lambda) f(y) - f(z) \in K. \tag{6.63}$$

Note that K-convexity $\to$ K-pre-invexity $\to$ K-convex-likeness $\to$ K-subconvexlikeness, see, e.g., [39]. The following properties of generalized convex functions defined above are from [39]:

(a) If f is K-pre-invex on ω and $k^* \in K^+$, then k^*f is R_+-pre-invex on ω, where $R_+ = \{\alpha \in R : \alpha \geq 0\}$.
(b) If f is K-subconvexlike on ω, then $f(\omega) + intK$ is convex.
(c) If f is K-subconvexlike on ω, then exactly one of the following holds:

$$(\exists x \in \omega) - f(x) \in intK; \tag{6.64}$$

$$(\exists k^* \in K^+ - \{0\})(k^*f)(\omega) \subset R_+. \tag{6.65}$$

(d) Suppose that $h : X \to R$ is locally Lipschitz and regular on ω. If h is R_+-pre-invex on ω, then there exists an $\eta : \omega \times \omega \to X$ such that for any $x, y \in \omega, \eta(y, x) \in F(\omega, x) = \{v \in X : (\exists t_0 > 0)(\forall t \in [0, t_0])x + tv \in \omega\}$ and

$$h(y) - h(x) \geq \langle \eta(y, x), x^* \rangle, \quad \forall x^* \in \partial^0 h(x). \tag{6.66}$$

The convex cone K defines a partial order on Y. We give the following solution concepts for (AVP).

Definition 6.78. *Let $\bar{x} \in \omega$.*

- *$\bar{x}$ is called a weak minimum of (AVP) if*

$$(\forall x \in \omega) f(\bar{x}) - f(x) \notin intK.$$

- *$\bar{x}$ is called a minimum of (AVP) if*

$$(\forall x \in \omega) f(\bar{x}) - f(x) \notin K - \{0\}.$$

- *$\bar{x}$ is called a strong minimum of (AVP) if*

$$(\forall x \in \omega) f(x) - f(\bar{x}) \in K.$$

The following lemma from [39] is needed for further discussion.

Lemma 6.79. *Suppose that $h : X \to R$ is locally Lipschitz and regular on ω. Then h gets a minimum over ω at x if h is R_+-pre-invex on ω and $\partial^0 h(x) \bigcap F(\omega, x)^+ \neq \phi$.*

For a weak minimum to the abstract vector minimization problem (AVP), Coladas and Wang [39] gave the following sufficient and necessary conditions:

Theorem 6.80. *If f is K-pre-invex on ω and if there exists a $k^* \in K^+ - \{0\}$ such that $\partial^0(k^*f)(x) \bigcap F(\omega, x)^+ \neq \phi$, then x is a weak minimum of (AVP).*

Proof. Suppose that f is K-pre-invex on ω then

$$\langle f(x) - f(y), k^* \rangle = k^*f(x) - k^*f(y) \leq 0, \quad \forall y \in \omega.$$

Hence there exists no $y \in \omega$ such that

$$f(x) - f(y) \in intK$$

because $k^* \in K^+ - \{0\}$ implies that $\langle k, k^* \rangle > 0, \forall k \in intK$. Therefore, x is a weak minimum to (AVP).

□

Theorem 6.81. *Suppose that ω is convex and that f is K-subconvexlike on ω. If x is a weak minimum for (AVP), then there exists a $k^* \in K^+ - \{0\}$ such that $\partial^0(k^*f)(x) \bigcap F(\omega, x)^+ \neq \phi$.*

For proof see [39].

The following corollary extends Theorem 2.1 from [229], the proof of the corollary is immediate from above discussions.

Corollary 6.82. *Suppose that ω is convex and that f is K-pre-invex on ω. Then x is a weak minimum for (AVP) if and only if there exists a $k^* \in K^+ - \{0\}$ such that $\partial^0(k^*f)(x) \bigcap F(\omega, x)^+ \neq \phi$.*

Coladas and Wang [39] have given a sufficient condition and a necessary condition for a minimum for (AVP) as well.

Theorem 6.83. *Let K be pointed and $x \in \omega$. if f is K-pre-invex on ω and if there exists a $k^* \in K^+ - \{0\}$ such that $\partial^0(k^*f)(x) \bigcap F(\omega, x)^{s+} \neq \phi$, then x is a minimum for (AVP).*

Coladas and Wang [39] extended Theorem 3.1 of [229] as the following corollary to Theorem 6.83.

Corollary 6.84. *Let $x \in \omega, Y = R^n, K = R^n_+$ and $f = (f_1, \dots, f_n)$. If f is R^n_+-pre-invex on ω and if there exists an i such that $\partial^0 f_i(x) \bigcap F(\omega, x)^{s+} \neq \phi$, then x is a minimum for (AVP).*

Theorem 6.85. *Suppose that ω is convex and that f is K-subconvexlike on ω. If x is a minimum of (AVP), then there exists a $k^* \in K^+ - \{0\}$ such that $\partial^0(k^*f)(x) \bigcap F(\omega, x)^+ \neq \phi$.*

These authors also proved a sufficient condition and a necessary condition for strong minimum for (AVP) in [39].

Another approach in nonsmooth vector optimization is composite (see [106]) multiobjective programming problem:
(CVOP)

$$\text{Minimize}(f_1(F_1(x)), \dots, f_p(F_p(x)))$$

$$\text{Subject to } x \in C, \quad g_j(G_j(x)) \leqq 0, \quad j = 1, \dots, m,$$

where C is a convex subset of a Banach space X, $f_i, i = 1, \dots, p$, $g_j, j = 1, \dots, m$, are real valued locally Lipschitz function on R^n, and $F_i, i = 1, \dots, p$ and $G_j, j = 1, \dots, m$ are locally Lipschitz and Gateaux differentiable functions from X into R^n with Gateaux derivatives F_i' and G_j', but are necessarily continuously Frechet differentiable, see Clarke [38].

The above composite model is broad and flexible enough to cover many common types of multiobjective problems, seen in the literature. Moreover, the model obviously includes the wide class of convex composite single objective problems, which is now recognized as fundamental for theory and computation in scalar nonsmooth optimization. To see the nature of the composite model, let us look at the following examples:

Example 6.86. Define $F_i, G_j : X^n \to R^{p+m}$ by

$$F_i(x) = (0, 0, \dots, l_i(x), 0, \dots, 0), \quad i = 1, \dots, p,$$

$$G_j(x) = (0, 0, \dots, h_j(x), 0, \dots, 0), \quad j = 1, \dots, m,$$

where $l_i(x)$ and $h_j(x)$ are locally Lipschitz and Gateaux differentiable functions on a Banach space X. Define $f_i, g_j : R^{p+m} \to R$ by $f_i(x) = x_i, i = 1, \dots, p$, and $g_j(x) = x_{p+j}, j = 1, \dots, m$. Let $C = X$. Then the composite multiobjective vector optimization problem (CVOP) is the problem:

$$\text{Minimize}(l_1(x), \dots, l_p(x))$$

$$\text{Subject to } x \in X^n, \quad h_j(x) \leqq 0, \quad j = 1, \dots, m,$$

which is a standard multiobjective differentiable nonlinear programming problem.

Example 6.87. Consider the vector approximation (model) problem:

$$V - \text{Minimize}(\|F_1(x)\|_1, \dots, \|F_p(x)\|_p)$$

$$\text{Subject to } x \in X,$$

where X is a Banach space, $\|\cdot\|_i, i = 1, \dots, p$, are norms in R^m, and for each $i = 1, \dots, p$, $F_i : X \to R^m$ is a Frechet differentiable (error) function. This problem is also the case of vector composite model, where for each $i = 1, \dots, p, f : R^n \to R$ is given by $f_i(x) = \|x\|_i$, and conditions on $G_j(x) = x$ and $g_j(x) = 0$. Various examples of vector approximation problems of this type that arise in simultaneous approximation are given in [97, 98].

The idea of Jeyakumar and Yang [106] is that by studying the composite model a unified framework can be given for the treatment of many questions of theoretical and computational interest in multiobjective optimization. Jeyakumar and Yang [106] presented new conditions under which the Kuhn–Tucker optimality conditions becomes sufficient for efficient and properly efficient solutions. The sufficient conditions in this section are significant even for scalar composite problems as these conditions are weaker than the conditions given by Jeyakumar [104].

Let $x, a \in X$. Define $K : X \to R^{n(p+m)}$ by

$$K(x) = (F_1(x), \dots, F_p(x), G_1(x), \dots, G_m(x)).$$

For each $x, a \in X$, the linear mapping

$$\begin{aligned} A_{x,a}(y) = (&\alpha_1(x,a)F_1'(a)y, \dots, \alpha_p(x,a)F_p'(a)y, \\ &\beta_1(x,a)G_1'(a)y, \dots, \beta_m(x,a)G_m'(a)y), \end{aligned}$$

where $\alpha_i(x,a), i = 1, \dots, p$ and $\beta_j(x,a), j = 1, \dots, m$ are real positive constants.

Equivalently, the null space condition means that for each $x, a \in X$, there exist real constants $\alpha_i(x,a) > 0, i = 1, \ldots, p$ and $\beta_j(x,a) > 0, j = 1, \ldots, m$ and $\mu(x,a) \in X$ such that $F_i(x) - F_i(a) = \alpha_i(x,a)F_i'(a)\mu(x,a)$ and $G_j(x) - G_j(a) = \beta_j(x,a)G_j'(a)\mu(x,a)$. For our composite vector optimization problem (CVOP), Jeyakumar and Yang [106] proposed the following generalized null space condition (GNC):
For each $x, a \in C$, there exist real constants $\alpha_i(x,a) > 0, i = 1, \ldots, p$, and $\beta_j(x,a) > 0, j = 1, \ldots, m$, and $\mu(x,a) \in (C - a)$ such that $F_i(x) - F_i(a) = \alpha_i(x,a)F_i'(a)\mu(x,a)$ and $G_j(x) - G_j(a) = \beta_j(x,a)G_j'(a)\mu(x,a)$. A condition of this type is called representation condition, has been used in the study of Chebyshev vector approximation problems in Jahn and Sachs [99].

Jeyakumar and Yang [106] proposed the following Kuhn–Tucker type optimality conditions (KT) for (CVOP):

$$\lambda \in R^p, \quad \lambda_i > 0, \quad i = 1, \ldots, p$$

$$\mu \in R^m, \quad \mu_j \geqq 0, \quad \mu_j g_j(G_j(a)) = 0,$$

$$0 \in \sum_{i=1}^{p} \lambda_i \partial f_i(F_i(a)) + \sum_{j=1}^{m} \mu_j \partial g_j(G_j(a))G_j' - (C - a)^+.$$

Jeyakumar and Yang [106] obtained the following result:

Theorem 6.88. *For the problem (CVOP), assume that f_i and g_j are convex functions, and F_i and G_j are locally Lipschitz and Gateaux differentiable functions. Let a be feasible for (CVOP). Suppose that the optimality conditions (KT) holds at a. If the generalized null space condition (GNC) holds at each feasible point x of (CVOP), then a is an efficient solution of (CVOP).*

For the proof, see [106].

The following example shows that the generalized null space condition (GNC) may not be sufficient for a feasible point which satisfies the optimality conditions (KT) to be a properly efficient solution for (CVOP).

Example 6.89. Consider the following simple multiobjective problem:

$$V - \text{Minimize} \left(\frac{x_1}{x_2}, \frac{x_2}{x_1} \right)$$

$$\text{Subject to } (x_1, x_2) \in R^2, \ \ 1 - x_1 \leq 0, \ \ 1 - x_2 \leq 0.$$

It is easy to check that $(1, 1)$ is an efficient solution for the problem, but it is not properly efficient. The generalized null space condition (GNC) holds at every feasible point (x_1, x_2) with $\alpha_1((x_1, x_2), (1, 1)) = \frac{1}{x_2}$, $\alpha_2((x_1, x_2), (1, 1)) = \frac{1}{x_1}$, $\beta_i((x_1, x_2), (1, 1)) = 1, i = 1, 2$.

The following example provides a nonsmooth convex composite problem for which the sufficient optimality condition given in above Theorem holds.

Example 6.90. Consider the multiobjective problem

$$\text{Minimize}\left(\frac{|2x_1 - x_2|}{|x_1 + x_2|}, \frac{x_1 + 2x_2}{x_1 + x_2}\right)$$

$$\text{Subject to } x_1 - x_2 \leq 0,\ \ 1 - x_1 \leq 0,\ \ 1 - x_2 \leq 0.$$

Let $F_1(x) = \frac{2x_1 - x_2}{x_1 + x_2}$, $F_2(x) = \frac{x_1 + 2x_2}{x_1 + x_2}$, $G_1(x) = x_1 - x_2$ $G_2(x) = 1 - x_1$, $G_3(x) = 1 - x_2$, $f_1(y) = |y|$, $f_2(y) = y$, and $g_1(y) = g_2(y) = g_3(y) = y$. Then, the problem becomes a convex composite problem with an efficient solution $(1, 2)$. It is easy to see that the null space condition holds at each feasible point of the problem with $\alpha_i(x, a) = 1$, for $i = 1, 2$, $\beta_j(x, a) = \frac{x_1 + x_2}{3}$, for $j = 1, 2, 3$ and $\mu(x, a) = \left(\frac{3x_1 - 3}{x_1 + x_2}, \frac{3x_2 - 6}{x_1 + x_2}\right)^T$. The optimality conditions (KT) hold with $v_1 = v_2 = 1, \lambda_1 = 1, \lambda_2 = 3$ and $\mu_j = 0, j = 1, 2, 3$.

The following example shows that the pseudolinear programming problems satisfy sufficient optimality conditions given in above theorem.

Example 6.91.

$$\text{Minimize}(l_1(x), \ldots, l_p(x))$$

$$\text{Subject to } x \in R^n,\ \ h_j(x) - b_j \leq 0,\ \ j = 1, \ldots, m,$$

where $l_i : R^n \to R$ and $h_j : R^n \to R$ are differentiable and pseudolinear, i.e., pseudo-convex and pseudo-concave, see Chap. 3, and $b_j \in R, j = 1, \ldots, m$. Define $F_i : R^n \to R^{p+m}$ by

$$F_i(x) = (0, \ldots, l_i(x), 0, \ldots, 0), \quad i = 1, \ldots, p,$$

and

$$G_j(x) = (0, \ldots, h_j(x) - b_j, 0, \ldots, 0), \quad j = 1, \ldots, m.$$

Define $f_i, g_j : R^{p+m} \to R$ by $f_i(x) = x_i, i = 1, \ldots, p$, $g_j(x) = x_{p+j}, j = 1, \ldots, m$. Then the above minimization problem can be written as:

$$\text{Minimize}\,(f_1(F_1(x)), \ldots, f_p(F_p(x)))$$

$$\text{Subject to } x \in R^n,\ \ g_j(G_j(x)) \leq 0,\ \ j = 1, \ldots, m.$$

The generalized null space condition (GNC) is verified at each feasible point by the pseudolinearity property of the functions involved. It follows from the above sufficiency theorem that is the optimality conditions

$$\sum_{i=1}^{p} \lambda_i l_i'(a) + \sum_{j=1}^{m} \mu_j g_j'(a) = 0,$$

$$\mu_j \left(g_j(a) - b_j\right) = 0,$$

holds with $\lambda_i > 0, i = 1, \ldots, p$ with $\mu_j \geq 0, j = 1, \ldots, m$, at the feasible point $a \in R^n$ of the pseudolinear programming problem, then a is an efficient solution for the pseudolinear programming problem.

Jeyakumar and Yang [106] further strengthened the generalized null space condition by constraining $\alpha_i(x,a) = \beta_j(x,a) = 1, \forall i,j$, in order to get sufficient conditions for properly efficient solutions for (CVOP). Jeyakumar and Yang [106] gave the following sufficient optimality for properly efficient solution.

Theorem 6.92. *Assume that the conditions on (CVOP) in Theorem 6.88 hold. Let a be feasible for (CVOP). Suppose that the optimality conditions (KT) hold at a. If the generalized null space condition (GNC) holds with $\alpha_i(x,a) = \beta_j(x,a) = 1, \forall i,j$, for each feasible x of (CVOP) then a is a properly efficient solution for (CVOP).*

It is well known that duality results have played a crucial role in the development of multiobjective programming [140, 217]. Following the success of multiobjective linear programming duality, various generalizations of the duality theory have been given for multiobjective nonlinear programming problems, see, e.g., [140,217]. For composite vector optimization model, Jeyakumar and Yang [106] gave the following Mond–Weir type of dual model:

(CVD)

$$\text{Maximize}\,(f_1(F_1(u)),\ldots,f_p(F_p(u)))$$

$$\text{Subject to } 0 \in \sum_{i=1}^{p} \lambda_i \partial f_i(F_i(u))F_i'(u) + \sum_{j=1}^{m} \mu_j \partial g_j(G_j(u))G_j'(u) - (C-u)^+,$$

$$\mu_j g_j(G_j(u)) \geq 0, \quad j = 1,\ldots,m,$$

$$u \in C, \quad \lambda \in R^P, \quad \lambda_i > 0, \quad \mu \in R^m, \quad \mu_j \geq 0.$$

Note that the problem (CVD) is considered as a dual to (CVOP) in the sense that

(a) (Zero duality gap) if $\bar{x}$ is a properly efficient solution for (CVOP) then, for some $\bar{\lambda} \in R^p$, $\bar{\mu} \in R^m$, $(\bar{x},\bar{\lambda},\bar{\mu})$ is a properly efficient solution for (CVD), and the objective values of (CVOP) and (CVD) at these points are equal.
(b) (Weak duality property) if x is feasible for (CVOP) and (u,λ,μ) is feasible for (CVD) then

$$(f_1(F_1(x)),\ldots,f_p(F_p(x)))^T,\ldots,(f_1(F_1(u),\ldots,f_p(F_p(u)))^T \notin -R_+^p - \{0\}.$$

Jeyakumar and Yang [106] obtained weak and strong duality results for (CVOP) and (CVD) under invexity hypothesis and generalized null space condition.

6.6 Vector Saddle Points

In this section we consider the vector valued optimization problem

(VP)

$$\text{Minimize} f(x), \quad \text{Subject to } g(x) \leqq 0,$$

where $f : S \to R^p, g : S \to R^m$ and the related vector saddle point.

By studying a natural generalization of the scalar Lagrangian

$$f(x) + \mu^T g(x)$$

Tanino and Sawaragi [234] have developed a saddle point and duality theory for convex (VP), generalizing that of the scalar case.

On the other hand Kuk et al. [134] studied weak vector saddle-point theorems for the nonsmooth multiobjective program (NVOP) in which functions are locally Lipschitz. For the problem (NVOP), a point (x, λ, μ) is said to be a critical point if, x is a feasible point for (NVOP), and

$$0 \in \partial \left(\sum_{i=1}^{p} \lambda_i f_i(x) + \sum_{j=1}^{m} \mu_j g_j(x) \right),$$

$$\mu_j g_j(x) = 0, \;\; j = 1, \ldots, m,$$
$$\lambda_i \geqq 0, \;\; i = 1, \ldots, p, \;\; \lambda^T e = 1.$$

Note that

$$\partial \left(\sum_{i=1}^{p} \lambda_i f_i(x) + \sum_{j=1}^{m} \mu_j g_j(x) \right) = \sum_{i=1}^{p} \partial \left(f_i(x) + \sum_{j=1}^{m} \mu_j g_j(x) \right).$$

Let $L(x, \lambda) = f(x) + \mu^T g(x)$, where $x \in R^m$ and $\mu \in R^m_+$. Then, a point $(\bar{x}, \bar{\lambda}) \in R^n \times R^m_+$ is said to be a weak vector saddle-point if

$$L(\bar{x}, \lambda) \leq L(\bar{x}, \bar{\lambda}) \leq L(x, \bar{\lambda}), \quad \forall x \in R^n, \;\; \mu \in R^m_+.$$

Kuk et al. [134] obtained the following saddle-point conditions:

Theorem 6.93. *Let $(\bar{x}, \bar{\Lambda}, \bar{\mu})$ be a critical point of (VP). Assume that $f(\cdot) + \bar{\mu}^T g(\cdot)$ is $V - \rho$-invex with respect to functions η and θ and $\sum_{i=1}^{p} \bar{\lambda} \rho_i \geqq 0$. Then $(\bar{x}, \bar{\lambda})$ is a weak vector saddle-point for (VP).*

For the proof, see Kuk et al. [134]. Further, they obtained:

Theorem 6.94. *If there exists $\bar{\lambda} \in R^m_+$ such that $(\bar{x}, \bar{\lambda})$ is a weak vector saddle-point, then $\bar{x}$ is a weak minimum for (VP).*

Li and Wang [139] presented several conditions for the existence of a Lagrange multipliers or a weak saddle point in multiobjective optimization. They established relationship between a Lagrange multiplier and a weak saddle point. Let K be a cone in R^m. Denote $K' = K \bigcup \{0\}$. K is said to be pointed if

$$K' \bigcap (-K)' = \{0\}.$$

K is said to be acute if the closure clK is pointed. Let K be a pointed convex cone of R^m with $intK \neq \phi$, let Y be a non-empty subset of R^m. For $y, z \in R^m$, the cone orders with respect to K is defined as follows:

$$y \leqq_K z \Leftrightarrow z - y \in K;$$

$$y \leq_K z \Leftrightarrow z - y \in K - \{0\};$$

$$y <_K z \Leftrightarrow z - y \in intK.$$

Li and Wang [139] considered the following multiobjective programming problem:

(VP)

$$K - \min f(x)$$

$$\text{Subject to } g(x) \leqq_K 0,$$

$$x \in X,$$

where X is a nonempty subset of R^n, $f : R^n \to R^m$, $g : R^n \to R^p$.

Definition 6.95. *$\bar{x} \in X_0 = \{x \in X : g(x) \leqq_K 0\}$ is called a weakly efficient solution for (VP) if $f(\bar{x}) \in W - \mathrm{Min}_K Y = \{\bar{y} \in Y :$ there is noy $\in Y$ such that $y <_K \bar{y}\}$; $\bar{x} \in X_0$ is called an efficient solution for (VP) if $f(\bar{x}) \in \mathrm{Min}_K f(X_0) = \{\bar{x} \in f(X_0) :$ there is no $y \in f(X_0$ such that $y \leq_K f(X_0))$.*

Denote by Γ the family of all $m \times p$ matrices Λ satisfying $\Lambda Q \subset K$. For any given $\eta \in K^0 - \{0\}$ and $\lambda \in Q^0$, one can easily verify that $\Lambda \in \Gamma$ if $\Lambda = e\lambda^T$, where $e \in K$ satisfying $\eta^T e = 1$. Here K^0 is positive dual cone of K.

Definition 6.96. *The vector-valued Lagrangian function for problem (VP) is defined as*

$$L(x, \Lambda) = f(x) + \Lambda g(x), \quad (x, \Lambda) \in X_0 \times \Gamma.$$

Here $X_0 \times \Gamma$ is the cartesian product of X_0 and Γ.

Definition 6.97. *A pair $(\bar{x}, \bar{\Lambda}) \in X \times T$ is called a saddle point of $L(x, \Lambda)$ if*

$$L(\bar{x}, \bar{\Lambda}) \in \mathrm{Min}_K\{L(x, \bar{\Lambda}) : x \in X\} \cap \mathrm{Max}_K\{L(\bar{x}, \Lambda) : \Lambda \in \Gamma\}.$$

Definition 6.98. *A pair $(\bar{x}, \bar{\Lambda}) \in X \times T$ is called a weak saddle point of $L(x, \Lambda)$ if*

$$L(\bar{x}, \bar{\Lambda}) \in W - \mathrm{Min}_K\{L(x, \bar{\Lambda}) : x \in X\} \cap W - \mathrm{Max}_K\{L(\bar{x}, \Lambda) : \Lambda \in \Gamma\}.$$

Li and Wang [139] used following result:

Theorem 6.99. *$(\bar{x}, \bar{\Lambda}) \in X \times \Gamma$ is a weak saddle point of $L(x, \Lambda)$ if and only if*

(a) $L(\bar{x}, \bar{\Lambda}) \in W - \mathrm{Max}_K \left\{L(x, \bar{\Lambda}) : x \in X\right\}$.
(b) $g(\bar{x}) \leqq_K 0$.
(c) $\bar{\Lambda} g(\bar{x}) \geq_K 0$.

Furthermore, Li and Wang [139] gave the following result to connect weak saddle point of the vector-valued Lagrangian function and weakly efficient solution of (VP).

Theorem 6.100. *If the pair* $(\bar{x}, \bar{\Lambda})$ *is a weak saddle point of the vector-valued Lagrangian function* L *and* $\bar{\Lambda} g(\bar{x}) = 0$, *then* $\bar{x}$ *is a weakly efficient solution of problem (VP).*

Li and Wang [139] have further discussed connection between properly efficient solution of (VP) and saddle point of the vector-valued Lagrangian function.

6.7 Linearization of Nonlinear Multiobjective Programming

Consider the following nonlinear multiobjective programming problem:
(CVOP)

$$\text{Minimize}\,(f_1(x), \ldots, f_p(x))$$

$$\text{Subject to } g_j(x) \leqq 0, \quad j = 1, \ldots, m.$$

Let x^* be a feasible solution for (CVOP), similar to Chew and Choo [36], Bector et al. [16] considered the linearized multiobjective program LP(x^*) given by
LP(x^*)

$$\text{Minimize}\left(x^T \nabla f_1(x^*), \ldots, x^T \nabla f_p(x^*)\right)$$

$$\text{Subject to } g_j(x) \leqq 0, \quad j = 1, \ldots, m.$$

Using the following extensions of results of Chew and Choo [36], Bector et al. [16] obtained connection of efficient solutions of the original problem to the efficient solutions of the linearized multiobjective programming problem.

Lemma 6.101. *Let* x^* *be efficient solution for (CVOP) and Slater type constraint qualification holds for (CVOP). Let* $g_j, j = 1, \ldots, m$ *be convex. Then* x^* *is efficient for LP*(x^*).

Lemma 6.102. *Let* $f_i, i = 1, \ldots, p$ *be pseudoconvex. Let* x^* *be efficient to LP*(x^*), *then* x^* *is efficient for (CVOP).*

In view of above lemmas, if f is pseudo-convex and g is convex then the constrained vector optimization problem (CVOP) and LP(x^*) are equivalent. Following Chew and Choo [36] and Bector et al. [16], Antczak [4] proposed the following equivalent multiobjective program:
$(CVOP_\eta(\bar{x}))$

$$\text{Minimize}\left(\eta(x, \bar{x})^T \nabla f_1(\bar{x}), \ldots, \eta(x, \bar{x})^T \nabla f_p(\bar{x})\right)$$

$$\text{Subject to } g_j(x) \leqq 0, \quad j = 1, \ldots, m,$$

where

$$X_0 = \{x \in X : g_j(x) \leqq 0, j = 1, \ldots, m\}, \eta : X_0 \times X_0 \to R^n.$$

Antczak [4] obtained the following result to connect (CVOP) and $(CVOP_\eta(\bar{x}))$. For proof see, Antczak [4].

Theorem 6.103. *Let $\bar{x}$ be (weak) efficient for (CVOP) and Slater type constraint qualification holds at $\bar{x}$ for (CVOP). Further, we assume that g is invex with respect to η at $\bar{x}$ on X_0 and $\eta(x, \bar{x}) = 0$. Then $\bar{x}$ is (weak) efficient for $(CVOP_\eta(\bar{x}))$.*

Theorem 6.104. *Let $\bar{x}$ be a feasible point for $(CVOP_\eta(\bar{x}))$. Further, we assume that f is invex with respect to η at $\bar{x}$ on X_0 and $\eta(\bar{x}, \bar{x}) = 0$. If $\bar{x}$ is efficient for $(CVOP_\eta(\bar{x}))$ then $\bar{x}$ is also efficient for (CVOP).*

Thus, if we assume that f and g are invex with respect to the same η at $\bar{x}$ on the set of feasible solution X_0 and $\eta(\bar{x}, \bar{x}) = 0$ then multiobjective programming problems (CVOP) and $(CVOP_\eta(\bar{x})$ are equivalent.

Example 6.105. Consider the following multiobjective programming problem

$$\text{Minimize} f(x) = \left(\frac{1}{3}x_1^3 - \frac{1}{2}x_1^2 + 5x_1 + \frac{1}{6}, 5x_1 + e^{x_2}\right)$$

$$\text{Subject to } g_1(x) = 1 - x_1 \leq 0,$$

$$g_2(x) = 1 - x_2 \leq 0.$$

Note that $\bar{x} = (1, 1)$ is an efficient point in the considered problem. Further, it can be proved that f and g are invex at $\bar{x}$ with respect to the same function η defined by

$$\eta(x, u) = \left(\frac{x_1 - u_1}{5}, \frac{x_2 - u_2}{e^{u_2}}\right)^T.$$

Then the problem $CVOP_\eta(\bar{x})$ for (CVOP) is the following linear multiobjective programming problem:

$$\text{Minimize}\,(x_1 - 1, x_1 + x_{22})$$

Subject to

$$g_1(x) = 1 - x_1 \leq 0,$$

$$g_2(x) = 1 - x_2 \leq 0.$$

It is not difficult to see, that $\bar{x} = (1, 1)$ is also efficient in the above optimization problem.

Antczak [4] introduced an η-Lagrange function for multiobjective programming problem $CVOP_\eta(\bar{x})$.

Definition 6.106. *An η-Lagrange function is said to be a Lagrange function for a multiobjective programming problem $CVOP_\eta(\bar{x})$*

$$L_\eta(x,\xi) = \left(\eta(x,\bar{x})^T \nabla f_1(\bar{x}) + \xi^T g(x), \ldots, \eta(x,\bar{x})^T \nabla f_1(\bar{x}) + \xi^T g(x)\right).$$

Definition 6.107. *A point $(\bar{x},\bar{\xi}) \in X_0 \times R_+^m$ is said to be a saddle point for the η-Lagrange function if*

$$L_\eta(\bar{x},\xi) \leqq L_\eta(\bar{x},\bar{\xi}) \leqq L_\eta(x,\bar{\xi}).$$

Antczak [4] obtained the following result to connect saddle point of the η-Lagrange function and the efficient solution of the (CVOP), for the proof, see Antczak [4].

Theorem 6.108. *We assume that f is (invex) strictly invex with respect to η at $\bar{x}$ on X_0 with $\eta(\bar{x},\bar{x}) = 0$ and some constraint qualification holds at $\bar{x}$ for (CVOP). If $(\bar{x},\bar{\xi})$ is saddle point for L_η, then $\bar{x}$ is a (weak) efficient solution for (CVOP).*

6.8 Multiobjective Symmetric Duality

Dorn [59] defined a nonlinear programming problem and its dual to be symmetric if the dual of the dual is the original problem; that is, if the dual program is recast in the form of the primal, its dual is the primal. A linear program and its dual are symmetric in this sense. Symmetric dual quadratic programs are given by Dorn [59] and Cottle [40]. Dantzig et al. [53] (see also [167]) first formulated a pair of symmetric dual nonlinear programming problems. The formulation given by Dantzig et al. [53] involves a scalar function $f(x,y), x \in R^n, y \in R^m$ that is required to be convex in x for fixed y and concave in y for fixed x. Mond and Weir [180] have given a different pair of symmetric dual nonlinear programs in which the convexity and concavity assumptions of [53] are weakened to pseudo-convexity and pseudo-concavity. Later Weir and Mond [246] presented multiobjective symmetric duality. Weir and Mond [246] proposed the following pair of nonlinear multiobjective symmetric dual:

(MSP)

$$\text{Minimize} f(x,y) - \left[y^T \left(\nabla_y \lambda^T f\right)(x,y)\right] e$$
$$\text{Subject to } \nabla_y \left(\lambda^T f\right)(x,y) \leqq 0,$$
$$x \geqq 0, \ \ \lambda \in \Lambda^+.$$

(MSD)

$$\text{Maximize} f(u,v) - \left[u^T \nabla_x \left(\lambda^T f\right)(u,v)\right] e$$
$$\text{Subject to } \nabla_x \left(\lambda^T f\right)(u,v) \geqq 0,$$
$$v \geqq 0, \ \ \lambda \in \Lambda^+,$$

where $f : R^n \times R^n \to R^p$; $\Lambda^+ = \{\lambda \in R^p : \lambda > 0, \sum_{i=1}^p \lambda_i = 1\}$ and $e = (1, \dots, 1)^T \in R^p$. Weir and Mond [246] obtained the following weak duality results.

Theorem 6.109. *Let $f(\cdot, y)$ be convex for fixed y and let $f(x, \cdot)$ be concave for fixed x. Let (x, y, λ) be feasible for (MSP) and let (u, v, λ) be feasible for (MSD). Then, the following cannot hold:*

$$f(x, y) - \left[y^T \nabla_y \left(\lambda^T f\right)\right] e \leq f(u, v) - \left[u^T \nabla_x \left(\lambda^T f\right)\right] e.$$

They also obtained strong duality result for (MSP) and (MSD). Weir and Mond [246] proposed another symmetric dual pair known as Mond–Weir type symmetric dual pair.

(MWSP)

$$\begin{aligned} &\text{Minimize} f(x, y) \\ &\text{Subject to } \nabla_y \left(\lambda^T f\right)(x, y) \leqq 0, \\ &y^T \nabla_y \left(\lambda^T f\right)(x, y) \geqq 0 \\ &x \geqq 0, \ \ \lambda \in \Lambda^+. \end{aligned}$$

(MWSD)

$$\begin{aligned} &\text{Maximize} f(u, v) \\ &\text{Subject to } \nabla_x \left(\lambda^T f\right)(u, v) \geqq 0, \\ &u^T \nabla_x \left(\lambda^T f\right)(u, v) \leqq 0 \\ &v \geqq 0, \ \ \lambda \in \Lambda^+, \end{aligned}$$

where $f : R^n \times R^n \to R^p$; $\Lambda^+ = \{\lambda \in R^p : \lambda > 0, \sum_{i=1}^p \lambda_i = 1\}$ and $e = (1, \dots, 1)^T \in R^p$. Weir and Mond [246] obtained the weak duality for (MWSP) and (MWSD):

Theorem 6.110. *Let (x, y) be feasible for (MWSP) and let (u, v) be feasible for (MWSD). Let $\lambda^T f(\cdot, \cdot, y)$ be pseudo-convex for fixed y and let $\lambda^T f(\cdot, \cdot, x)$ be pseudo-convex for fixed x. Then $f(x, y) \nleq f(u, v)$.*

Gulati et al. [76] extended the results of Weir and Mond [246] to the case of invex functions, but this was not a direct approach, they proposed a slightly modified dual models for Mond–Weir dual models (MWSP) and (MWSP) (however, the Wolfe type dual model (MSP) and (MSD) remains the same)and an additional restriction was imposed on the kernel function η. Weak duality for (MSP) and (MSD) under invexity given by Gulati et al. [76] is (for proof, see [76]):

Theorem 6.111. *Let $f(\cdot, y)$ be invex in x with respect to η and $-f(x, \cdot)$ be invex in y with respect to ξ, with $\eta(x, u) + u \geq 0$ and $\xi(v, y) + y \geq 0$, whenever (x, y, λ) feasible for (MSP) and (u, v, λ) feasible for (MSD). Then*

$$\begin{aligned} &\lambda^T \left(f(x, y) - \left(y^T \nabla_y \left(\lambda^T K\right)(x, y) e\right)\right) \\ &\geq \lambda^T \left(f(u, v) - \left(u^T \nabla_x \left(\lambda^T K\right)(u, v) e\right)\right). \end{aligned}$$

Gulati et al. [76] established strong duality for (MSP) and (MSD) under invexity conditions. Further, these authors proposed a slightly different Mond–Weir type symmetric dual as compared to Weir and Mond [246].

(MWSP1)

$$\begin{aligned} &\text{Minimize} f(x,y) \\ &\text{Subject to } \nabla_y \left(\lambda^T f\right)(x,y) \leqq 0, \\ &y^T \nabla_y \left(\lambda^T f\right)(x,y) \geqq 0, \\ &\lambda > 0. \end{aligned}$$

(MWSD1)

$$\begin{aligned} &\text{Maximize} f(u,v) \\ &\text{Subject to } \nabla_x \left(\lambda^T f\right)(u,v) \geqq 0, \\ &u^T \nabla_x \left(\lambda^T f\right)(u,v) \leqq 0, \\ &\lambda > 0. \end{aligned}$$

The weak duality relation for (MWSP1) and (MWSD1) under invexity assumption given by Gulati et al. [76]:

Theorem 6.112. *Let (x,y,λ) be feasible for (MWSP1) and (u,v,λ) feasible for (MWSD1). Let $\lambda^T f(\cdot,y)$ be pseudo-invex with respect to η for fixed y, and $-\lambda^T f(x,\cdot)$ be pseudo-invex with respect to ξ for fixed x, with $\eta(x,u)+u \geqq 0$ and $\eta(v,y)+y \geqq 0$. Then*

$$\lambda^T f(x,y) \geq \lambda^T f(u,v).$$

Gulati et al. [76] also obtained strong and converse duality for (MWSP1) and (MWSD1) under pseudo-invexity assumptions. These authors also discussed self-duality under the generalized invexity assumptions. Later Kim et al. [123] formulated a pair of multiobjective symmetric dual programs for pseudo-invex functions and arbitrary cones. The model proposed by Kim et al. [123] unifies the Wolfe vector symmetric dual model and the Mond–Weir vector symmetric dual model.

(KSP)

$$\begin{aligned} &\text{Minimize} f(x,y) - \left[y^T \nabla_y \left(\lambda^T f\right)(x,y)\right] e \\ &\text{Subject to } (x,y) \in C_1 \times C_2, \\ &\nabla_y \left(\lambda^T f\right)(x,y) \in C_2^*, \\ &\lambda \geqq 0, \quad \lambda^T e = 1. \end{aligned}$$

(KSD)

$$\begin{aligned} &\text{Maxmize} f(u,v) - \left[u^T \nabla_x \left(\lambda^T f\right)(u,v)\right] e \\ &\text{Subject to } (x,y) \in C_1 \times C_2, \\ &-\nabla_x \left(\lambda^T f\right)(u,v) \in C_1^*, \\ &\lambda \geqq 0, \quad \lambda^T e = 1, \end{aligned}$$

where $C_i, i = 1, 2$ are convex cones with nonempty interior in R^n. Recall that $C_i^*, i = 1, 2$ is called the polar of $C_i, i = 1, 2$ if

$$C_i^* = \{z : x^T z \leqq 0, \forall x \in C_i\}.$$

Kim et al. [123] established the following weak duality relation for (KSP) and (KSD).

Theorem 6.113. *Let (x, y, λ) be feasible for (KSP) and (u, v, λ) be feasible for (KSD). Assume that either*

(a) $x \neq u, (\lambda^T f)(\cdot, y)$ is strictly pseudo-invex for fixed y with respect to η_1 on C_1 and $-(\lambda^T f)(x, \cdot)$ is pseudo-invex for fixed x with respect to η_2 on C_2; or

(b) $y \neq v$ $(\lambda^T f)(\cdot, y)$ is pseudo-invex for fixed y with respect to η_1 on C_1 and $-(\lambda^T f)(x, \cdot)$ is strictly pseudo-invex for fixed x with respect to η_2 on C_2; or

(c) $\lambda > 0$, $(\lambda^T f)(\cdot, y)$ is pseudo-invex for fixed y with respect to η_1 on C_1 and $-(\lambda^T f)(x, \cdot)$ is pseudo-invex for fixed x with respect to η_2 on C_2.

Then

$$f(x, y) - \left[y^T \nabla_y \left(\lambda^T f\right)(x, y)\right] e \nleq f(u, v) - \left[u^T \nabla_x \left(\lambda^T f\right)(u, v)\right] e.$$

Kim et al. [123] also obtained strong and self-duality relations under pseudo-invexity assumptions. For more results on symmetric duality, reader may see, Suneja et al. [230], Mishra [156], Mishra and Lai [157].

7

Variational and Control Problems Involving Invexity

7.1 Scalar Variational Problems with Invexity

The relationship between mathematical programming and classical calculus of variation was explored and extended by Hanson [82]. Duality results are obtained for scalar valued variational problems in Mond and Hanson [170] under convexity. Mond and Hanson [170] considered the following problems as continuous analogue of the usual primal and dual problems in nonlinear programming problems:
Consider the determination of a piece-wise smooth extremal $x = x(t)$, $t \in I$ for the following modified Lagrange problem:

Problem I. (Primal) $\equiv$ P

$$\text{Minimize} \int_a^b f(t, x, \dot{x}) dt$$

$$\text{Subject to } x(a) = \alpha, \quad x(b) = \beta, \tag{7.1}$$

$$g(t, x, \dot{x}) \geq 0, \quad t \in I. \tag{7.2}$$

Consider also the determination of $(m+n)$-dimensional extremal $(x, \lambda) \equiv (x(t), \lambda(t)), a \leq t \leq b$, for the following maximization problem:

Problem II. (Dual) $\equiv$ D

$$\text{Maximize} \int_a^b \{f(t, x, \dot{x}) - \lambda(t) g(t, x, \dot{x})\} dt$$

$$\text{Subject to } x(a) = \alpha, \quad x(b) = \beta, \tag{7.3}$$

$$f_x(t, x, \dot{x}) - \lambda(t) g(t, x, \dot{x}) = \frac{d}{dt} f(t, x, \dot{x}) - \lambda(t) g(t, x, \dot{x}), \tag{7.4}$$

$$\lambda(t) \geq 0. \tag{7.5}$$

Here $I = [a, b]$ be a real interval and $f : I \times R^n \times R^n \to R$ be a continuously differentiable function. In order to consider $f(t, x, \dot{x})$, where $x : I \to R^n$ is differentiable with derivative $\dot{x}$. Let $g(t, x, \dot{x})$ be an m-dimensional function which similarly has continuous derivatives up to and including second order, $x(t)$ is an n-dimensional piecewise smooth function and $\lambda(t)$ is an m-dimensional function continuous except possibly for values of t corresponding to corners of $x(t)$. For values of t corresponding to corners of $x(t)$, (7.4) must be satisfied for right and left hand limits.

No notational distinction is made between row and column vectors. Subscripts denote partial derivatives, superscripts denote vector components. Thus,

$$f_{x^1} - \left(\sum_{i=1}^{m} \lambda_i g^i\right)_{x^1} = \frac{d}{dt}\left\{f_{\dot{x}^1} - \left(\sum_{i=1}^{m} \lambda_i g^i\right)_{\dot{x}^1}\right\},$$

$$f_{x^2} - \left(\sum_{i=1}^{m} \lambda_i g^i\right)_{x^2} = \frac{d}{dt}\left\{f_{\dot{x}^2} - \left(\sum_{i=1}^{m} \lambda_i g^i\right)_{\dot{x}^2}\right\},$$

$$f_{x^n} - \left(\sum_{i=1}^{m} \lambda_i g^i\right)_{x^n} = \frac{d}{dt}\left\{f_{\dot{x}^n} - \left(\sum_{i=1}^{m} \lambda_i g^i\right)_{\dot{x}^n}\right\}.$$

Remark 7.1. It was pointed out by Mond and Hanson [170] that if all the functions are independent of t, the problems P and D reduce to nonlinear programs,

$$\text{Minimize } f(x)$$
$$\text{Subject to } g(x) \geq 0,$$

and its dual

$$\text{Maximize } f(x) - \lambda g(x)$$
$$\text{Subject to } f_x(x) - \lambda g_x(x) = 0,$$
$$\lambda \geq 0.$$

Mond et al. [174] extended the work of Mond and Hanson [170] to the class of invex functions by extending the concept of invex functions due to Hanson [83] to continuous case as follows:

Definition 7.2. *$\int_a^b f(t, x, \dot{x})$ will be said to be invex with respect to η if there exists a vector function $\eta(t, x^*, x)$ with $\eta(t, x, x) = 0$ such that*

$$\int_a^b f(t, x^*, \dot{x}^*)dt - \int_a^b f(t, x, \dot{x})dt$$
$$\geq \int_a^b \left\{\eta(t, x^*, x) f_x(t, x, \dot{x}) + \left(\frac{d}{dt}\eta(t, x^*, x)\right) f_{\dot{x}}(t, x, \dot{x})\right\} dt.$$

Remark 7.3. Mond and others pointed out that if the function f is independent of t, the above definition of invex function reduces to the following:

$$f(x^*) - f(x) \geq \eta(x^*, x) f_x(x),$$

which is the original definition of invexity given by Hanson [83].

Definition 7.4. *$\int_a^b f(t, x, \dot{x})$ will be said to be pseudoinvex with respect to η if there exists a vector function $\eta(t, x^*, x)$ with $\eta(t, x, x) = 0$ such that*

$$\int_a^b \left\{ \eta(t, x^*, x) f_x(t, x, \dot{x}) + \left(\frac{d}{dt} \eta(t, x^*, x) \right) f_x(t, x, \dot{x}) \right\} dt \geq 0$$
$$\Rightarrow \int_a^b f(t, x^*, \dot{x}^*) dt \geq \int_a^b f(t, x, \dot{x}) dt,$$

Or equivalently,

$$\int_a^b f(t, x^*, \dot{x}^*) dt < \int_a^b f(t, x, \dot{x}) dt$$
$$\Rightarrow \int_a^b \left\{ \eta(t, x^*, x) f_x(t, x, \dot{x}) + \left(\frac{d}{dt} \eta(t, x^*, x) \right) f_x(t, x, \dot{x}) \right\} dt < 0.$$

Definition 7.5. *$\int_a^b f(t, x, \dot{x})$ will be said to be quasiinvex with respect to η if there exists a vector function $\eta(t, x^*, x)$ with $\eta(t, x, x) = 0$ such that*

$$\int_a^b f(t, x^*, \dot{x}^*) dt \leq \int_a^b f(t, x, \dot{x}) dt$$
$$\Rightarrow \int_a^b \left\{ \eta(t, x^*, x) f_x(t, x, \dot{x}) + \left(\frac{d}{dt} \eta(t, x^*, x) \right) f_x(t, x, \dot{x}) \right\} dt \leq 0.$$

However, Mond et al. [174] established only weak duality theorem and discussed natural boundary values. We will report the weak duality of Mond and others as follows.

Theorem 7.6 (Weak duality). *If f and $-g$ are invex for some function η, then* $\inf(P) \geq \sup(D)$.

□

Assume that the necessary constraints (see, e.g., Valentine [237]) for the existence of multipliers at an extremal of (P) are satisfied. Thus for every minimizing arc $x = x^*(t)$ of (P), there exists a function of the form

$$F = \lambda_0^* f - \lambda^*(t) g$$

such that

$$F_x = \frac{d}{dt}(F_{\dot{x}}) \tag{7.6}$$

$$\lambda^{*i} g^i = 0, \quad i = 1, \dots, m \tag{7.7}$$

$$\lambda^*(t) \geq 0 \tag{7.8}$$

holds throughout $a \leq t \leq b$ (except at corners of $x^*(t)$, where (7.6) holds for unique left- and right-hand limits). Here λ_0^* is a constant, $\lambda^*(t)$ is continuous except possibly for values of t corresponding to the corners of $x^*(t)$, and $(\lambda_0^*, \lambda^*(t))$ cannot vanish for any t, $a \leq t \leq b$.

Assuming that the minimizing arc $x^*(t)$ is normal, i.e., λ_0^* can be taken equal to 1, the following strong duality theorem is given in Mond et al. [174].

Theorem 7.7 (Strong duality). *Assume that f and $-g$ are invex for some function η. If the function $x^*(t)$ minimizes the primal problem (P), then there exists a $\lambda^*(t)$, such that $x^*(t), \lambda^*(t)$ maximizes the dual problem (D) and the extreme values of (P) and (D) are equal.*

Proof. See Mond et al. [174].

□

Mond and others have further discussed the variational problems with natural boundary values rather than fixed end points, however, we will confine ourselves to fixed end point variational problems.

Later Mond and Husain [172] considered the following minimization problem:
Consider the determination of a piece-wise smooth extremal $x = x(t)$, $t \in I$ for the following modified Lagrange problem:

Problem I. (Primal) ≡ P

$$\text{Minimize} \int_a^b f(t, x, \dot{x})dt$$

$$\text{Subject to } x(a) = \alpha, \quad x(b) = \beta,$$

$$g(t, x, \dot{x}) \leq 0, \quad t \in I.$$

Consider also the determination of $(m+n)$-dimensional extremal $(x, \lambda) \equiv (x(t), \lambda(t)), a \leq t \leq b$, for the following maximization problem:

Problem II. (Dual) ≡ MWD

$$\text{Maximize} \int_a^b f(t, u, \dot{u})dt$$

$$\text{Subject to } x(t_0) = \alpha, \quad x(t_f) = \beta,$$

$$f_u(t, u, \dot{u}) + y(t)g_u(t, u, \dot{u})$$

$$= \frac{d}{dt}\{f_{\dot{u}}(t, u, \dot{u}) + y(t)g_{\dot{u}}(t, u, \dot{u})\}, \tag{7.9}$$

$$\int_a^b y(t)g(t, u, \dot{u})dt \geq 0,$$

$$y(t) \geq 0, \quad t \in I.$$

Here $I = [a, b]$ be a real interval and $f : I \times R^n \times R^n \to R$ be a continuously differentiable function. In order to consider $f(t, x, \dot{x})$ where $x : I \to R^n$ is differentiable with derivative $\dot{x}$. Let $g(t, x, \dot{x})$ be an m-dimensional function which similarly has continuous derivatives up to and including second order, $x(t)$ is an n-dimensional piecewise smooth function and $\lambda(t)$ is an m-dimensional function continuous except possibly for values of t corresponding to corners of $x(t)$. For values of t corresponding to corners of $x(t)$, (7.9) must be satisfied for right- and left-hand limits. Notice that the dual problem is now Mond–Weir type of dual and this allows further weakening of the invexity assumptions in the duality results. Mond and Husain [172] obtained several Kuhn–Tucker type sufficient optimality results and duality theorems under pseudo-invexity and quasi-invexity assumptions.

Theorem 7.8 (Sufficient optimality conditions). *Let x^* be feasible for (P) and assume that f is pseudoinvex at x^* with respect to η and that for each $i \in I(x^*), g^i$ is quasiinvex at x^* with respect to η. If there exists a piecewise smooth $y^* : I \to R^m$ such that (x^*, y^*) satisfies the conditions*

$$f_x(t, x^*, \dot{x}^*) + y^*(t)g_x(t, x^*, \dot{x}^*) = \frac{d}{dt}\{f_{\dot{x}}(t, x^*, \dot{x}^*) + y^*(t)g_{\dot{x}}(t, x^*, \dot{x}^*)\}$$

$$y^*(t)g(t, x^*, \dot{x}^*) = 0, \quad t \in I,$$

$$y^*(t) \geq 0, \quad t \in I,$$

then x^* is a global optimal solution of the problem (P).

Proof. From feasibility and quasiinvexity assumption, we get

$$\eta(t, x, x^*)g_x^i(t, x^*, \dot{x}^*) + \left(\frac{d}{dt}\eta(t, x, x^*)\right) g_{\dot{x}}^i(t, x^*, \dot{x}^*) \leq 0,$$

and hence, by taking $y^{i^*} = 0$, for $i \notin M(x^*)$,

$$\int_a^b \left\{\eta(t, x, x^*)g_x^i(t, x^*, \dot{x}^*) + \left(\frac{d}{dt}\eta(t, x, x^*)\right) g_{\dot{x}}^i(t, x^*, \dot{x}^*)\right\} dt \leq 0.$$

On the other hand from the first necessary optimality condition, integration by parts and the above inequality, one gets

$$\int_a^b \left\{\eta(t, x, x^*)f_x(t, x^*, \dot{x}^*) + \left(\frac{d}{dt}\eta(t, x, x^*)\right) f_{\dot{x}}(t, x^*, \dot{x}^*)\right\} dt \geq 0.$$

This inequality together with pseudoinvexity gives

$$\int_a^b f(t, x, \dot{x})dt \geq \int_a^b f(t, x^*, \dot{x}^*)dt,$$

that is x^* is a global optimal solution for (P).

□

Theorem 7.9 (Sufficient optimality conditions). *Let x^* be a feasible point for (P). If there exists a piecewise smooth $y^* : I \to R^m$ such that (x^*, y^*) satisfies the necessary optimality conditions given in Theorem 7.8, and if the Lagrangian function $\Psi(x, y^*) : X \to R$, defined by*

$$\Psi(x, y^*) = \int_a^b \{f(t, x, \dot{x}) + y^*(t)g(t, x, \dot{x})\}dt$$

is pseudoinvex at x^* with respect to η, then x^* is a global optimal solution for (P).

Proof. The proof is left as a simple exercise to the reader. □

The following theorem is slight generalization of Theorem 7.6

Theorem 7.10 (Weak duality). *If for all feasible x for (P) and (u, y) for (MWD), there exists a differentiable vector function η with $\eta(t, x, x) = 0$ such that, for all feasible (x, u, y), f is pseudoinvex at u with respect to η and that for each $i = 1, 2, \ldots, m$, g^i is quasiinvex at u with respect to η. Then $\int_a^b f(t, x, \dot{x})dt \geq \int_a^b f(t, u, \dot{u})dt$.*

Proof. The proof is very similar to the proof of Theorem 7.8. □

Theorem 7.11 (Strong duality). *Let x^* be a normal solution for (P). Assume that a differentiable vector function η, with $\eta(t, x, x) = 0$, exists such that the invexity hypothesis of Theorem 7.10 are satisfied. Then there exists a piecewise smooth $y^* : I \to R^m$ such that (x^*, y^*) solves (MWD) and* Minimum(P) = Maximum(MWD).

Mond and Husain [172] have also given strict converse duality for (P) and (MWD) under several generalized invexity assumptions.

As discussed in Chaps. 5 and 6, the symmetric duality is one of the important problems of interest for researchers. Mond and Hanson [171] extended the symmetric duality to variational problems, giving continuous analogues of the previous results in the literature on symmetric duality.

Consider the real scalar function $f(t, x, \dot{x}, y, \dot{y})$, where $I = [a, b]$, x and y are functions of t with $x(t) \in R^n$ and $y(t) \in R^m$, and $\dot{x}$ and $\dot{y}$ denote derivatives of x and y, respectively, with respect to t. assume that f has continuous fourth-order partial derivatives with respect to $x, \dot{x}, y$ and $\dot{y}$.

In symmetric duality, we consider the problem of finding functions $x : [a, b] \to R^n$ and $y : [a, b] \to R^m$, with $(\dot{x}(t), \dot{y}(t))$ piecewise smooth on $[a, b]$, to solve the following pair of optimization problems.

(SP)

$$\text{Minimize} \int_a^b [f(t,x,\dot{x},y,\dot{y}) - y(t)^T f_y(t,x,\dot{x},y,\dot{y}) + y(t)^T \frac{d}{dt} f_{\dot{y}}(t,x,\dot{x},y,\dot{y})]dt,$$

$$\text{Subject to } x(a)=\alpha,\ \ x(b)=\beta,\ \ y(a)=\alpha,\ \ y(b)=\beta \tag{7.10}$$

$$\frac{d}{dt} f_{\dot{y}}(t,x,\dot{x},y,\dot{y}) \geq f_y(t,x,\dot{x},y,\dot{y}),\quad t\in I, \tag{7.11}$$

(SD)

$$\text{Maximize} \int_a^b [f(t,x,\dot{x},y,\dot{y}) - x(t)^T f_x(t,x,\dot{x},y,\dot{y}) + x(t)^T \frac{d}{dt} f_{\dot{x}}(t,x,\dot{x},y,\dot{y})]dt,$$

$$\text{Subject to } x(a)=\alpha,\ \ x(b)=\beta,\ \ y(a)=\alpha,\ \ y(b)=\beta \tag{7.12}$$

$$\frac{d}{dt} f_{\dot{x}}(t,x,\dot{x},y,\dot{y}) \leq f_x(t,x,\dot{x},y,\dot{y}),\quad \in I, \tag{7.13}$$

where (7.11) and (7.13) may fail to hold at corners of $(\dot{x}(t),\dot{y}(t))$, but must be satisfied for unique right- and left-hand limits.

The above (SP) and (SD) are *problemsI* and *problemII* stated in Mond and Hanson [171] with the constraint $x(t) \geq 0$ removed from *ProblemI* and $y(t) \geq 0$ removed from *ProblemII*.

Remark 7.12. If the time dependency of problems (SP) and (SD) is removed and is considered to have domain $R^n \times R^m$, we obtain the symmetric dual pair as follows:

(SP)

$$\text{Minimize } f(x,y) - y^T f_y(x,y)$$
$$\text{Subject to } f_y(x,y) \leq 0$$

(SD)

$$\text{Minimize } f(x,y) - x^T f_x(x,y)$$
$$\text{Subject to } f_x(x,y) \geq 0.$$

These are the programs considered in Chap. 5 (see also Dantzing et al. [53]), except that the positivity constraints have been omitted.

The concept of invexity for symmetric dual problems needs a little extension as follows:

Definition 7.13. *The functional $\int_a^b f(t,x,\dot{x},y,\dot{y})dt$ is invexinvex functional in x and $\dot{x}$ if for each $y:[a,b]\to R^m$, with $\dot{y}$ piecewise smooth, there exists a function $\eta:[a,b]\times R^n\times R^n\times R^n\times R^n\to R^n$ such that*

$$\int_a^b [f(t,x,\dot{x},y,\dot{y})-f(t,u,\dot{u},y,\dot{y})]dt$$
$$\geq \int_a^b \eta(t,x,\dot{x},y,\dot{y})^T\times[f_x(t,u,\dot{u},y,\dot{y})-\frac{d}{dt}f_{\dot{x}}(t,u,\dot{u},y,\dot{y})]dt$$

for all $x:[a,b]\to R^n, u:[a,b]\to R^n$ with $(\dot{x}(t),\dot{u}(t))$ piecewise smooth on $[a,b]$.

Similarly, the functional $-\int_a^b f(t,x,\dot{x},y,\dot{y})dt$ is invex in y and $\dot{y}$ if for each $x:[a,b]\to R^n$, with $\dot{x}$ piecewise smooth, there exists a function $\xi:[a,b]\times R^m\times R^m\times R^m\times R^m\to R^m$ such that

$$-\int_a^b [f(t,x,\dot{x},v,\dot{v})-f(t,x,\dot{x},y,\dot{y})]dt$$
$$\geq -\int_a^b \xi(t,v,\dot{v},y,\dot{y})^T\times[f_y(t,x,\dot{x},y,\dot{y})-\frac{d}{dt}f_{\dot{y}}(t,x,\dot{x},y,\dot{y})]dt$$

for all $v:[a,b]\to R^m, y:[a,b]\to R^m$ with $(\dot{v}(t),\dot{y}(t))$ piecewise smooth on $[a,b]$.

For the sake of simplicity, in the sequel, we will write $\eta(x,u)$ for $\eta(t,x,\dot{x},u,\dot{u})$ and $\xi(v,y)$ for $\xi(t,v,\dot{v},y,\dot{y})$.

The following weak duality relating (SP) and (SD) is established in Smart and Mond [226].

Theorem 7.14 (Weak duality). *If $\int_a^b f$ is invex in x and $\dot{x}$ and $-\int_a^b f$ is invex in y and $\dot{y}$ with $\eta(x,u)+u(t)\geq 0$ and $\xi(v,y)+y(t)\geq 0$ for all $t\in[a,b]$ (except perhaps at corners of $(\dot{x}(t),\dot{y}(t))$ or $(\dot{u}(t),\dot{v}(t))$ whenever (x,y) is feasible for (SP) and (u,v) is feasible for (SD), then $\inf(SP)\geq\sup(SD)$.*

Proof. Let (x,y) be feasible for (SP) and (u,v) be feasible for (SD). Then using the invexity assumptions, we get

$$\int_a^b \left[\left\{f(t,x,\dot{x},y,\dot{y})-y(t)^Tf_y(t,x,\dot{x},y,\dot{y})+y(t)^T\frac{d}{dt}f_{\dot{y}}(t,x,\dot{x},y,\dot{y})\right\}\right.$$
$$\left.-\left\{f(t,u,\dot{u},v,\dot{v})-u(t)^Tf_x(t,u,\dot{u},v,\dot{v})+u(t)^T\frac{d}{dt}f_{\dot{x}}(t,u,\dot{u},v,\dot{v})\right\}\right]dt$$
$$\geq \int_a^b \eta(x,u)^T\left[f_x(t,u,\dot{u},v,\dot{v})-\frac{d}{dt}f_{\dot{x}}(t,u,\dot{u},v,\dot{v})\right]dt$$
$$-\int_a^b \xi(v,y)^T\left[f_y(t,x,\dot{x},y,\dot{y})-\frac{d}{dt}f_{\dot{y}}(t,x,\dot{x},y,\dot{y})\right]dt$$

$$
\begin{aligned}
&+\int_a^b u(t)^T\left[f_x(t,u,\dot{u},v,\dot{v})-\frac{d}{dt}f_{\dot{x}}(t,u,\dot{u},v,\dot{v})\right]dt\\
&-\int_a^b y(t)^T\left[f_y(t,x,\dot{x},y,\dot{y})-\frac{d}{dt}f_{\dot{y}}(t,x,\dot{x},y,\dot{y})\right]dt\\
&=\int_a^b (\eta(x,u)+u(t))^t\left[f_x(t,u,\dot{u},v,\dot{v})-\frac{d}{dt}f_{\dot{x}}(t,u,\dot{u},v,\dot{v})\right]dt\\
&-\int_a^b (\xi(v,y)+y(t))^T\left[f_y(t,x,\dot{x},y,\dot{y})-\frac{d}{dt}f_{\dot{y}}(t,x,\dot{x},y,\dot{y})\right]dt\\
&\geq 0,
\end{aligned}
$$

by (7.11) and (7.13) with $\eta(x,u)+u(t)\geq 0$ and $\xi(v,y)+y(t)\geq 0$. Hence, $\inf(SP)\geq\sup(SD)$.

□

Remark 7.15. If the invexity assumptions of Theorem 7.8 are replaced by convexity and concavity, then the conditions $\eta(x,u)+u(t)\geq 0$ and $\xi(v,y)+y(t)\geq 0$ become $x(t)\geq 0$ and $y(t)\geq 0$. These constraints may be added to problems (SP) and (SD), respectively, to obtain the dual pair of Mond and Hanson [171].

Smart [224] further developed strong and converse duality results for the pair of problems (SP) and (SD). Another important discussion on duality is self duality:

Assume that $m=n, f(t,x(t),\dot{x}(t),y(t),\dot{y}(t))$ is said to be skew symmetric (i.e., $f(t,x(t),\dot{x}(t),y(t),\dot{y}(t))=-f(t,y(t),\dot{y}(t),x(t),\dot{x}(t))$ for all $x(t)$ and $y(t)$ in the domain of f, such that $(\dot{x}(t),\dot{y}(t))$ is piecewise smooth on $[a,b]$ and that $x(a)=y(a), x(b)=y(b)$.

It follows that (SD) may be written as a minimization problem:

(SD)

$$
\begin{aligned}
\text{Minimize} \int_a^b \Big[&f(t,y,\dot{y},x,\dot{x})-x(t)^T f_x(t,y,\dot{y},x,\dot{x})\\
&+x(t)^T\frac{d}{dt}f_{\dot{x}}(t,x,\dot{x},y,\dot{y})\Big]dt,
\end{aligned}
$$

$$
\text{Subject to } x(a)=\alpha,\quad x(b)=\beta,\quad y(a)=\alpha,\quad y(b)=\beta
$$

$$
\frac{d}{dt}f_{\dot{x}}(t,y,\dot{y},x,\dot{x})\geq f_x(t,y,\dot{y},x,\dot{x}).
$$

(SD) is formally identical to (SP); that is, the objective and constraint functions and initial conditions of (SP) and (SD) are identical. This problem is said to be self-dual.

It is easy to see that whenever (x,y) is feasible for (SP), then (y,x) is feasible for (SD), and vice-versa.

Theorem 7.16 (Self-duality). *Assume that (SP) is self-dual and that the invexity conditions of Theorem 7.14 are satisfied. If (x^*, y^*) is optimal for (SP) and the system*

$$p(t)^T \left(f_{yy}(t, x^*, \dot{x}^*, y^*, \dot{y}^*) - \frac{d}{dt} f_{y\dot{y}}(t, x^*, \dot{x}^*, y^*, \dot{y}^*) \right) + \frac{d}{dt}\left[p(t)^T \frac{d}{dt} f_{y\dot{y}}(t, x^*, \dot{x}^*, y^*, \dot{y}^*) \right] + \frac{d^2}{dt^2}\left[-p(t)^T \frac{d}{dt} f_{\dot{y}\dot{y}}(t, x^*, \dot{x}^*, y^*, \dot{y}^*) \right] = 0.$$

only has the solution $p(t) = 0, a \le t \le b$, then (y^*, x^*) is optimal for both (SP) and (SD) and the common optimal value is zero.

Proof. We leave the proof for the reader as a simple exercise. □

Natural boundary (free boundary) conditions may be discussed as in Mond and Hanson [171], since the extra transversality conditions required for the formulation of (SP) and (SD) are independent of any positivity constraints on x and y.

Kim and Lee [121] formulated the following pair of symmetric dual variational problems different from the one presented by Smart and Mond [226], that is different from the (SP) and (SD) above in this section:

(SP1)

$$\text{Minimize} \int_a^b f(t, x, \dot{x}, y, \dot{y})dt$$

$$\text{Subject to } x(a) = \alpha, \quad x(b) = \beta, \quad y(a) = \alpha, \quad y(b) = \beta$$

$$f_y(t, x, \dot{x}, y, \dot{y}) - \frac{d}{dt} f_{\dot{y}}(t, x, \dot{x}, y, \dot{y}) \le 0, \quad t \in I,$$

$$y^T \left[f_y(t, x, \dot{x}, y, \dot{y}) - \frac{d}{dt} f_{\dot{y}}(t, x, \dot{x}, y, \dot{y}) \right] \ge 0.$$

(SD1)

$$\text{Maximize} \int_a^b f(t, x, \dot{x}, y, \dot{y})dt$$

$$\text{Subject to } x(a) = \alpha, \quad x(b) = \beta, \quad y(a) = \alpha, \quad y(b) = \beta$$

$$f_x(t, x, \dot{x}, y, \dot{y}) - \frac{d}{dt} f_{\dot{x}}(t, x, \dot{x}, y, \dot{y}) \ge 0, \quad t \in I,$$

$$x^T \left[f_x(t, x, \dot{x}, y, \dot{y}) - \frac{d}{dt} f_{\dot{x}}(t, x, \dot{x}, y, \dot{y}) \right] \le 0.$$

If $(SP1)$ and $(SD1)$ are independent of t and f is considered to have domain $R^n \times R^m$, we obtain the following symmetric dual pair

(MWSP)

$$\begin{aligned} \text{Minimize } & f(x,y) \\ \text{Subject to } & f_y(x,y) \leq 0 \\ & y^T f_y(x,y) \geq 0. \end{aligned}$$

(MWSD)

$$\begin{aligned} \text{Maximize } & f(x,y) \\ \text{Subject to } & f_x(x,y) \geq 0 \\ & x^T f_x(x,y) \leq 0. \end{aligned}$$

Notice that $(MWSP)$ and $(MWSD)$ are the pair of problems given in Chap. 5, Sect. 5.4.

Definition 7.17. *The functional $\int_a^b f(t,x,\dot{x},y,\dot{y})dt$ is pseudoinvex in x and $\dot{x}$ if for each $y:[a,b]\to R^m$, with $\dot{y}$ piecewise smooth, there exists a function $\eta:[a,b]\times R^n\times R^n\times R^n\times R^n\to R^n$ such that*

$$\int_a^b \eta(t,x,\dot{x},u,\dot{u})^T \times \left[f_x(t,u,\dot{u},y,\dot{y}) - \frac{d}{dt}f_{\dot{x}}(t,u,\dot{u},y,\dot{y})\right] dt \geq 0$$
$$\Rightarrow \int_a^b [f(t,x,\dot{x},y,\dot{y}) - f(t,u,\dot{u},y,\dot{y})]dt \geq 0$$

for all $x:[a,b]\to R^n, u:[a,b]\to R^n$ with $(\dot{x}(t),\dot{y}(t))$ piecewise smooth on $[a,b]$.

Similarly, the functional $-\int_a^b f(t,x,\dot{x},y,\dot{y})dt$ is pseudoinvex in y and $\dot{y}$ if for each $x:[a,b]\to R^n$, with $\dot{x}$ piecewise smooth, there exists a function $\xi:[a,b]\times R^m\times R^m\times R^m\times R^m\to R^m$ such that

$$-\int_a^b \xi(t,v,\dot{v},y,\dot{y})^T \times \left[f_y(t,x,\dot{x},y,\dot{y}) - \frac{d}{dt}f_{\dot{y}}(t,x,\dot{x},y,\dot{y})\right] dt \geq 0$$
$$\Rightarrow -\int_a^b [f(t,x,\dot{x},v,\dot{v}) - f(t,x,\dot{x},y,\dot{y})]dt \geq 0$$

for all $v:[a,b]\to R^m, y:[a,b]\to R^m$ with $(\dot{v}(t),\dot{y}(t))$ piecewise smooth on $[a,b]$.

For the sake of simplicity, in the sequel, we will write $\eta(x,u)$ for $\eta(t,x,\dot{x},u,\dot{u})$ and $\xi(v,y)$ for $\xi(t,v,\dot{v},y,\dot{y})$.

Kim and Lee [121] established following duality results:

Theorem 7.18 (Weak duality). *If $\int_a^b f$ is pseudoinvex in x and $\dot{x}$ and $-\int_a^b f$ is pseudoinvex in y and $\dot{y}$, with $\eta(x,u)+u(t) \geq 0$ and $\xi(v,y)+y(t) \geq 0$ for all $t \in [a,b]$ (except perhaps at corners of $(\dot{x}(t), \dot{y}(t))$ or $(\dot{u}(t), \dot{v}(t))$ whenever (x,y) is feasible for $(SP1)$ and (u,v) is feasible for $(SD1)$, then* $\inf(SP1) \geq \sup(SD1)$.

□

Notice that if f is independent of t and have the domain f is considered to be $R^n \times R^m$, then the above Theorem 7.18 is Theorem 5.30 of Chap. 5.

Theorem 7.19 (Strong duality). *Let (x^*, y^*) be a minimizing function for $(SP1)$. Suppose that*

$$\left(p(t)^T \left(f_{\dot{y}y}(t,x^*,\dot{x}^*,y^*,\dot{y}^*) - \frac{d}{dt} f_{y\dot{y}}(t,x^*,\dot{x}^*,y^*,\dot{y}^*)\right)\right) + \frac{d}{dt}\left[p(t)^T \frac{d}{dt} f_{y\dot{y}}(t,x^*,\dot{x}^*,y^*,\dot{y}^*)\right] + \frac{d^2}{dt^2}\left[-p(t)^T \frac{d}{dt} f_{\dot{y}\dot{y}}(t,x^*,\dot{x}^*,y^*,\dot{y}^*)\right] = 0.$$

Only has the solution $p(t) = 0$ for all $t \in [a,b]$ and $f_y(t,x^*,\dot{x}^*,y^*,\dot{y}^*) \neq \frac{d}{dt} f_{\dot{y}}(t,x^*,\dot{x}^*,y^*,\dot{y}^*)$.

Then, (x^*, y^*) is feasible for $(SD1)$. If in addition, the pseudoinvexity conditions of Theorem 7.18 are satisfied, then (x^*, y^*) is a maximizing function for $(SD1)$, and the extreme values of $(SP1)$ and $(SD1)$ are equal.

Proof. For proof we refer the reader to Kim and Lee [121].

□

The static case of the above Theorem is Theorem 5.31. Kim and Lee [121] have also obtained converse duality result. However, Kim and Lee [121] did not discuss self duality results. Later, Nahak and Nanda [188] obtained symmetric duality results for the Wolfe type dual pair constraints in cone and using pseudoinvexity assumptions.

7.2 Multiobjective Variational Problems with Invexity

Let $I = [a,b]$ be a real interval and $f : I \times R^n \times R^n \to R$ and $g : I \times R^n \times R^n \to R^m$ be continuously differentiable functions. In order to consider $f(t, x(t), \dot{x}(t))$, where $x : I \to R^n$ with derivative $\dot{x}$. Let $C(I, R^n)$ denote the space of piecewise smooth functions x with norm $\|x\| = \|x\|_\infty + \|Dx\|_\infty$, where the differential operator D is given by

$$u^i = Dx^i \Leftrightarrow x^i(t) = \alpha + \int_a^t u^i(s)ds,$$

in which α is a given boundary value. Therefore, $D = \frac{d}{dt}$ except at discontinuities.

For a multiobjective continuous-time programming:

(CMP)

$$\text{Minimize} \int_a^b f(t,x,\dot{x})dt = \left(\int_a^b f_1(t,x,\dot{x})dt, \ldots, \int_a^b f_p(t,x,\dot{x})dt\right)$$

$$\text{Subject to } x(a) = \alpha, \quad x(b) = \beta, \tag{7.14}$$

$$g(t,x,\dot{x}) \leq 0, \quad t \in I, \quad x \in C(I, R^n), \tag{7.15}$$

where $f_i : I \times R^n \times R^n \to R, i \in P = \{1, \ldots, p\}, g : I \times R^n \times R^n \to R^m$ are assumed to be continuously differentiable functions. Let K denote the set of all feasible solutions for (CMP), that is $K = \{x \in C(I, R^n) : x(a) = \alpha, x(b) = \beta, g(t, x(t), \dot{x}(t)) \leq 0, t \in I\}$.

Craven [46] obtained Kuhn–Tucker type necessary conditions for the above problem and proved that the necessary conditions are also sufficient if the objective functions are pseudoconvex and constraints are quasiconvex.

Definition 7.20. *A point $x^* \in K$ is said to be an efficient solution of (CMP) if for all $x \in K$*

$$\int_a^b f_i(t, x^*(t), \dot{x}^*(t))dt \geq \int_a^b f_i(t, x(t), \dot{x}(t))dt \quad \forall i \in P$$

$$\Rightarrow \int_a^b f_i(t, x^*(t), \dot{x}^*(t))dt = \int_a^b f_i(t, x(t), \dot{x}(t))dt \quad \forall i \in P.$$

Definition 7.21. *A point x^* in K is said to be a weak minimum for (CMP) if there exists no other $x \in K$ for which*

$$\int_a^b f(t, x^*(t), \dot{x}^*(t))dt = \int_a^b f(t, x(t), \dot{x}(t))dt.$$

Remark 7.22. From the Definitions 7.20 and 7.21 it follows that if x^* in K is efficient for (CMP) then it is also a weak minimum for (CMP).

Definition 7.23 (Geoffrion [67]). *A point x^* in K is said to be properly efficient solution for (CMP) if there exists a scalar $M > 0$ such that, $\forall i \in \{1, \ldots, p\}$*

$$\int_a^b f_i(t, x^*(t), \dot{x}^*(t))dt - \int_a^b f_i(t, x(t), \dot{x}(t))dt$$

$$\leq M \left(\int_a^b f_j(t, x^*(t), \dot{x}^*(t))dt - \int_a^b f_j(t, x(t), \dot{x}(t))dt\right)$$

for some j such that

$$\int_a^b f_j(t, x(t), \dot{x}(t))dt > \int_a^b f_j(t, x^*(t), \dot{x}^*(t))dt$$

whenever $x \in K$ and

$$\int_a^b f_j(t, x^*(t), \dot{x}^*(t))dt > \int_a^b f_j(t, x(t), \dot{x}(t))dt.$$

An efficient solution that is not properly efficient is said to be improperly efficient. Thus for x^* to be improperly efficient means that to every sufficiently large $M > 0$, there is an $x \in K$ and an $i \in \{1, \ldots, p\}$, such that

$$\int_a^b f_i(t, x(t), \dot{x}(t))dt < \int_a^b f_i(t, x^*(t), \dot{x}^*(t))dt$$

and

$$\int_a^b f_i(t, x^*(t), \dot{x}^*(t))dt - \int_a^b f_i(t, x(t), \dot{x}(t))dt$$
$$\geq M \left(\int_a^b f_j(t, x^*(t), \dot{x}^*(t))dt - \int_a^b f_j(t, x(t), \dot{x}(t))dt \right), \quad \forall j \in \{1, \ldots, p\},$$

such that

$$\int_a^b f_j(t, x(t), \dot{x}(t))dt > \int_a^b f_j(t, x^*(t), \dot{x}^*(t))dt.$$

Now we consider the following Geoffrion type parametric Variational problem for predetermined Geoffrion's parameter $\lambda \in \Lambda^+$, where

$$\Lambda^+ = \{\lambda \in R^p : \lambda > 0, \lambda^T e = 1, e = (1, \ldots, 1) \in R^p\}.$$

(CP$_\lambda$)

$$\text{Minimize} \sum_{i=1}^{p} \lambda_i \int_a^b f_i(t, x, \dot{x})dt = \int_a^b \lambda^T f(t, x, \dot{x})dt.$$

Subject to (7.14) and (7.15). The problems (CMP) and (CP_λ) are equivalent in the sense of Geoffrion [67]. The following theorems are valid when R^n is replaced by some normed space of functions, as the proofs of these theorems do not depend on the dimensionality of the space in which the feasible set of (CMP) lies. For problem (CMP) the feasible set K lies in the normed space $C(I, R^n)$. For proofs of the following Theorems we refer the reader to Nahak and Nanda [187].

Theorem 7.24. *Let $\lambda \in \Lambda^+$ be fixed. If x^* is optimal for (CP_λ), then x^* is properly efficient for (CMP).*

Proof. First we must prove that x^* is efficient. Suppose contrary, i.e., there exists $x \in K$ such that

$$\int_a^b f^i(t, x^*, \dot{x}^*)dt \geq \int_a^b f^i(t, x, \dot{x})dt, \quad i = 1, \ldots, p$$

and

$$\int_a^b f^{i_0}(t, x^*, \dot{x}^* dt \geq \int_a^b f^{i_0}(t, x, \dot{x})dt, \quad \text{for at least one } i_0.$$

Since $\lambda_i > 0 \; \forall i = 1, \ldots, p$, we have

$$\int_a^b \lambda_i f^i(t, x^*, \dot{x}^*)dt \geq \int_a^b \lambda_i f^i(t, x, \dot{x})dt, \quad i = 1, \ldots, p.$$

That is,

$$\int_a^b \lambda_i f^i(t, x, \dot{x})dt \leq \int_a^b \lambda_i f^i(t, x^*, \dot{x}^*)dt, \quad i = 1, \ldots, p,$$

which contradicts the optimality of x^*. Next we have to prove that x^* is properly efficient for (CMP) with

$$M = (p-1)\text{Max}_{i,j} \frac{\lambda_j}{\lambda_i}$$

(we may assume that $p \geq 2$). Suppose to the contrary that for every sufficiently large $M > 0$ there is an $x \in K$ and an $i \in \{1, \ldots, p\}$, such that

$$\int_a^b f^i(t, x, \dot{x})dt < \int_a^b f^i(t, x^*, \dot{x})^* dt$$

and

$$\int_a^b f^i(t, x^*, \dot{x}^*)dt - \int_a^b f^i(t, x, \dot{x})dt$$
$$> M \left(\int_a^b f^j(t, x, \dot{x})dt - \int_a^b f^j(t, x^*, \dot{x}^*)dt \right), \quad \forall j \in \{1, \ldots, p\}$$

such that

$$\int_a^b f^j(t, x, \dot{x})dt > \int_a^b f^j(t, x^*, \dot{x}^*)dt.$$

It follows that

$$\int_a^b f^i(t, x^*, \dot{x}^*)dt - \int_a^b f^i(t, x, \dot{x})dt$$
$$> \frac{p-1}{\lambda_i} \left(\int_a^b f^j(t, x, \dot{x})dt - \int_a^b f^j(t, x^*, \dot{x}^*)dt \right), \quad j \neq i.$$

Multiplying both sides by $\frac{\lambda_i}{p-1}$ and summing over $j \neq i$ yields

$$\lambda_i \left(\int_a^b f^i(t, x^*, \dot{x}^*)dt - \int_a^b f^i(t, x, \dot{x})dt \right) > \sum_{j \neq i}^{p} \left(\int_a^b f^j(t, x, \dot{x})dt - \int_a^b f^j(t, x^*, \dot{x}^*)dt \right)$$

which contradicts the optimality of x^* in (CP_λ).

□

Theorem 7.25. *Let f and g be convex in $(x, \dot{x})$ on K. Then x^* is properly efficient for (CMP) if and only if x^* is optimal for (CP_λ) for some $\lambda \in \Lambda^+$.*

The following necessary optimality conditions for (CP_λ) from Bector and Husain [14] will be needed in the sequel.

Proposition 7.26. *If x is optimal for(CP_λ) and is normal [31], there exists a piecewise smooth $y : I \to R^m$ such that for $t \in I$*

$$\begin{aligned} &\lambda^T f_x(t, x(t), \dot{x}(t)) + y(t)^T g_x(t, x(t), \dot{x}(t)) \\ &\quad = D\left[\lambda^T f_{\dot{x}}(t, x(t), \dot{x}(t)) + y(t)^T g_{\dot{x}}(t, x(t), \dot{x}(t))\right] \\ &\quad y(t)^T g_{\dot{x}}(t, x(t), \dot{x}(t)) > 0 \\ &\quad y(t) > 0. \end{aligned}$$

A Wolfe type dual to (CP_λ) as suggested by Geoffrion [67] is:

(CD$_\lambda$)

$$\text{Maximize} \int_a^b \left\{ \lambda^T f(t, u, \dot{u}) + y(t)^T g(t, u, \dot{u}) \right\} dt$$

$$\text{Subject to } x(a) = \alpha, \quad x(b) = \beta,$$

$$\lambda^T f(t, u, \dot{u}) + y(t)^T g(t, u, \dot{u})e = \frac{d}{dt}\left[\lambda^T f_{\dot{x}}(t, u, \dot{u}) + y(t)^T g_{\dot{x}}(t, u, \dot{u})\right]$$

$$y \geq 0, \quad t \in I.$$

In view of Theorems 7.24 and 7.25, Bector and Husain [14] proposed the following Wolfe type dual to (CMP):

(CWMD)

$$\text{Maximize} \int_a^b \left\{ f(t, u, \dot{u}) + y(t)^T g(t, u, \dot{u}) \right\} dt$$

$$\text{Subject to } x(a) = \alpha, \quad x(b) = \beta,$$

$$\lambda^T f(t,u,\dot{u}) + y(t)^T g(t,u,\dot{u})e = \frac{d}{dt}\left[\lambda^T f_{\dot{x}}(t,u,\dot{u}) + y(t)^T g_{\dot{x}}(t,u,\dot{u})\right]$$

$$y \geq 0, \quad t \in I,$$

$$\lambda \in \Lambda^+.$$

In problems (CP_λ) and (CD_λ) the vector $0 < \lambda < R^p$ is predetermined. Note that if $p = 1$ problems (CMP) and $(CWMD)$ become the pair of Wolfe type dual Variational problems studied by Mond and Hanson [170]. However, Bector and Husain [14] obtained duality results for convex functions involved in the problem. These authors also obtained Mond–Weir type duality results for (CMP) for convex case. Nahak and Nanda [187] extended the results of Bector and Husain [14] to invex functions.

Theorem 7.27 (Weak duality). *Let $x(t)$ be feasible for (CMP) and $(u(t), \lambda, y(t))$ be feasible for $(CWMD)$. Let f and g be invex with respect to the same function η at $(u, \dot{u})$ over K. Then the following cannot hold:*

$$\int_a^b f^i(t,x,\dot{x}) \leq \left\{f^i(t,u,\dot{u}) + y(t)^T g^i(t,u,\dot{u})\right\} dt, \quad \forall i \in \{1,\ldots,p\},$$

$$\int_a^b f^j(t,x,\dot{x}) < \left\{f^i(t,u,\dot{u}) + y(t)^T g^j(t,u,\dot{u})\right\} dt, \quad \textit{for at least one } j.$$

Proof. The proof follows from the invexity assumptions and integration by part.

□

Proposition 7.28. *Let $u^*(t)$ be feasible for (CMP) and $(u^*(t), \lambda^*, y^*(t))$ be feasible for $(CWMD)$. Let f and g be invex with respect to the same function η and if ${y^*}^T g(t,u^*,\dot{u}^*) = 0, \quad t \in I$, then $u^*(t)$ is properly efficient for (CMP) and $(u^*(t), \lambda^*, y^*(t))$ is properly efficient for (CMWD).*

Proof. We leave the proof as an exercise.

□

Assuming the constraint conditions for the existence of multipliers at the extrema of (CMP) hold, Nahak and Nanda [187] established usual strong duality for (CMP) and $(CWMD)$ under invexity assumptions. These authors also stated duality results for (CMP) and the following Mond–Weir type dual, without proof:

(CMWMD)

$$\text{Maximize} \int_a^b f(t,u,\dot{u})dt$$

$$\text{Subject to } x(a) = \alpha, \quad x(b) = \beta,$$

$$\lambda^T f_u(t,u,\dot{u}) + y(t)^T g_u(t,u,\dot{u}) = \frac{d}{dt}\left[f_{\dot{u}}(t,u,\dot{u}) + y(t)^T g_{\dot{u}}(t,u,\dot{u})\right]$$

$$y(t)^T g_u(t, u, \dot{u}) \geq 0,$$
$$y \geq 0,$$
$$\lambda \in \Lambda = \{\lambda \in R^p : \lambda^T e = 1\}.$$

We state the Theorem 6 of Nahak and Nanda [187] with few corrections.

Theorem 7.29 (Weak duality). *Let $x(t)$ be feasible for (CMP) and $(u(t), \lambda, y(t))$ be feasible for $(CMWMD)$. Let f and g be invex with respect to the same function η at $(u, \dot{v})$ over K. Then the following cannot hold:*

$$\int_a^b f^i(t, x, \dot{x})dt \leq \int_a^b f^i(t, u, \dot{u})dt, \quad \forall i \in \{1, \ldots, p\},$$
$$\int_a^b f^j(t, x, \dot{x})dt < \int_a^b f^j(t, u, \dot{u})dt, \quad \textit{for at least one } j.$$

□

In Theorem 6 [187] have taken $(u(t), \lambda, y(t))$ to be feasible for $(CWMD)$ instead of $(CMWWD)$. Strong and converse duality theorems are also stated by Nahak and Nanda [187].

On the other hand, Mukherjee and Mishra [183] obtained Wolfe and Mond–Weir type duality results for (CMP) and corresponding dual problems involving pseudoinvex and quasiinvex functions. In relation to (CMP) Mukherjee and Mishra [183] considered the following Wolfe type multiobjective continuous-time dual problem:

(CWMD)

$$\text{Maximize} \int_a^b \left\{ f(t, u, \dot{u}) + y(t)^T g(t, u, \dot{u}) e \right\} dt$$
$$\text{Subject to } x(a) = \alpha, \quad x(b) = \beta,$$
$$f_u(t, u, \dot{u}) + y(t)^T g_u(t, u, \dot{u}) e = \frac{d}{dt} \left[f_{\dot{u}}(t, u, \dot{u}) + y(t)^T g_{\dot{u}}(t, u, \dot{u}) e \right]$$
$$y \geq 0,$$
$$\lambda \in \Lambda = \{\lambda \in R^p : \lambda^T e = 1\}.$$

Mukherjee and Mishra [183] established several weak duality results relating (CMP) and $(NWMD)$.

Theorem 7.30 (Weak duality). *If for all feasible (x, u, y, λ),*

(a) $\int_a^b \left\{ f(t, \cdot, \cdot) + y(t)^T g(t, \cdot, \cdot) e \right\} dt$ *is pseudoinvex; or*

(b) $\int_a^b \left\{ \lambda^T f(t, \cdot, \cdot) + y(t)^T g(t, \cdot, \cdot) e \right\} dt$ *is pseudoinvex.*

Then $\int_a^b f(t, x, \dot{x})dt \geq \int_a^b \left\{ f(t, u, \dot{u}) + y(t)^T g(t, u, \dot{u}) e \right\} dt.$

Proof. For the proof we refer to Mukherjee and Mishra [183].

□

Mukherjee and Mishra [183] have also established strong duality between (CMP) and $(CWMD)$. These authors have considered Mond–Weir type of multiobjective dual for (CMP):

(CMWMD)

$$\text{Maximize} \int_a^b f(t, u, \dot{u})dt$$

$$\text{Subject to } x(a) = \alpha, \quad x(b) = \beta,$$

$$\lambda^T f_u(t, u, \dot{u}) + y(t)^T g_u(t, u, \dot{u}) = \frac{d}{dt}\left[f_{\dot{u}}(t, u, \dot{u}) + y(t)^T g_{\dot{u}}(t, u, \dot{u})\right]$$

$$y(t)^T g_u(t, u, \dot{u}) \geq 0,$$

$$y \geq 0,$$

$$\lambda \in \Lambda = \{\lambda \in R^p : \lambda^T e = 1\}.$$

Various weak duality between (CMP) and $(CMWMD)$ are also obtained by Mukherjee and Mishra [183].

Theorem 7.31 (Weak duality). *If for all feasible* (x, u, y, λ),

(a) $\int_a^b f(t, \cdot, \cdot)dt$ *is pseudoinvex and* $\int_a^b y(t)^T g(t, \cdot, \cdot)dt$ *is quasiinvex; or*

(b) $\int_a^b \lambda^T f(t, \cdot, \cdot)dt$ *is pseudoinvex and* $\int_a^b y(t)^T g(t, \cdot, \cdot)dt$ *is quasiinvex; or*

(c) $\int_a^b f(t, \cdot, \cdot)dt$ *is quasiinvex and* $\int_a^b y(t)^T g(t, \cdot, \cdot)dt$ *is strictly pseudoinvex; or*

(d) $\int_a^b \lambda^T f(t, \cdot, \cdot)dt$ *is quasiinvex and* $\int_a^b y(t)^T g(t, \cdot, \cdot)dt$ *is strictly pseudoinvex,*

then $\int_a^b f(t, x, \dot{x})dt \geq \int_a^b f(t, u, \dot{u})dt$.

□

Theorem 7.32 (Strong duality). *Let* x_0 *be a weak minimum for* (CMP) *at which the Kuhn–Tucker constraint qualification is satisfied. Then there exists* (y, λ) *such that* $x_0, y, \lambda)$ *is feasible for* $(CMWMD)$ *and the objective values of* (CMP) *and* $(CMWMD)$ *are equal. If, also, the any one of (a)–(d) of the invexity assumptions of Theorem 7.31 is satisfied then* $x_0, y, \lambda)$ *is a weak minimum for* $(CMWMD)$.

□

Jeyakumar and Mond [105] introduced a new class of invex functions which preserves the sufficient optimality and duality results in the scalar case and avoids major difficulty of verifying that the inequality holds for the same kernel function. Mukherjee and Mishra [183] extended this concept to the continuous case and obtained sufficient optimality and duality results under the assumptions of this new class of functions. The following definitions and examples have appeared in Mukherjee and Mishra [184].

Let $F_i : X \to R$ defined by $F_i(x) = \int_a^b f_i(t, x, \dot{x})dt, \ i = 1, \ldots, p$ be differentiable.

Definition 7.33 (V-invex). *A vector function $F = (F_1, \ldots, F_p)$ is said to be V-invex if there exists differentiable vector function $\eta : I \times R^n \times R^n \to R^n$ with $\eta(t, x, x) = 0$ and $\beta_i : I \times X \times X \to R_+ \backslash \{0\}$ such that for each $x, \bar{x} \in X$ and for $i = 1, \ldots, p$*

$$F_i(x) - F_i(\bar{x}) \geq \int_a^b \Big[\alpha_i(t, x(t), \bar{x}(t)) f_x^i(t, x(t), \dot{x}(t)) \eta(t, x(t), \bar{x}(t)) + \frac{d}{dt}\eta(t, x(t), \bar{x}(t)) \alpha_i(t, x(t), \bar{x}(t)) f_{\dot{x}}^i(t, x(t), \dot{x}(t))\Big] dt.$$

Definition 7.34 (V-pseudo-invex). *A vector function $F = (F_1, \ldots, F_p)$ is said to be V-pseudo-invex if there exists differentiable vector function $\eta : I \times R^n \times R^n \to R^n$ with $\eta(t, x, x) = 0$ and $\beta_i : I \times X \times X \to R_+ \backslash \{0\}$ such that for each $x, \bar{x} \in X$ and for $i = 1, \ldots, p$*

$$\int_a^b \left[\sum_{i=1}^p \eta(t, x, \bar{x}) f_x^i(t, x, \dot{x}) + \frac{d}{dt}\eta(t, x, \bar{x}) f_{\dot{x}}^i(t, x, \dot{x})\right] dt \geq 0$$
$$\Rightarrow \int_a^b \sum_{i=1}^p \beta_i(t, x(t), \bar{x}(t)) f_i(t, x(t), \dot{x}(t)) dt$$
$$\geq \int_a^b \sum_{i=1}^p \beta_i(t, x(t), \bar{x}(t)) f_i(t, \bar{x}(t), \dot{\bar{x}}(t)) dt$$

or equivalently

$$\int_a^b \sum_{i=1}^p \beta_i(t, x(t), \bar{x}(t)) f_i(t, x(t), \dot{x}(t)) dt$$
$$< \int_a^b \sum_{i=1}^p \beta_i(t, x(t), \bar{x}(t)) f_i(t, \bar{x}(t), \dot{\bar{x}}(t)) dt$$
$$\Rightarrow \int_a^b \left[\sum_{i=1}^p \eta(t, x, \bar{x}) f_x^i(t, x, \dot{x}) + \frac{d}{dt}\eta(t, x, \bar{x}) f_{\dot{x}}^i(t, x, \dot{x})\right] dt < 0.$$

Definition 7.35 (V-Quasi-invex). *A vector function $F = (F_1, \ldots, F_p)$ is said to be V-Quasi-invex if there exists differentiable vector function $\eta : I \times R^n \times R^n \to R^n$ with $\eta(t, x, x) = 0$ and $\beta_i : I \times X \times X \to R_+ \backslash \{0\}$ such that for each $x, \bar{x} \in X$ and for $i = 1, \ldots, p$*

$$\int_a^b \sum_{i=1}^p \beta_i(t, x(t), \bar{x}(t)) f_i(t, x(t), \dot{x}(t)) dt$$
$$\leq \int_a^b \sum_{i=1}^p \beta_i(t, x(t), \bar{x}(t)) f_i(t, \bar{x}(t), \dot{\bar{x}}(t)) dt$$
$$\Rightarrow \int_a^b \left[\sum_{i=1}^p \eta(t, x, \bar{x}) f_x^i(t, x, \dot{x}) + \frac{d}{dt}\eta(t, x, \bar{x}) f_{\dot{x}}^i(t, x, \dot{x})\right] dt \leq 0.$$

or equivalently

$$\int_a^b \left[\sum_{i=1}^p \eta(t,x,\bar{x}) f_x^i(t,x,\dot{\bar{x}}) + \frac{d}{dt}\eta(t,x,\bar{x}) f_{\dot{x}}^i(t,x,\dot{\bar{x}}) \right] dt > 0$$

$$\Rightarrow \int_a^b \sum_{i=1}^p \beta_i(t,x(t),\bar{x}(t)) f_i(t,\bar{x}(t),\dot{x}(t))dt$$

$$> \int_a^b \sum_{i=1}^p \beta_i(t,x(t),\bar{x}(t)) f_i(t,\bar{x}(t),\dot{\bar{x}}(t))dt.$$

It is to be noted here that, if the function f is independent of t, Definitions 7.33–7.35 reduce to the definitions of V-invexity, V-pseudo-invexity and V-quasi-invexity of Jeyakumar and Mond [105], respectively. It is apparent that every V-invex function is V-pseudo-invex and V-quasi-invex. The following example shows that V-invexity is wider than that of invexity:

Example 7.36 (Mukherjee and Mishra [184]). Consider

$$\text{Min}_{x_1,x_2\in R}\left(\int_a^b \frac{x_1^2(t)}{x_2(t)}dt, \int_a^b \frac{x_2(t)}{x_1(t)}dt \right)$$

Subject to

$$1 - x_1(t) \le 0, \quad 1 - x_2(t) \le 0.$$

Then for

$$\alpha_1(x,u) = \frac{u_2(t)}{x_2(t)}, \quad \alpha_2(x,u) = \frac{u_1(t)}{x_1(t)},$$

$$\beta_i(x,u) = 1 \quad \text{for} \quad i = 1,2 \quad \text{and} \quad \eta(x,u) = x(t) - u(t),$$

we shall show that

$$\int_a^b f_i(t,x,\dot{u}) - f(t,u,\dot{u})$$
$$-\alpha_i(t,x(t),u(t)) f_x^i(t,x(t),u(t))\eta(t,x(t),u(t))dt \ge 0, \quad i = 1,2.$$

Now

$$\int_a^b \frac{x_1^2(t)}{x_2(t)}dt - \int_a^b \frac{u_1^2(t)}{u_2(t)}dt$$
$$- \int_a^b \frac{u_2(t)}{x_2(t)} \left(\frac{2u_1(t)}{u_2(t)}, \frac{-u_1^2(t)}{u_2^2(t)} \right)(x_1 - 1)(x_2 - 1)dt$$
$$= \int_a^b \frac{x_1^2(t)}{x_2(t)}dt - \int_a^b 1dt - \int_a^b \frac{1}{x_2(t)}(2,-1)(x_1-1)(x_2-1)dt$$
$$= \int_a^b \frac{x_1^2(t)}{x_2(t)}dt - \int_a^b 1dt - \int_a^b \frac{1}{x_2(t)}(2x_1 - 2 - x_2 + 1)dt$$

$$= \int_a^b \frac{x_1^2(t)}{x_2(t)} dt - \int_a^b 1dt - \int_a^b \left\{ \frac{2x_1}{x_2(t)} - 1 - \frac{1}{x_2(t)} \right\}$$
$$= \int_a^b \left\{ \frac{x_1^2(t)}{x_2(t)} - \frac{2x_1}{x_2(t)} + \frac{1}{x_2(t)} \right\} dt$$
$$= \int_a^b \left\{ \frac{(x_1(t)-1)^2}{x_2(t)} \right\} dt \geq 0.$$

The following example shows that V-invex functions can be formed from certain nonconvex functions.

Example 7.37 (Mukherjee and Mishra [183]). Consider the function $h : I \times X \times X \to R^p$

$$h(t, x(t), \dot{x}(t)) = \left(\int_a^b f_1(t, x(t), \dot{x}(t))dt, \ldots, \int_a^b f_p(t, x(t), \dot{x}(t))dt \right)$$

where $f_i : I \times X \times X \to R, i = 1, \ldots, p$ are strongly pseudo-convex functions with real positive functions $\alpha_i(t, x, u), \psi : I \times X \times X \to R^n$ is surjective with $\psi'(t, u, \dot{u})$ onto for each $u \in R^n$. Then the function $h : I \times X \times X \to R^p$ is V-invex. To see this, let $x, u \in X, v = \psi(t, x, \dot{x}), w = \psi(t, u, \dot{u})$. Then, by strong-pseudo-convexity, we get

$$\int_a^b \{f_i(\psi(t, x, \dot{x})) - f_i(\psi(t, u, \dot{u}))\}dt = \int_a^b \{f_i(v) - f_i(w)\}dt$$
$$> \int_a^b \alpha_i(t, v, w) f_i'(w)(v - w)\psi_x(t, x, \dot{x})dt$$
$$+ \frac{d}{dt}\alpha_i(t, v, w)(v - w) f_i'(w)\psi_x(t, x, \dot{x})dt.$$

Since $\psi(t, u, \dot{u})$ is onto, $v - w = \psi'(t, u, \dot{u})\eta(t, x, u)$ is solvable for some $\eta(t, x, u)$.
Hence

$$\int_a^b \{f_i(\psi(t, x, \dot{x})) - f_i(\psi(t, u, \dot{u}))\}dt \geq \int_a^b \alpha_i(t, v, w)(f_i \circ \psi)_x dt$$
$$+ \int_a^b \frac{d}{dt}\alpha_i(t, v, w)(f_i \circ \psi)_x dt.$$

Mukherjee and Mishra [184] have shown that the following necessary optimality conditions will be sufficient for optimality under generalized V-invexity assumptions. There exists a piecewise smooth $\lambda^* : I \to R^m$ such that

$$\sum_{i=1}^p \tau_i f_x^i(t, x^*, \dot{x}^*) + \sum_{j=1}^m \lambda_j^* g_x^j(t, x^*, \dot{x}^*)$$
$$= \frac{d}{dt}\left(\sum_{i=1}^p f_{\dot{x}}^i(t, x^*, \dot{x}^*) + \sum_{j=1}^m \lambda_j^*(t) g_{\dot{x}}^j(t, x^*, \dot{x}^*) \right), \tag{7.16}$$

$$\lambda_j^*(t) g_x^j(t, x^*, \dot{x}^*) = 0, \quad t \in I, \quad j = 1, \ldots, m, \tag{7.17}$$

$$\tau \in R^p, \quad \tau \neq 0, \quad \tau \geq 0, \quad \lambda^*(t) \geq 0, \quad t \in I. \tag{7.18}$$

Theorem 7.38 (Sufficient optimality conditions). *Let x^* be a feasible solution for (VCP) and assume that $\left(\int_a^b \tau_1 f_1(t,\cdot,\cdot)dt, \ldots, \int_a^b \tau_p f_p(t,\cdot,\cdot)dt\right)$ is V-pseudo-invex and $\left(\int_a^b \lambda_1 g_1(t,\cdot,\cdot)dt, \ldots, \int_a^b \tau_m g_m(t,\cdot,\cdot)dt\right)$ is V-quasi-invex with respect to η. If there exists a piecewise smooth $\lambda^* : I \to R^m$ such that $(x^*(t), \lambda^*(t))$ satisfies the conditions (7.16)–(7.18), then x^* is a global weak minimum for (VCP).*

Proof. We refer the reader to Mukherjee and Mishra [184].

□

Mukherjee and Mishra [184] also established Mond–Weir type duality results under generalized V-invexity assumption on the functions involved.

Theorem 7.39 (Weak duality). *Let x be feasible for (VCP) and let (u, τ, λ) be feasible for (VCD). If $\left(\int_a^b \tau_1 f_1(t,\cdot,\cdot)dt, \ldots, \int_a^b \tau_p f_p(t,\cdot,\cdot)dt\right)$ is V-pseudo-invex and $\left(\int_a^b \lambda_1 g_1(t,\cdot,\cdot)dt, \ldots, \int_a^b \tau_m g_m(t,\cdot,\cdot)dt\right)$ is V-quasi-invex with respect to η, then $\left(\int_a^b \tau_1 f_1(t,x,\dot{x})dt, \ldots, \int_a^b \tau_p f_p(t,x,\dot{x})dt\right)^T - \left(\int_a^b \tau_1 f_1(t,u,\dot{u})dt, \ldots, \int_a^b \tau_p f_p(t,u,\dot{u})dt\right)^T \notin -intR_+^p$.*

Proof. Follows the lines of the proof of sufficient optimality conditions.

□

Theorem 7.40 (Strong duality). *Assume that u is a weak minimum for (VCP) and that a suitable constraint qualification is satisfied at u. Then there exist (τ, λ) such that (u, τ, λ) is feasible for (VCD) and the objective functions of (VCP) and (VCD) are equal at these points. If, also for all feasible (u, τ, λ), $\left(\int_a^b \tau_1 f_1(t,\cdot,\cdot)dt, \ldots, \int_a^b \tau_p f_p(t,\cdot,\cdot)dt\right)$ is pseudo-invex and $\left(\int_a^b \lambda_1 g_1(t,\cdot,\cdot)dt, \ldots, \int_a^b \tau_m g_m(t,\cdot,\cdot)dt\right)$ is quasi-invex, then (u, τ, λ) is weak maximum for (VCD).*

Proof. We leave the proof as an easy exercise.

□

Kim and Kim [120] extended the concept of invexity studied by Mukherjee and Mishra [184] to the V-type I invex functions and other generalizations, thus extending the V-type I and related functions of Hanson et al. [85] to the continuous case. However, Kim and Kim [120] did not compare their results with that of Mukherjee and Mishra [184].

Definition 7.41. *We say the problem (VCP) to be V-type I invex at $u \in C(I, R^n)$ with respect to η, α_i and β_j if there exist vector function $\eta\colon I \times R^n \times R^n \to R^n$ with $\eta(t,x,x) = 0$ and real valued functions $\alpha_i \in R_+\backslash\{0\}$ and $\beta_j \in R_+\backslash\{0\}$ such that*

$$\int_a^b f_i(t,x,\dot{x})dt - \int_a^b f_i(t,u,\dot{u})dt$$
$$\geq \int_a^b \left[\alpha_i(x,u,\dot{x},\dot{u})\eta(t,x,u)\left\{f_x^i(t,u,\dot{u}) - \frac{d}{dt}f_{\dot{x}}^i(t,u,\dot{u})\right\}\right] dt$$

and

$$-\int_a^b g(t,u,\dot{u})dt$$
$$\geq \int_a^b \left[\beta_j(x,u,\dot{x},\dot{u})\eta(t,x,u)\left\{g_x^j(t,u,\dot{u}) - \frac{d}{dt}g_{\dot{x}}^j(t,u,\dot{u})\right\}\right] dt,$$

$\forall x \in K$, and for all $i \in P = \{1,2,\ldots,p\}, j \in M = \{1,2,\ldots,m\}$. If the first inequality is strict (whenever $x \neq x^*$) we say that (VCP) is semi strictly V-type I invex at x^*.

Definition 7.42. *We say the problem (VCP) is quasi V-type I at $u \in C(I, R^n)$ with respect to η, α_i and β_j if there exist vector function $\eta\colon I \times R^n \times R^n \to R^n$ with $\eta(t,x,x) = 0$ and real valued functions $\alpha_i \in R_+\backslash\{0\}$ and $\beta_j \in R_+\backslash\{0\}$ such that for some vector $\tau \in R^p, \tau \geq 0$, and piecewise smooth function $\lambda : I \to R^m, \lambda(t) \geq 0$,*

$$\int_a^b \sum_{i=1}^p \tau_i\alpha_i(x,u,\dot{x},\dot{u})f_i(t,x,\dot{x})dt \leq \int_a^b \sum_{i=1}^p \tau_i\alpha_i(x,u,\dot{x},\dot{u})f_i(t,u,\dot{u})dt$$
$$\Rightarrow \int_a^b \sum_{i=1}^p \tau_i\eta(t,x,u)\left\{f_x^i(t,u,\dot{u}) - \frac{d}{dt}f_{\dot{x}}^i(t,u,\dot{u})\right\} dt \leq 0,$$

and

$$-\int_a^b \sum_{j=1}^m \lambda_j(t)\beta_j(x,u,\dot{x},\dot{u})g_j(t,u,\dot{u})dt \leq 0$$
$$\Rightarrow \int_a^b \sum_{j=1}^m \lambda_j(t)\eta(t,x,u)\left\{g_x^j(t,u,\dot{u}) - \frac{d}{dt}g_{\dot{x}}^j(t,u,\dot{u})\right\} dt \leq 0,$$

whenever $x \neq x^*, \forall x \in K$, for all $i \in P = \{1,2,\ldots,p\}, j \in M = \{1,2,\ldots,m\}$.

Definition 7.43. *We say the problem (VCP) is pseudo V-type I at $u \in C(I, R^n)$ with respect to η, α_i and β_j if there exist vector function $\eta : I \times R^n \times R^n \to R^n$ with $\eta(t,x,x) = 0$ and real valued functions $\alpha_i \in R_+\backslash\{0\}$*

and $\beta_j \in R_+\backslash\{0\}$ *such that for some vector* $\tau \in R^p, \tau \geq 0$, *and piecewise smooth function* $\lambda : I \to R^m, \lambda(t) \geq 0$,

$$\int_a^b \sum_{i=1}^p \tau_i \eta(t,x,u) \left\{ f_x^i(t,u,\dot{u}) - \frac{d}{dt} f_{\dot{x}}^i(t,u,\dot{u}) \right\} dt \geq 0$$
$$\Rightarrow \int_a^b \sum_{i=1}^p \tau_i \alpha_i(x,u,\dot{x},\dot{u}) f_i(t,x,\dot{x}) dt \geq \int_a^b \sum_{i=1}^p \tau_i \alpha_i(x,u,\dot{x},\dot{u}) f_i(t,u,\dot{u}) dt$$

and

$$\int_a^b \sum_{j=1}^m \lambda_j(t) \eta(t,x,u) \left\{ g_x^j(t,u,\dot{u}) - \frac{d}{dt} g_{\dot{x}}^j(t,u,\dot{u}) \right\} dt \geq 0$$
$$\Rightarrow \int_a^b \sum_{j=1}^m \lambda_j(t) \beta_j(x,u,\dot{x},\dot{u}) g_j(t,u,\dot{u}) dt \leq 0,$$

hold $\forall x \in K$, *for all* $i \in P = \{1,2,\ldots,p\}, j \in M = \{1,2,\ldots,m\}$.

Definition 7.44. *We say the problem (VCP) is pseudo quasi* V*-type I at* $u \in C(I,R^n)$ *with respect to* η, α_i *and* β_j *if there exist vector function* $\eta : I \times R^n \times R^n \to R^n$ *with* $\eta(t,x,x) = 0$ *and real valued functions* $\alpha_i \in R_+\backslash\{0\}$ *and* $\beta_j \in R_+\backslash\{0\}$ *such that for some vector* $\tau \in R^p, \tau \geq 0$, *and piecewise smooth function* $\lambda : I \to R^m, \lambda(t) \geq 0$,

$$\int_a^b \sum_{i=1}^p \tau_i \eta(t,x,u) \left\{ f_x^i(t,u,\dot{u}) - \frac{d}{dt} f_{\dot{x}}^i(t,u,\dot{u}) \right\} dt \geq 0$$
$$\Rightarrow \int_a^b \sum_{i=1}^p \tau_i \alpha_i(x,u,\dot{x},\dot{u}) f_i(t,x,\dot{x}) dt \geq \int_a^b \sum_{i=1}^p \tau_i \alpha_i(x,u,\dot{x},\dot{u}) f_i(t,u,\dot{u}) dt$$

and

$$-\int_a^b \sum_{j=1}^m \lambda_j(t) \beta_j(x,u,\dot{x},\dot{u}) g_j(t,u,\dot{u}) dt \leq 0$$
$$\Rightarrow \int_a^b \sum_{j=1}^m \lambda_j(t) \eta(t,x,u) \left\{ g_x^j(t,u,\dot{u}) - \frac{d}{dt} g_{\dot{x}}^j(t,u,\dot{u}) \right\} dt \leq 0,$$

$\forall x \in K$, *for all* $i \in P = \{1,2,\ldots,p\}, j \in M = \{1,2,\ldots,m\}$.

Definition 7.45. *We say the problem (VCP) is pseudo quasi* V*-type I at* $u \in C(I,R^n)$ *with respect to* η, α_i *and* β_j *if there exist vector function* $\eta : I \times R^n \times R^n \to R^n$ *with* $\eta(t,x,x) = 0$ *and real valued functions* $\alpha_i \in R_+\backslash\{0\}$ *and* $\beta_j \in R_+\backslash\{0\}$ *such that for some vector* $\tau \in R^p, \tau \geq 0$, *and piecewise* *smooth function* $\lambda : I \to R^m, \lambda(t) \geq 0,$

$$\int_a^b \sum_{i=1}^p \tau_i \alpha_i(x,u,\dot{x},\dot{u}) f_i(t,x,\dot{x})dt \leq \int_a^b \sum_{i=1}^p \tau_i \alpha_i(x,u,\dot{x},\dot{u}) f_i(t,u,\dot{u})dt$$

$$\Rightarrow \int_a^b \sum_{i=1}^p \tau_i \eta(t,x,u) \left\{ f_x^i(t,u,\dot{u}) - \frac{d}{dt} f_{\dot{x}}^i(t,u,\dot{u}) \right\} dt \leq 0$$

and

$$\int_a^b \sum_{j=1}^m \lambda_j(t) \eta(t,x,u) \left\{ g_x^j(t,u,\dot{u}) - \frac{d}{dt} g_{\dot{x}}^j(t,u,\dot{u}) \right\} dt \geq 0$$

$$\Rightarrow \int_a^b \sum_{j=1}^m \lambda_j(t) \beta_j(x,u,\dot{x},\dot{u}) g_j(t,u,\dot{u})dt \leq 0,$$

hold $\forall x \in K$, *for all* $i \in P = \{1,2,\ldots,p\}, j \in M = \{1,2,\ldots,m\}$.

Theorem 7.46 (Sufficient optimality conditions). *Let x^* be a feasible solution for (VCP) and assume that (VCP) is V-pseudo quasi type I with respect to η. If there exists a piecewise smooth $\lambda^* : I \Rightarrow R^m$ such that $(x^*(t), \lambda^*(t))$ satisfies the conditions (7.16)–(7.18), then x^* is an efficient solution for (VCP).*

Proof. We refer the reader to Kim and Kim [120].

□

Further Kim and Kim [120] obtained Wolfe and Mond–Weir type duality results under generalized V-type I and related functions. One can see the results of Kim and Kim [120] as an extension of the results of Mukherjee and Mishra [184] to generalized V-type I and related functions as well as the results of Hanson et al. [85] to continuous case.

Now we turn towards multiobjective symmetric duality. Gulati et al. [76] are the first to propose the symmetric duality for multiobjective variational problems. However, these authors obtained Mond–Weir type symmetric duality results for convex and concave functions, they did not consider invexity of the functions involved in the multiobjective symmetric dual models. Kim and Lee [122] are the first one to discuss symmetric duality for multiobjective variational problems with invexity assumptions on the functions involved in the symmetric dual models. We state the problem considered by these authors as follows:

(CMSP)

$$\text{Minimize} \int_a^b \{ f(t,x,\dot{x},y,\dot{y}) - [y(t)^T(\lambda^T f_y(t,x,\dot{x},y,\dot{y})$$

$$- \frac{d}{dt} \lambda^T f_{\dot{y}}(t,x,\dot{x},y,\dot{y}))]e\} dt$$

$$\text{Subject to } x(a) = x_0, \ \ x(b) = x_1, \ \ y(a) = y_0, \ \ y(b) = y_1$$

$$\lambda^T f_y(t,x,\dot{x},y,\dot{y}) - \frac{d}{dt}\lambda^T f_{\dot{y}}(t,x,\dot{x},y,\dot{y}) \leq 0$$

$$\lambda \geq 0, \ \ \lambda^T e = 1,$$

(CMSD)

$$\text{Maximize} \int_a^b \{f(t,u,\dot{u},v,\dot{v}) - [u(t)^T(\lambda^T f_x(t,u,\dot{u},v,\dot{v}) - \frac{d}{dt}\lambda^T f_{\dot{x}}(t,u,\dot{u},v,\dot{v}))]e\}dt$$

$$\text{Subject to } u(a) = x_0, \ \ u(b) = x_1, \ \ v(a) = y_0, \ \ v(b) = y_1$$

$$\lambda^T f_x(t,u,\dot{u},v,\dot{v}) - \frac{d}{dt}\lambda^T f_{\dot{x}}(t,u,\dot{u},v,\dot{v}) \geq 0$$

$$\lambda \geq 0, \ \ \lambda^T e = 1,$$

where first inequalities in constraints of (CMSP) and (CMSD) hold at corners of $(\dot{x}(t),\dot{y}(t))$ and $(\dot{u}(t),\dot{v}(t))$, respectively, but must be satisfied for unique right- and left-hand limits, $\lambda \in R^p$, and $e = (1,\dots,1)^T \in R^p$.

Remark 7.47. If $p = 1$ in $(CMSP)$ and $(CMSD)$ above, then we get (SP) and (SD) discussed in Sect. 7.1.

Theorem 7.48 (Weak duality). *Let (x,y,λ) be feasible for $(CMSP)$ and (u,v,λ) be feasible for $(CMSD)$. Assume that either for all $t \in [a,b]$*

(a) $x \neq u$, $\int_a^b f$ is strictly invex in x and $\dot{x}$, and $-\int_a^b f$ is invex in y and $\dot{y}$, with $\eta(x,u) + u(t) \geq 0$ and $\xi(v,y) + y(t) \geq 0$; or

(b) $y \neq v$, $\int_a^b f$ is invex in x and $\dot{x}$, and $-\int_a^b f$ is strictly invex in y and $\dot{y}$, with $\eta(x,u) + u(t) \geq 0$ and $\xi(v,y) + y(t) \geq 0$; or

(c) $\lambda > 0$, $\int_a^b f$ is invex in x and $\dot{x}$, and $-\int_a^b f$ is invex in y and $\dot{y}$, with $\eta(x,u) + u(t) \geq 0$ and $\xi(v,y) + y(t) \geq 0$ (except perhaps at the corners of $(\dot{x}(t),\dot{y}(t))$ and $(\dot{u}(t),\dot{v}(t))$).

Then,

$$\int_a^b \{f(t,x,\dot{x},y,\dot{y}) - [y(t)^T(\lambda^T f_y(t,x,\dot{x},y,\dot{y}) - \frac{d}{dt}\lambda^T f_{\dot{y}}(t,x,\dot{x},y,\dot{y}))]e\}dt$$

$$\nleq \int_a^b \{f(t,u,\dot{u},v,\dot{v}) - [u(t)^T(\lambda^T f_x(t,u,\dot{u},v,\dot{v}) - \frac{d}{dt}\lambda^T f_{\dot{x}}(t,u,\dot{u},v,\dot{v}))]e\}dt$$

Proof. For the proof we refer the reader to Kim and Lee [122].

□

Theorem 7.49 (Strong duality). *Let (x^*, y^*, λ^*) be an efficient solution for $(CMSP)$. Suppose that the system*

$$\Big(p(t)^T\Big(\lambda^{*T} f_{yy}(t,x^*,\dot{x}^*,y^*,\dot{y}^*) - \frac{d}{dt}\lambda^{*T} f_{y\dot{y}}(t,x^*,\dot{x}^*,y^*,\dot{y}^*)\Big)$$
$$+\frac{d}{dt}\Big[p(t)^T\frac{d}{dt}\lambda^{*T} f_{y\dot{y}}(t,x^*,\dot{x}^*,y^*,\dot{y}^*)\Big]$$
$$+\frac{d^2}{dt^2}\Big[-p(t)^T\lambda^{*T} f_{\dot{y}\dot{y}}(t,x^*,\dot{x}^*,y^*,\dot{y}^*)\Big]\Big)\,p(t)=0.$$

only has the solution $p(t)=0$ for all $t\in[a,b]$ and

$$f_y^i(t,x^*,\dot{x}^*,y^*,\dot{y}^*) - \frac{d}{dt}f_{\dot{y}}^i(t,x^*,\dot{x}^*,y^*,\dot{y}^*)$$

is linearly independent. Assume that $\lambda > 0$ and the invexity hypothesis of Theorem 7.48. Then (x^*, y^*, λ^*) is an efficient solution for (CMSP), and the optimal values of (CMSP) and (CMSD) are equal.

□

Kim and Lee [122] also established converse duality theorem. Further, these authors have discussed self duality result as well.

Assume that $m = n$, $f(t,x(t),\dot{x}(t),y(t),\dot{y}(t))$ is said to be skew-symmetric (i.e., $f(t,x(t),\dot{x}(t),y(t),\dot{y}(t)) = -f(t,y(t),\dot{y}(t),x(t),\dot{x}(t))$) for all $x(t)$ and $y(t)$ in the domain of f, such that $(\dot{x}(t),\dot{y}(t))$ is piecewise smooth on $[a,b]$ and that $x(a)=y(a), x(b)=y(b)$.

It follows that $(CMSD)$ may be written as a minimization problem:

(CMSD)′

$$\text{Minimize} \int_a^b [f(t,y,\dot{y},x,\dot{x}) - x(t)^T(\lambda^T f_x)(t,y,\dot{y},x,\dot{x})$$
$$+\,x(t)^T\frac{d}{dt}(\lambda^T f_{\dot{x}})(t,y,\dot{y},x,\dot{x})]dt$$
$$\text{Subject to } x(a)=\alpha,\quad x(b)=\beta,\quad y(a)=\alpha,\quad y(b)=\beta$$
$$\frac{d}{dt}(\lambda^T f_{\dot{x}})(t,y,\dot{y},x,\dot{x}) \geq (\lambda^T f_x)(t,y,\dot{y},x,\dot{x}),$$
$$\lambda\geq 0,\quad \lambda^T e = 1.$$

$(CMSD)'$ is formally identical to $(CMSP)$; that is, the objective and constraint functions and initial conditions of $(CMSP)$ and $(CMSD)'$ are identical. This problem is said to be self-dual.

It is easy to see that whenever (x,y,λ) is feasible for $(CMSP)$, then (y,x,λ) is feasible for $(CMSD)$, and vice versa.

Theorem 7.50 (Self-duality). *Assume that (CMSP) is self-dual and that the invexity conditions of Theorem 7.48 are satisfied. If (x^*, y^*, λ^*) is an efficient solution for (CMSP) and the system*

$$\Bigg(p(t)^T \left(\lambda^{*T} f_{yy}(t, x^*, \dot{x}^*, y^*, \dot{y}^*) - \frac{d}{dt}\lambda^{*T} f_{y\dot{y}}(t, x^*, \dot{x}^*, y^*, \dot{y}^*)\right) + \frac{d}{dt}\left[p(t)^T \frac{d}{dt}\lambda^{*T} f_{y\dot{y}}(t, x^*, \dot{x}^*, y^*, \dot{y}^*)\right] + \frac{d^2}{dt^2}\left[-p(t)^T \lambda^{*T} f_{\dot{y}\dot{y}}(t, x^*, \dot{x}^*, y^*, \dot{y}^*)\right]\Bigg) p(t) = 0.$$

only has the solution $p(t) = 0, \; a \le t \le b$, and

$$f_y^i(t, x^*, \dot{x}^*, y^*, \dot{y}^*) - \frac{d}{dt} f_{\dot{y}}^i(t, x^*, \dot{x}^*, y^*, \dot{y}^*)$$

is linearly independent, then (y^*, x^*, λ^*) is an efficient solution for both (CMSP) and (CMSD) and the common optimal value is zero.

Proof. We refer the reader to Kim and Lee [122].

□

Kim and Song [126] presented two pair of nonlinear Multiobjective mixed integer programs for the polars of arbitrary cones. Chen [34] formulated a pair of multiobjective variational mixed integer programs for arbitrary cones and established weak, strong, converse and self-duality theorems under partial-invexity and separability assumptions on the functions involved. To discuss minimax mixed integer variational problems we need the following notations and preliminaries:
Let $I = [a, b]$ be a real interval, $x : I \to R^n$ and $y : I \to R^m$. We constrain some of the components of x and y which belong to arbitrary sets of integers. Suppose that the first n_1 components of x and the first m_1 components of $y (0 \le n_1 \le n, 0 \le m_1 \le m)$ are constrained to be integers and the following notations are introduced:

$$(x, y) = (x_1, x_2, y_1, y_2) \quad x_1 \in U, \;\; y_1 \in V, \;\; x_2 \in C_1, \;\; y_2 \in C_2,$$

where U and V are two arbitrary sets of integers in R^{n_1} and R^{m_1}, respectively. Let C_1 and C_2 be closed convex cones in R^{n_2} and R^{m_2} with nonempty interiors, respectively. $n = n_1 + n_2$ and $m = m_1 + m_2$.

Let $f_i(t, x_1(t), x_2(t), \dot{x}_2(t), y_1(t), y_2(t), \dot{y}_2(t)), i = (1, \ldots, p)$ be twice continuously differentiable functions at $x_2, \dot{x}_2, y_2$ and $\dot{y}_2$ where $x_1(t) \in U, \; y_1(t) \in V, \; x_2(t) \in C_1, \; y_2(t) \in C_2, \; t \in I, \; x_2 : I \to S_1$ and $y_2 : I \to S_2$ with derivatives $\dot{x}_2$ and $\dot{y}_2$, respectively; $S_1 \subset R^{n_2}$ and $S_2 \subset R^{m_2}$ be open and $C_1 \subset S_1, C_2 \subset S_2$. Denote by f_{ix_2} and $f_{i\dot{x}_2}$ the first partial derivatives of f_i with respect to $x_2(t)$ and $\dot{x}_2(t)$, respectively; $f_{ix_2x_2}$ the Hessian matrix of f_i with respect to $x_2(t)$. Similarly, $f_{iy_2}, f_{i\dot{y}_2}$ denote the first partial derivatives of f_i with respect to $y_2(t)$ and $\dot{y}_2(t)$, respectively; $f_{ix_2\dot{x}_2}$, $f_{ix_2y_2}$, $f_{ix_2\dot{y}_2}$, $f_{i\dot{x}_2x_2}$, $f_{i\dot{x}_2\dot{x}_2}$, $f_{i\dot{x}_2y_2}$, $f_{i\dot{x}_2\dot{y}_2}$, $f_{iy_2x_2}$, $f_{iy_2\dot{x}_2}$, $f_{iy_2y_2}$, $f_{iy_2\dot{y}_2}$, $f_{i\dot{y}_2x_2}$, $f_{i\dot{y}_2\dot{x}_2}$, $f_{i\dot{y}_2y_2}$ and $f_{i\dot{y}_2\dot{y}_2}$ denote the other Hessian matrices of with respect to $x_2(t), \dot{x}_2(t), y_2(t)$ and $\dot{y}_2(t)$, respectively, for $i = 1, \ldots, p$.

Let $C(I, R^s)$ denote the space of piecewise smooth functions $z : I \to R^s$ with norm $\|z\| = \|z\|_\infty + \|Dz\|_\infty$, where the differentiation operator D is given by

$$u = Dz \Leftrightarrow z(t) = u(\alpha) + \int_a^t u(s)ds,$$

where $u(a)$ is a given boundary value. Therefore $d/dt = D$ except at discontinuities. Consider the following multiobjective programming problem:

(MP)

$$\begin{aligned} \text{Minimize} \quad & g(x) \\ \text{Subject to} \quad & h(x) \leq 0, \end{aligned}$$

where $g : C \to R^p, h : C \to R^m$ and C is a closed convex cone with nonempty interior in R^n. Denote by Q the set of feasible solutions of (MP).

Definition 7.51. *A point $\bar{x} \in Q$ is said to be an efficient solution of (MP) if there exists no other $x \in Q$ such that $g(x) \leq g(\bar{x})$.*

Definition 7.52. *A cone C_i^* is said to be polar of C_i for $i = 1, 2$ if*

$$C_i^* = \{z : z^i x \leq 0, \ \text{for all } x \in C_i\}.$$

Definition 7.53. *A real valued function $\phi(t, x_1, \dot{x}_1, x_2, \dot{x}_2, \ldots, x_p, \dot{x}_p)$ is said to be separable with respect to x_1 and $\dot{x}_1$ if there exist real valued functions $\phi(t, x_1, \dot{x}_1)$ and $\theta(t, x_2, \dot{x}_2, \ldots, x_p, \dot{x}_p)$ such that*

$$\phi(t, x_1, \dot{x}_1, x_2, \dot{x}_2, \ldots, x_p, \dot{x}_p) = \phi(t, x_1, \dot{x}_1) + \theta(t, x_2, \dot{x}_2, \ldots, x_p, \dot{x}_p).$$

Definition 7.54 (Chen and Yang [35]). *Let $S \subset R^n$ be an open set. If there exists a vector function $\eta_1(t, x(t), \dot{x}(t), u(t), \dot{u}(t)) \in S$ with $\eta_1(t, x(t), \dot{x}(t), x(t), \dot{x}(t)) = 0$, such that for the scalar function $h(t, x(t), \dot{x}(t), y(t), \dot{y}(t))$ the functional*

$$H(x, \dot{x}, y, \dot{y}) = \int_a^b h(t, x(t), \dot{x}(t), y(t), \dot{y}(t))dt$$

satisfies

$$\begin{aligned} H(x, \dot{x}, y, \dot{y}) - H(u, \dot{u}, y, \dot{y}) \geq \int_a^b & \left\{ \eta_1^T h_x(t, u(t), \dot{u}(t), y(t), \dot{y}(t)) \right. \\ & \left. + (D\eta_1)^T h_{\dot{x}}(t, u(t), \dot{u}(t), y(t), \dot{y}(t)) \right\} dt \end{aligned}$$

then $H(x,\dot{x},y,\dot{y})$ *is said to be partially invex in* x *and* $\dot{x}$ *on* I *with respect to* η_1 *for fixed* y. *If* $H(x,\dot{x},y,\dot{y})$ *satisfies*

$$H(x,\dot{x},y,\dot{y}) - H(x,\dot{x},v,\dot{v}) \geq \int_a^b \left\{\eta_2^T h_y(t,x(t),\dot{x}(t),v(t),\dot{v}(t)) + (D\eta_2)^T h_{\dot{y}}(t,x(t),\dot{x}(t),v(t),\dot{v}(t))\right\} dt,$$

where $\eta_2(t,y(t),\dot{y}(t),v(t),\dot{v}(t)) \in S$ *then* $H(x,\dot{x},y,\dot{y})$ *is said to be partially invex in* y *and* $\dot{y}$ *on* I *with respect to* η_2 *for fixed* x. *If* $-H(x,\dot{x},y,\dot{y})$ *is partially invex in* x *and* $\dot{x}$ *on* I *with respect to* η_1 *for fixed* y, *then* $H(x,\dot{x},y,\dot{y})$ *is said to be partially incave in* x *and* $\dot{x}$ *on* I *with respect to* η_1 *for fixed* y *and if* $-H(x,\dot{x},y,\dot{y})$ *is partially invex in* y *and* $\dot{y}$ *on* I *with respect to* η_2 *for fixed* x, *then* $H(x,\dot{x},y,\dot{y})$ *is said to be partially incave in* y *and* $\dot{y}$ *on* I *with respect to* η_2 *for fixed* x.

The following definition is a natural extension to the Definition 7.54.

Definition 7.55. *Let* $s \subset R^n$ *be an open set. If there exists a vector function* $\eta_1(t,x(t),\dot{x}(t),u(t),\dot{u}(t)) \in S$ *with* $\eta_1(t,x(t),\dot{x}(t),x(t),\dot{x}(t)) = 0$, *such that for the scalar function* $h(t,x(t),\dot{x}(t),y(t),\dot{y}(t)) \in S$ *the functional*

$$H(x,\dot{x},y,\dot{y}) = \int_a^b h(t,x(t),\dot{x}(t),y(t),\dot{y}(t))dt$$

satisfies

$$\int_a^b \left\{\eta_1^T h_x(t,u(t),\dot{u}(t),y(t),\dot{y}(t)) + (D\eta_1)^T h_{\dot{x}}(t,u(t),\dot{u}(t),y(t),\dot{y}(t))\right\} dt \geq 0 \Rightarrow H(x,\dot{x},y,\dot{y}) \geq H(u,\dot{u},y,\dot{y}),$$

then $H(x,\dot{x},y,\dot{y})$ *is said to be partially pseudo-invex in* x *and* $\dot{x}$ *on* I *with respect to* η_1 *for fixed* y. *If* $H(x,\dot{x},y,\dot{y})$ *satisfies*

$$\int_a^b \left\{\eta_2^T h_y(t,x(t),\dot{x}(t),v(t),\dot{v}(t)) + (D\eta_2)^T h_{\dot{y}}(t,x(t),\dot{x}(t),v(t),\dot{v}(t))\right\} dt \geq 0 \Rightarrow H(x,\dot{x},y,\dot{y}) \geq H(x,\dot{x},v,\dot{v}),$$

where $\eta_2(t,y(t),\dot{y}(t),v(t),\dot{v}(t)) \in S$ *then* $H(x,\dot{x},y,\dot{y})$ *is said to be partially pseudo invex in* y *and* $\dot{y}$ *on* I *with respect to* η_2 *for fixed* x. *If* $-H(x,\dot{x},y,\dot{y})$ *is partially pseudoinvex in* x *and* $\dot{x}$ *on* I *with respect to* η_1 *for fixed* y, *then* $H(x,\dot{x},y,\dot{y})$ *is said to be partially pseudoincave in* x *and* $\dot{x}$ *on* I *with respect to* η_1 *for fixed* y *and if* $-H(x,\dot{x},y,\dot{y})$ *is partially pseudoinvex in* y *and* $\dot{y}$ *on* I *with respect to* η_2 *for fixed* x, *then* $H(x,\dot{x},y,\dot{y})$ *is said to be partially incave in* y *and* $\dot{y}$ *on* I *with respect to* η_2 *for fixed* x, *then* $H(x,\dot{x},y,\dot{y})$ *is said to be partially pseudoincave in* y *and* $\dot{y}$ *on* I *with respect to* η_2 *for fixed* x.

The following Lemma, which is said to be the generalized form of the Fritz-John condition for the vector-valued functions proposed by Bazaraa and Goode [13], play a main role in the proof of the strong duality.

Lemma 7.56. *Let K be a convex set with nonempty interior in R^n and let C be a closed convex cone in R^n having a nonempty interior. Let F and G be two vector-valued functions defined on K. If z_0 is an efficient solution of the following problem:*

$$\text{Minimize } F(z)$$

$$\text{Subject to } G(z) \in C, \quad z \in K,$$

then there exists a nonzero vector (r_0, r) such that

$$(r_0^T F_z(z_0) + r^T G_z(z_0))(z - z_0) \geq 0,$$

for all $z \in K$, and $r_0 \geq 0$, $r \in C^*$, and $r^T G(z_0) = 0$.

We consider the following Mond–Weir type symmetric dual minimax multiobjective variational mixed integer programs for arbitrary cones given by Chen and Yang [35]:

(MSP)

$$\underset{x_1}{\text{Max}}\ \underset{x_2, y}{\text{Min}} \left(\int_a^b f(t, x_1(t), x_2(t), \dot{x}_2(t), y_1(t), y_2(t)\dot{y}_2(t))dt \right.$$

$$= \left(\int_a^b f^1(t, x_1(t), x_2(t), \dot{x}_2(t), y_1(t), y_2(t)\dot{y}_2(t))dt, \ldots, \right.$$

$$\left.\left. \int_a^b f^p(t, x_1(t), x_2(t), \dot{x}_2(t), y_1(t), y_2(t)\dot{y}_2(t))dt \right)\right)$$

$$\text{Subject to } x(a) = 0 = x(b); \quad y(a) = 0 = y(b), \tag{7.19}$$

$$\dot{x}_2(a) = 0 = \dot{x}_2(b); \quad \dot{y}_2(a) = 0 = \dot{x}_2(b), \tag{7.20}$$

$$x_1(t) \in U, \quad x_2(t) \in C_1, \quad y_1(t) \in V, \quad y_2(t) \in C_2, \quad \forall t \in I, \tag{7.21}$$

$$f_{iy_2}(t, x_1(t), x_2(t), \dot{x}_2(t), y_1(t), y_2(t)\dot{y}_2(t))$$

$$-Df_{i\dot{y}_2}(t, x_1(t), x_2(t), \dot{x}_2(t), y_1(t), y_2(t)\dot{y}_2(t)) \in C_2^*, \tag{7.22}$$

$$i = 1, \ldots, p, \quad \forall t \in I,$$

$$(y_2(t))^T [f_{iy_2}(t, x_1(t), x_2(t), \dot{x}_2(t), y_1(t), y_2(t)\dot{y}_2(t))$$

$$-Df_{i\dot{y}_2}(t, x_1(t), x_2(t), \dot{x}_2(t), y_1(t), y_2(t)\dot{y}_2(t))] \geq 0, \tag{7.23}$$

$$i = 1, \ldots, p, \quad \forall t \in I,$$

(MSD)

$$\underset{v_1}{\text{Min}}\ \underset{u, v_2}{\text{Max}} \int_a^b f(t, u_1(t), u_2(t), \dot{u}_2(t), v_1(t) v_2(t) \dot{v}_2(t))dt$$

$$= \left(\int_a^b f^1(t, u_1(t), u_2(t), \dot{u}_2(t), v_1(t) v_2(t) \dot{v}_2(t))dt, \ldots, \right.$$

$$\left. \int_a^b f^p(t, u_1(t), u_2(t), \dot{u}_2(t), v_1(t) v_2(t) \dot{v}_2(t))dt \right)$$

$$\text{Subject to } u(a) = 0 = u(b); \ \ v(a) = 0 = v(b), \tag{7.24}$$

$$\dot{u}_2(a) = 0 = \dot{u}_2(b); \ \ \dot{v}_2(a) = 0 = \dot{v}_2(b), \tag{7.25}$$

$$u_1(t) \in U, \ \ u_2(t) \in C_1, \ \ v_1(t) \in V, \ \ v_2(t) \in C_2, \ \ \forall t \in I, \tag{7.26}$$

$$\begin{aligned} & f_{ix_2}(t, u_1(t), u_2(t), \dot{u}_2(t), v_1(t) v_2(t) \dot{v}_2(t)) \\ & - D f_{i\dot{x}_2}(t, u_1(t), u_2(t), \dot{u}_2(t), v_1(t) v_2(t) \dot{v}_2(t)) \in C_1^*, \\ & i = 1, \ldots, p, \ \ \forall t \in I, \end{aligned} \tag{7.27}$$

$$\begin{aligned} & (u_2(t))^T [-f_{ix_2}(t, u_1(t), u_2(t), \dot{u}_2(t), v_1(t) v_2(t) \dot{v}_2(t)) \\ & + D f_{i\dot{x}_2}(t, u_1(t), u_2(t), \dot{u}_2(t), v_1(t) v_2(t) \dot{v}_2(t))] \leq 0, \\ & i = 1, \ldots, p, \ \ \forall t \in I, \end{aligned} \tag{7.28}$$

where the function $f(t, x_1(t), x_2(t), \dot{x}_2(t), y_1(t), y_2(t)\dot{y}_2(t))$ is a p-dimensional vector-valued function. In (MSP) and (MSD), if f_i does not depend explicitly for t for $i = 1, \ldots, p$, then our problems become the pair of problems given by Kim and Song [126]:

(MSP1)

$$\underset{x_1}{\text{Max}}\ \underset{x_2, y}{\text{Min}} \ f(x, y)$$

$$\begin{aligned} \text{Subject to } & x_1 \in U, \ \ x_2 \in C_1, \ \ y_1 \in V, \ \ y_2 \in C_2, \\ & (\lambda^T f)_{y_2}(x, y) \in C_2^*, \\ & (y_2)^T (\lambda^T f)_{y_2}(x, y) \geq 0, \\ & \lambda > 0. \end{aligned}$$

(MSD1)

$$\underset{v_1}{\text{Min}}\ \underset{u, v_2}{\text{Max}} \ f(u, v)$$

$$\text{Subject to } u_1 \in U, \ \ u_2 \in C_1, \ \ v_1 \in V, \ \ v_2 \in C_2,$$

$$(\lambda^T f)_{x_2}(u,v) \in C_1^*,$$
$$(u_2(t))^T(\lambda^T f)_{x_2}(u,v) \leq 0,$$
$$\lambda > 0.$$

The pair of problems ($MSP1$) and ($MSD1$) are Mond–Weir type mixed integer symmetric dual programs studied by Kim and Song [126] under pseudo-invexity/pseudo-incavity assumptions and duality results are established. Furthermore, if $p = 1, C_1 = R_+^{n_2}$ and $C_2 = R_+^{m_2}$, then our problems again become the pair of problems considered by Mishra and Das [151]. Denote by X and Y the set of feasible solutions of (MSP) and (MSD), respectively. In order to obtain the duality theorems in next section, suppose that

(a) The functions $f_i(t, x_1(t), x_2(t), \dot{x}_2(t), y_1(t), y_2(t), \dot{y}_2(t))$ are separable with respect to x_1 or $y_1, i = 1, 2, \ldots, p$. Without loss of generality, we suppose that $f_i(t, x_1(t), x_2(t), \dot{x}_2(t), y_1(t), y_2(t), \dot{y}_2(t))$ are separable with respect to x_1, that is, f_i can be expressed as

$$f_i(t, x_1(t), x_2(t), \dot{x}_2(t), y_1(t), y_2(t), \dot{y}_2(t))$$
$$= f_i^1(t, x_1(t)) + f_i^2(t, x_2(t), \dot{x}_2(t), y_1(t), y_2(t), \dot{y}_2(t));$$

(b) The sets of feasible solutions X and Y having the properties that if $(x(t), y(t)) \in X$ and $(u(t), v(t)) \in Y$ then

$$\eta_1(t, x_2(t), \dot{x}_2(t), u_2(t), \dot{u}_2(t)) \in C_1 \text{ and}$$

$$\eta_2(t, y_1(t), y_2(t), \dot{y}_2(t), v_1(t), v_2(t), \dot{v}_2(t)) \in C_2.$$

Remark 7.57 (Chen and Yang [35]). Under the above assumptions, (MSP) can be expressed as the following form:

$$\underset{x_1}{\text{Max}}\ \underset{x_2, y}{\text{Min}} \int_a^b \{f^1(t, x_1(t)) + f^2(t, x_2(t), \dot{x}_2(t), y_1(t), y_2(t), \dot{y}_2(t))\}$$

Subject to (7.19)–(7.23)

where $f^1(t, x_1(t), \dot{x}_1(t)) = \left(f_1^1(t, x_1(t), \dot{x}_1(t)), \ldots, f_1^p(t, x_1(t), \dot{x}_1(t))\right)^T$

and $f_2(t, x_2(t), \dot{x}_2(t), y_1(t), y_2(t), \dot{y}_2(t))$

$$= \left(f_2^1(t, x_2(t), \dot{x}_2(t), y_1(t), y_2(t), \dot{y}_2(t)), \ldots,\right.$$
$$\left.f_2^p(t, x_2(t), \dot{x}_2(t), y_1(t), y_2(t), \dot{y}_2(t))\right)^T.$$

Let

(MP)

$$\phi_1(y_1) = \underset{x_2, y_2}{\text{Min}} \int_a^b \{f_2(t, x_2(t), \dot{x}_2(t), y_1(t), y_2(t), \dot{y}_2(t))\}dt$$

$$\text{Subject to (7.19)–(7.20),} \tag{7.29}$$

$$f^2_{iy_2}(t, x_2(t), \dot{x}_2(t), y_1(t), y_2(t), \dot{y}_2(t)) - Df^2_{i\dot{y}_2}(t, x_2(t), \dot{x}_2(t), y_1(t), y_2(t), \dot{y}_2(t)) \in C_2^* \tag{7.30}$$

$$i = 1, \ldots, p, \quad \forall t \in I,$$

$$(y_2(t))^T \left[f^2_{iy_2}(t, x_2(t), \dot{x}_2(t), y_1(t), y_2(t), \dot{y}_2(t)) - Df^2_{i\dot{y}_2}(t, x_2(t), \dot{x}_2(t), y_1(t), y_2(t), \dot{y}_2(t))\right] \geq 0 \tag{7.31}$$

$$i = 1, \ldots, p, \quad \forall t \in I,$$

So, (MSP) may be written as

$$\underset{x_1}{\text{Max}}\ \underset{x_2, y}{\text{Min}} \left\{ \left[\int_a^b f_1(t, x_1(t))dt + \phi_1(y_1) \right] : x_1 \in U, \quad y_1 \in V \right\}$$

Similarly, (MSD) can also be written as

$$\underset{v_1}{\text{Min}}\ \underset{u, v_2}{\text{Max}} \left\{ \left[\int_a^b f_1(t, u_1(t))dt + \psi_1(v_1) \right] : u_1 \in U, \quad v_1 \in V \right\},$$

where

(MD)

$$\psi_1(v_1) = \underset{u_2, v_2}{\text{Max}} \int_a^b \{f^2(t, u_2(t), \dot{u}_2(t), v_1(t), v_2(t), \dot{v}_2(t))\}dt$$

$$\text{Subject to (7.19)–(7.20)}, \ u_2(t) \in R^n, \quad v_2(t) \in C_2, \tag{7.32}$$

$$- f^2_{ix_2}(t, u_2(t), \dot{u}_2(t), v_1(t), v_2(t), \dot{v}_2(t)) + Df^2_{i\dot{x}_2}(t, u_2(t), \dot{u}_2(t), v_1(t), v_2(t), \dot{v}_2(t)) \in C_1^* \tag{7.33}$$

$$i = 1, \ldots, p, \quad \forall t \in I,$$

$$(u_2(t))^T \left[-f^2_{ix_2}(t, u_2(t), \dot{u}_2(t), v_1(t), v_2(t), \dot{v}_2(t)) + Df^2_{i\dot{x}_2}(t, u_2(t), \dot{u}_2(t), v_1(t), v_2(t), \dot{v}_2(t))\right] \leq 0 \tag{7.34}$$

$$i = 1, \ldots, p, \quad \forall t \in I,$$

Remark 7.58. Let $\psi : I \times R^n \times R^n \to R$ be a continuously differentiable function with respect to each of its arguments. Let $x, u : I \to R^n$ be differentiable with $x(a) = u(a) = \alpha$ and $x(b) = u(b) = \beta$. Then

$$\int_a^b \frac{d}{dt}(\eta(t,x,u))^T \psi_{\dot{x}}(t,u,\dot{u})dt = -\int_a^b \eta(t,x,u)^T \frac{d}{dt}(\psi_{\dot{x}}(t,u,\dot{u}))dt$$

Theorem 7.59 (Weak duality). *Let $(x(t), y(t))$ and $(u(t), v(t))$ be feasible solutions of (MSP) and (MSD), respectively. Assume that the function $f_i(t, x_1(t), x_2(t), \dot{x}_2(t), y_1(t), y_2(t), \dot{y}_2(t))$ is separable with respect to $x_1(t)$, and $\int_a^b f_i^2(t, \cdot, \cdot, y_1(t), y_2(t), \dot{y}_2(t))dt$ is partially pseudo-invex in x_2 and $\dot{x}_2$ for each $y_1(t)$ and $y_2(t)$ on I with respect to*

$$\eta_1 = \eta_1(t, x_2(t), \dot{x}_2(t), u_2(t), \dot{u}_2(t)) \in C_1, \quad i = 1, \dots, p;$$

and $\int_a^b f_i^2(t, x_2(t), \dot{x}_2(t), \cdot, \cdot)dt$ is partially pseudo-incave in y_1, y_2 and $\dot{y}_2$ for each $x_1(t), x_2(t)$ on I with respect to

$$\eta_2 = \eta_2(t, y_1(t), y_2(t), \dot{y}_2(t), v_1(t), v_2(t), \dot{v}_2(t)) \in C_2, \quad i = 1, \dots, p;$$

and

$$\eta_1(t, x_2, \dot{x}_2, u_2, \dot{u}_2)^T \xi + (u_2)^T \xi \geq 0, \quad \forall x_2, \dot{x}_2, u_2, \dot{u}_2 \in C_1, \ \ \xi \in C_1^*$$
$$\eta_2(t, y_1(t), y_2(t), \dot{y}_2(t), v_1(t), v_2(t), \dot{v}_2(t))^T \zeta + (y_2)^T \zeta \geq 0,$$
$$\forall v_2, \dot{v}_2, y_1, y_2, \dot{y}_2 \in C_2, \ \ \zeta \in C_2^*.$$

Then the following cannot hold:

$$\int_a^b f_i(t, x_1(t), x_2(t), \dot{x}_2(t), y_1(t), y_2(t), \dot{y}_2(t))dt$$
$$\leq \int_a^b f_i(t, u_1(t), u_2(t), \dot{u}_2(t), v_1(t), v_2(t), \dot{v}_2(t))dt,$$

$\forall i = 1, \dots, p$, and

$$\int_a^b f_j(t, x_1(t), x_2(t), \dot{x}_2(t), y_1(t), y_2(t), \dot{y}_2(t))dt$$
$$< \int_a^b f_j(t, u_1(t), u_2(t), \dot{u}_2(t), v_1(t), v_2(t), \dot{v}_2(t))dt,$$

for some $j \in \{1, \dots, p\}$.

Proof. Let $(x(t), y(t))$ and $(u(t), v(t))$ be feasible solutions of (MSP) and (MSD), respectively. In order to prove the Theorem, it suffices to prove that the following cannot hold:

$$\int_a^b f_i(t, x_1(t), x_2(t), \dot{x}_2(t), y_1(t), y_2(t), \dot{y}_2(t))dt$$
$$\leq \int_a^b f_i(t, u_1(t), u_2(t), \dot{u}_2(t), v_1(t), v_2(t), \dot{v}_2(t))dt,$$

$\forall i = 1, \ldots, p$, and

$$\int_a^b f_j(t, x_1(t), x_2(t), \dot{x}_2(t), y_1(t), y_2(t), \dot{y}_2(t))dt$$
$$< \int_a^b f_j(t, u_1(t), u_2(t), \dot{u}_2(t), v_1(t), v_2(t), \dot{v}_2(t))dt,$$

for some $j \in \{1, \ldots, p\}$. Now taking,

$$\xi = f^2_{ix_2}(t, u_2(t), \dot{u}_2(t), v_1(t), v_2(t), \dot{v}_2(t))$$
$$- Df^2_{i\dot{x}_2}(t, u_2(t), \dot{u}_2(t), v_1(t), v_2(t), \dot{v}_2(t)).$$

By

$$\eta_1(t, x_2, \dot{x}_2, u_2, \dot{u}_2)^T\xi + (u_2)^T\xi \geq 0, \quad \forall x_2, \dot{x}_2, u_2, \dot{u}_2 \in C_1, \ \ \xi \in C_1^*$$

and inequality (7.34), we have,

$$\eta_1(t, x_2, \dot{x}_2, u_2, \dot{u}_2)^T(f^2_{ix_2}(t, u_2(t), \dot{u}_2(t), v_1(t), v_2(t), \dot{v}_2(t))$$
$$- Df^2_{i\dot{x}_2}(t, u_2(t), \dot{u}_2(t), v_1(t), v_2(t), \dot{v}_2(t)))$$
$$> -(u_2)^T(f^2_{ix_2}(t, u_2(t), \dot{u}_2(t), v_1(t), v_2(t), \dot{v}_2(t))$$
$$- Df^2_{i\dot{x}_2}(t, u_2(t), \dot{u}_2(t), v_1(t), v_2(t), \dot{v}_2(t))),$$

which implies that

$$\eta_1(t, x_2, \dot{x}_2, u_2, \dot{u}_2)^T(f^2_{ix_2}(t, u_2(t), \dot{u}_2(t), v_1(t), v_2(t), \dot{v}_2(t))$$
$$- Df^2_{i\dot{x}_2}(t, u_2(t), \dot{u}_2(t), v_1(t), v_2(t), \dot{v}_2(t))) \geq 0.$$

By partial pseudo-invexity of $\int_a^b f_i^2(t, \cdot, \cdot, y_1(t), y_2(t), \dot{y}_2(t))dt$ in $(x_2, \dot{x}_2)$, we get

$$\int_a^b f_i^2(t, x_2, \dot{x}_2, v_1, v_2, \dot{v}_2)dt \geq \int_a^b f_i^2(t, u_2, \dot{u}_2, v_1, v_2, \dot{v}_2)dt,$$

which along with

$$\lambda \in \Lambda = \left\{ \lambda = (\lambda_1, \ldots, \lambda_p) : \lambda_i \geq 0, i = 1, \ldots, p, \sum_{i=1}^{p} \lambda_i = 1 \right\},$$

implies that

$$\sum_{i=1}^{p} \lambda_i f_i^2(t, x_2, \dot{x}_2, v_1, v_2, \dot{v}_2)dt \geq \sum_{i=1}^{p} \lambda_i \int_a^b f_i^2(t, u_2, \dot{u}_2, v_1, v_2, \dot{v}_2)dt. \quad (7.35)$$

By

$$\eta_2(t, y_1(t), y_2(t), \dot{y}_2(t), v_1(t), v_2(t), \dot{v}_2(t))^T \zeta + (y_2)^T \zeta \geq 0,$$

with

$$\begin{aligned} \zeta = &- \left(f_{iy_2}^2(t, x_2(t), \dot{x}_2(t), v_1(t), v_2(t), \dot{v}_2(t))\right. \\ &\left. - D f_{i\dot{y}_2}^2(t, x_2(t), \dot{x}_2(t), v_1(t), v_2(t), \dot{v}_2(t))\right) \end{aligned}$$

and inequality (7.31), and partial of pseudo-incavity

$$\int_a^b f_i^2(t, x_2(t), \dot{x}_2(t), \cdot, \cdot)dt$$

in y_1, y_2 and $\dot{y}_2$ for each $x_1(t), x_2(t)$ on I with respect to

$$\eta_2 = \eta_2(t, y_1(t), y_2(t), \dot{y}_2(t), v_1(t), v_2(t), \dot{v}_2(t)) \in C_2, \quad i = 1, \ldots, p,$$

we have,

$$\sum_{i=1}^{p} \lambda_i f_i^2(t, x_2, \dot{x}_2, v_1, v_2, \dot{v}_2)dt \leq \sum_{i=1}^{p} \lambda_i \int_a^b f_i^2(t, x_2, \dot{x}_2, y_1, y_2, \dot{y}_2)dt. \quad (7.36)$$

From inequalities (7.35) and (7.36), we obtain

$$\sum_{i=1}^{p} \lambda_i f_i^2(t, x_2, \dot{x}_2, y_1, y_2, \dot{y}_2)dt \geq \sum_{i=1}^{p} \lambda_i \int_a^b f_i^2(t, u_2, \dot{u}_2, v_1, v_2, \dot{v}_2)dt.$$

□

Remark 7.60. The above theorem is established under the assumption of partial invexity/partial incavity assumptions by Chen and Yang [35]. The strong duality theorem (Theorem 3.5) given in Chen and Yang [35] can be extended to partial pseudo-invexity/partial pseudo-incavity assumptions as in the proof of Strong duality, one needs the Weak duality result. Therefore, we omit the strong duality theorem.

A mathematical programming problem is said to be self-dual, if when the dual is recast in the form of the primal, the new problem so obtained is the same as the primal problem. From Chen [34], we recall the following definition.

Definition 7.61. *The function $h : I \times R^{n_1} \times R^{n_2} \times R^{n_2} \times R^{n_1} \times R^{n_2} \times R^{n_2} \to R$ is said to be skew symmetric if for all x and y in the domain of h where*

$$h(t, x_1, x_2, \dot{x}_2, y_1, y_2, \dot{y}_2) = -h(t, y_1, y_2, \dot{y}_2, x_1, x_2, \dot{x}_2), \quad t \in I,$$

where $x_1, x_2 \in U,\ \ y_1, y_2 \in C,\ \ t \in I,\ \ U$ is an arbitrary set of integers in R^{n_1} and C is a closed convex cone in R^{n_2} with nonempty interior, $n_1 + n_2 = n$.

One can easily establish the following self-duality theorem.

Theorem 7.62 (Self-duality). *If f_i is skew symmetric for $i = 1, 2, \ldots, p$, then (MSP) is self-dual. Moreover, the feasibility of $(x(t), y(t))$ for (MSP) implies the feasibility of $(y(t), x(t))$ for (MSD) and vise-versa.*

□

Theorem 7.63. *Under the conditions of Theorem 7.59 if $(x(t), y(t))$ is an efficient solution for (MSP), then $(y(t), x(t))$ is an efficient solution for (MSD) and the common optimal value is zero and the converse.*

□

Remark 7.64. Theorem 7.63 is obtained in Chen and Yang [35] under the assumptions of partial invexity/partial incavity.

7.3 Scalar Control Problems

The control problem is to transfer the state variable from an initial state x_0 at a to a final state x_f at b so as to minimize a given functional, subject to constraints on the control and state variables, that is:

(P)

$$\text{Minimize} \int_a^b f(t, x, u)dt$$

$$\text{Subject to } x(a) = \alpha, \quad x(b) = \beta, \tag{7.37}$$

$$G(t, x, u) = \dot{x}, \tag{7.38}$$

$$R(t, x, u) \geq 0. \tag{7.39}$$

$x(t)$ and $u(t)$ are required to be piecewise smooth functions on $[a, b] = I$; their derivatives are continuous except perhaps at points of discontinuity of $u(t)$, which has piecewise continuous first and second derivatives. The constraints (7.38) and (7.39) may fail to hold at these points of discontinuities of $u(t)$.

Throughout this section $x \in R^n, u \in R^m, t$ is the independent variable, $u(t)$ is the control variable and $x(t)$ is the state variable; $u(t)$ is related to $x(t)$ via the state equations $G(t, x, u) = \dot{x}$ where $\dot{x}$ denotes the derivative with respect to t. If $x = (x^1, x^2, \ldots, x^n)^T$, the gradient vector of f with respect to x is denoted by $f_x = \left(\frac{\partial f}{\partial x^1}, \ldots, \frac{\partial f}{\partial x^n}\right)^T$. For an r-dimensional vector function $R(t, x, u)$ the gradient with respect to x is denoted as

$$R_x = \begin{pmatrix} \frac{\partial R^1}{\partial x^1} & \cdots & \frac{\partial R^r}{\partial x^1} \\ \vdots & \vdots & \vdots \\ \frac{\partial R^1}{\partial x^n} & \cdots & \frac{\partial R^r}{\partial x^n} \end{pmatrix}$$

Remark 7.65. If f, G and R are independent of time (without loss of generality, assume $t_f - t_0 = 1$), then the above control problem reduces to a static mathematical programming problem:

$$\begin{aligned} &\text{Minimize } f(z) \\ &\text{Subject to } G(z) = 0, \\ &R(z) \geq 0. \end{aligned}$$

In this section, we recall the concepts of invex and generalized invex functionals instead of functions following Mond and Smart [176] and we will show that invexity of a functional is necessary as well as sufficient for its critical points to be global minima, which coincides with the concept of an invex function.

Definition 7.66. *If there exist vector functions $\eta(t, x, x^*, \dot{x}, \dot{x}^*, u, u^*) \in R^n$ with $\eta = 0$ at t if $x(t) = x^*(t)$ and $\xi(t, x, x^*, \dot{x}, \dot{x}^*, u, u^*) \in R^m$ such that for the scalar function $h(t, x, \dot{x}, u)$ the functional $H(x, \dot{x}, u) = \int_a^b h(t, x, \dot{x}, u)dt$ satisfies*

$$\begin{aligned} &H(x, \dot{x}, u) - H(x^*, \dot{x^*}, u^*) \\ &\geq \int_a^b \left(\eta^T h_x(t, x^*, \dot{x}^*, u^*) + \frac{d\eta^T}{dt} h_{\dot{x}}(t, x^*, \dot{x}^*, u^*) + \xi^T h_u(t, x^*, \dot{x}^*, u^*) \right) dt. \end{aligned}$$

Then H is said to be invex in $x, \dot{x}$ and u on I with respect to η and ξ.

Remark 7.67. (x^*, u^*) is a critical point of H if $h_x(t, x^*, \dot{x}^*, u^*) = \frac{d}{dt} h_{\dot{x}}(t, x^*, \dot{x}^*, u^*)$ and $h_u(t, x^*, \dot{x}^*, u^*) = 0$ almost everywhere in the interval I. If $x(t_0)$ and $x(t_f)$ are free, the transversality conditions $h_{\dot{x}}(t, x^*, \dot{x}^*, u^*) = 0$ at t_0 and t_f are included.

Lemma 7.68. *$H(x, \dot{x}, u) = \int_a^b h(t, x, \dot{x}, u)dt$ is invex iff every critical point of H is a global minimum.*

Proof. Assume that H is invex with respect to η and ξ. Let (x^*, u^*) be a critical point of H. Then

$$\begin{aligned} &H(x, \dot{x}, u) - H(x^*, \dot{x^*}, u^*) \\ &\geq \int_a^b \left(\eta^T h_x(t, x^*, \dot{x}^*, u^*) + \frac{d\eta^T}{dt} h_{\dot{x}}(t, x^*, \dot{x}^*, u^*) + \xi^T h_u(t, x^*, \dot{x}^*, u^*) \right) dt \\ &= \int_a^b \left(\eta^T h_x(t, x^*, \dot{x}^*, u^*) - \eta^T \frac{d}{dt} h_{\dot{x}}(t, x^*, \dot{x}^*, u^*) + \xi^T h_u(t, x^*, \dot{x}^*, u^*) \right) dt \\ &\quad + \eta^T h_{\dot{x}}(t, x^*, \dot{x}^*, u^*)|_a^b \qquad \text{by integration by parts} \\ &= 0. \end{aligned}$$

As (x^*, u^*) is a critical point, and either fixed boundary conditions imply that $\eta = 0$ at t_0 and t_f, or free boundary conditions imply $h_{\dot{x}} = 0$ at t_0 and t_f.

Therefore, (x^*, u^*) is a global minimum of H. Conversely, assume that every critical point is global minimum. If (x^*, u^*) is critical point, put $\eta = 0 = \xi$. If (x^*, u^*) is not a critical point, then if $h_x \neq \frac{d}{dt} h_{\dot{x}}$ at (x^*, u^*) put

$$\eta = \frac{h(t,x,\dot{x},u) - h(t,x^*,\dot{x}^*,u^*)}{2\left(h_x - \frac{d}{dt}h_{\dot{x}}\right)^T \left(h_x - \frac{d}{dt}h_{\dot{x}}\right)} \left(h_x - \frac{d}{dt}h_{\dot{x}}\right)$$

Or, if $h_x = \frac{d}{dt} h_{\dot{x}}$, put $\eta = 0$; and if $h_u \neq 0$, put

$$\xi = \frac{h(t,x,\dot{x},u) - h(t,x^*,\dot{x}^*,u^*)}{2h_u^T h_u} h_u$$

Or, if $h_u = 0$ put $\xi = 0$. Then H is invex on I with respect to η and ξ.

□

Definition 7.69. *If there exist vector functions $\eta(t,x,x^*,\dot{x},\dot{x}^*,u,u^*) \in R^n$ with $\eta = 0$ at t if $x(t) = x^*(t)$ and $\xi(t,x,x^*,\dot{x},\dot{x}^*,u,u^*) \in R^m$ such that for the scalar function $h(t,x,\dot{x},u)$ the functional $H(x,\dot{x},u) = \int_a^b h(t,x,\dot{x},u)dt$ such that*

$$\begin{aligned}
&\int_a^b [\eta^T h_x(t,x^*,\dot{x}^*,u^*) \\
&\qquad + \frac{d\eta^T}{dt} h_{\dot{x}}(t,x^*,\dot{x}^*,u^*) + \xi^T h_u(t,x^*,\dot{x}^*,u^*)]dt \geq 0 \\
&\Rightarrow H(x,\dot{x},u) \geq H(x^*,\dot{x}^*,u^*)(>),
\end{aligned}$$

then H is said to be pseudo-invex (strictly pseudo-invex) in $x, \dot{x}$ and u on I with respect to η and ξ.

Definition 7.70. *If there exist vector functions $\eta(t,x,x^*,\dot{x},\dot{x}^*,u,u^*) \in R^n$ with $\eta = 0$ at t if $x(t) = x^*(t)$ and $\xi(t,x,x^*,\dot{x},\dot{x}^*,u,u^*) \in R^m$ such that for the scalar function $h(t,x,\dot{x},u)$ the functional $H(x,\dot{x},u) = \int_a^b h(t,x,\dot{x},u)dt$ such that*

$$\begin{aligned}
&H(x,\dot{x},u) \leq H(x^*,\dot{x}^*,u^*) \\
&\Rightarrow \int_a^b [\eta^T h_x(t,x^*,\dot{x}^*,u^*) \\
&\qquad + \frac{d\eta^T}{dt} h_{\dot{x}}(t,x^*,\dot{x}^*,u^*) + \xi^T h_u(t,x^*,\dot{x}^*,u^*)]dt \leq 0
\end{aligned}$$

then H is said to be quasi-invex in $x, \dot{x}$ and u on I with respect to η and ξ.

Remark 7.71. If h is independent of t, the above Definitions 7.66–7.70 reduce to the definitions of invexity, (strict) pseudo-invexity and quasi-invexity, respectively of the static case, see Hanson [83] and Kaul and Kaur [114]. Mond and Smart [176] have shown that, for invex functions, the necessary conditions of Berkovitz [19] are also sufficient with normality of the constraints. The Berkovitz's necessary optimality theorem for (P) is:

Theorem 7.72. *Assuming the constraint conditions for the existence of multipliers $\lambda(t)$ and $\mu(t)$ at extrema of (P) hold, the necessary conditions for (x^*, u^*) to be optimal for (P) are: there exist $\lambda_0 \in R, \lambda(t), \mu(t)$ such that*

$$F = \lambda_0 f - \lambda(t)^T[G - \dot{x}] - \mu(t)^T R$$

$$\text{satisfies } F_x = \frac{d}{dt} F_{\dot{x}}, \tag{7.40}$$

$$F_u = 0, \tag{7.41}$$

$$\mu(t)^T R(t, x^*, u^*) = 0, \tag{7.42}$$

$$\mu(t) \geq 0. \tag{7.43}$$

hold throughout the interval I (except that at t corresponding to discontinuities of u^*, (7.40) holds for right and left limits). Here λ_0 is a nonnegative constant, $\lambda(t)$ is continuous on I, and $\lambda_0, \lambda(t)$ and $\mu(t)$ can not vanish simultaneously for any $t \in I$.

It is assumed from now on that the minimizing solution (x^*, u^*) of (P) is normal; that is, λ_0 is non-zero, so that without loss of generality, we can take $\lambda_0 = 1$.

Proof. We refer the reader to Mond and Smart [176].

□

Theorem 7.73 (Sufficient optimality conditions). *If there exists $(x^*, u^*, \lambda^*, \mu^*)$ such that conditions (7.40)–(7.43) hold, with (x^*, u^*) feasible for (P), and $\int_a^b f dt$, $\int_a^b -(\lambda^*)^T(G - \dot{x})dt$ and $\int_a^b -(\mu^*)^T R dt$ are all invex with respect to the same functions η and ξ, then (x^*, u^*) is optimal for (P).*

Proof. Assume (x^*, u^*) is not optimal for (P). Then there exists $(x, u) \neq (x^*, u^*), (x, u)$ feasible for (P), such that

$$\int_a^b f(t, x, u)dt < \int_a^b f(t^*, x^*, u^*)dt.$$

As $\int_a^b f dt$ is invex with respect to η and ξ, it follows that

$$\int_a^b \left(\eta^T f_x(t, x^*, u^*) + \xi^T f_u(t, x^*, u^*)\right) dt < 0. \tag{7.44}$$

Now

$$(\lambda^*)^T[G(t, x, u) - \dot{x}] = 0 = (\lambda^*)^T[G(t, x^*, u^*) - \dot{x}^*]$$

implies

$$\int_a^b -(\lambda^*)^T \left(G(t,x,u) - \dot{x} - G(t,x^*,u^*) + \dot{x}^T\right) dt = 0 \leq 0.$$

Thus, by invexity of $\int_a^b -(\lambda^*)^T (G - \dot{x})\, dt$, we get

$$-\int_a^b \left(\eta^T G_x(t,x^*,u^*)\lambda^*(t) - \frac{d\eta^T}{dt}\lambda^*(t) + \xi^T G_u(t,x^*,u^*)\lambda^*(t)\right) dt \leq 0. \tag{7.45}$$

Also, $\lambda^*(t)R(t,x^*,u^*) = 0 \leq \lambda^*(t)R(t,x,u)$ implies that

$$\int_a^b - (\mu^*(t))^T (R(t,x,u) - R(t,x^*,u^*))\, dt \leq 0.$$

By invexity of $\int_a^b -(\mu^*)^T R dt$, we get

$$-\int_a^b \left(\eta^T R_x(t,x^*,u^*)\mu^*(t) + \xi^T R_u(t,x^*,u^*)\right) dt \leq 0. \tag{7.46}$$

Combining (7.44), (7.45) and (7.46), we get

$$\begin{aligned}\int_a^b & \left[\eta^T f_x(t,x^*,u^*) + \xi^T f_u(t,x^*,u^*)\right.\\ & - \left(\eta^T G_x(t,x^*,u^*)\lambda^*(t) - \frac{d\eta^T}{dt}\lambda^*(t) + \xi^T G_u(t,x^*,u^*)\lambda^*(t)\right)\\ & \left. - \left(\eta^T R_x(t,x^*,u^*)\mu^*(t) + \xi^T R_u(t,x^*,u^*)\right)\right] dt < 0. \end{aligned} \tag{7.47}$$

Now, premultiplying (7.40) by η^T and (7.41) by ξ^T, adding and integrating gives

$$\begin{aligned}\int_a^b & [\eta^T f_x(t,x^*,u^*) + \xi^T f_u(t,x^*,u^*)\\ & - (\eta^T G_x(t,x^*,u^*)\lambda^*(t) + \xi^T G_u(t,x^*,u^*)\lambda^*(t))\\ & - (\eta^T R_x(t,x^*,u^*)\mu^*(t) + \xi^T R_u(t,x^*,u^*)) - \eta^T \frac{d}{dt}\lambda^*(t)]dt = 0. \end{aligned} \tag{7.48}$$

But,

$$\int_a^b \eta^T \frac{d}{dt}\lambda^*(t)dt = \eta^T \lambda^*(t)|_a^b - \int_a^b \frac{d\eta^T}{dt}\lambda^*(t)dt = -\int_a^b \frac{d\eta^T}{dt}\lambda^*(t)dt,$$

as fixed boundary conditions give $\eta = 0$ at a and b. This contradicts (7.47). Hence (x^*, u^*) is optimal for (P).

□

Later Kim et al. [119] extended the above Theorem to pseudoinvex and quasiinvex functions.

Theorem 7.74 (Sufficiency). *If there exists $(x^*, u^*, \lambda^*, \mu^*)$ such that conditions (7.40)–(7.43) hold, with (x^*, u^*) feasible for (P), and $\int_a^b f dt$, is pseudo-invex with respect to η and ξ, and*

$$\int_a^b \left(-(\lambda^*)^T(G - \dot{x}) - (\mu^*)^T R\right) dt$$

is quasi-invex with respect to the same functions η and ξ, then (x^*, u^*) is optimal for (P).

Proof. The proof of this theorem is an exercise to the reader.

□

The following dual problem to primal problem (P) is from Mond and Hanson [171]:

(D)

$$\begin{gathered}
\text{Maximize} \int_a^b \{f(t,x,u) - \lambda(t)^T[G(t,x,u) - \dot{x}]dt - \mu(t)^T R(t,x,u)\}dt \\
\text{Subject to } x(a) = \alpha, \ \ x(b) = \beta, \\
f_x(t,x,u) - G_x(t,x,u)\lambda(t) - R_x(t,x,u) = \dot{\lambda}(t), \\
f_u(t,x,u) - G_u(t,x,u)\lambda(t) - R_u(t,x,u) = 0, \\
\mu(t) \geq 0,
\end{gathered}$$

where $\lambda : I \to R^n$ and $\mu : I \to R^k$.

Remark 7.75. If f, G and R are independent of time (without loss of generality, assume $b - a = 1$), then the above control problem reduces to a static mathematical programming problem:

$$\begin{gathered}
\text{Maximize } f - \lambda^T G(z) - \mu^T R(z) \\
\text{Subject to } f_z(z) - G_z(z)\lambda - R_z(z) = 0 \\
\mu(t) \geq 0,
\end{gathered}$$

where $\lambda \in R^n$ and $\mu \in R^k$.

Mond and Smart [176] proved that the problems (P) and (D) are a dual pair subject to invexity conditions on the objective and constraint functions.

Theorem 7.76 (Weak duality). *If $\int_a^b f dt, \int_a^b -(\lambda^*)^T(G - \dot{x})dt$ and $\int_a^b -(\mu^*)^T R dt$ are all invex with respect to the same functions η and ξ, for any $\lambda(t) \in R^n$ and $\mu(t) \in R^k$, then $\inf(P) \geq \sup(D)$.*

Proof. The proof of this theorem is very similar to the proof of Theorem 7.73. □

In the following Theorem, it is assumed that the minimizing solution (x^*, u^*) for (P) is normal; that is, λ_0 is non-zero, so that without loss of generality, we can take $\lambda_0 = 1$.

Theorem 7.77 (Strong duality). *Under the invexity conditions of Theorem 7.76, if (x^*, u^*) is an optimal solution for (P), then there exists $\lambda(t)$ and $\mu(t)$ such that (x^*, u^*, λ, μ) is optimal for (D), and the corresponding objective values are equal.*

Proof. Left to the reader as an exercise. □

Similarly, converse duality holds if we further assume that f, G and R have continuous third derivatives, and writing () as $P(t, x, u, \lambda, \mu) = 0$ the matrix $\frac{dP_i}{dz_i}, i = 1, \ldots, m, \ \ j = 1, \ldots, n + m$ where $z = \begin{pmatrix} x \\ u \end{pmatrix}$ has rank m.

Theorem 7.78 (Converse duality). *If $(x^*, u^*, \lambda^*, \mu^*)$ is optimal for (P), and if*

$$\begin{pmatrix} f_{xx} - (G_x\lambda)_x - (R_x\mu)_x & f_{ux} - (G_x\lambda)_u - (R_x\mu)_u \\ f_{xu} - (G_u\lambda)_x - (R_u\mu)_x & f_{uu} - (G_u\lambda)_u - (R_u\mu)_u \end{pmatrix}$$

is non-singular for all $t \in I$, then (x^*, u^*) is optimal for (P), and the corresponding objective values are equal.

Proof. See Mond and Hanson [171]. □

Kim et al. [119] introduced the Mond–Weir type dual problem for the problem (P) given in this section as follows:

(MWD)

$$\text{Maximize} \int_a^b f(t, x, u)dt$$

$$\text{Subject to } x(a) = \alpha, \ \ x(b) = \beta,$$

$$f_x(t, x, u) - G_x(t, x, u)\lambda(t) - R_x(t, x, u) = \dot{\lambda}(t),$$

$$f_u(t, x, u) - G_u(t, x, u)\lambda(t) - R_u(t, x, u) = 0,$$

$$\int_a^b \{\lambda(t)^T[G(t, x, u) - \dot{x}]dt + \mu(t)^T R(t, x, u)\}dt \leq 0$$

$$\mu(t) \geq 0, \ \ t \in I$$

where $\lambda : I \to R^n$ and $\mu : I \to R^k$.

Remark 7.79. If f, G and R are independent of time (without loss of generality, assume $b - a = 1$), then the above control problem reduces to a static mathematical programming problem:

$$\text{Maximize } f(z)$$
$$\text{Subject to } f_z(z) - G_z(z)\lambda - R_z(z) = 0$$
$$\mu(t) \geq 0,$$

where $\lambda \in R^n$ and $\mu \in R^k$.

As it is well known that Mond–Weir type of dual allows further weakening of the invexity conditions, therefore, Kim et al. [119] established some duality theorems for (P) and (MWD) under generalized invexity conditions.

Theorem 7.80 (Weak duality). *If $\int f dt$, is pseudo-invex with respect to η and ξ, and $\int_a^b (-\lambda^T (G - \dot{x}) - \mu^T R) dt$ for any $\lambda(t) \in R^n$ and $\mu(t) \in R^m$ is quasi-invex with respect to the same functions η and ξ, then* $\inf(P) \geq \sup(MWD)$.

Proof. See Kim et al. [119].

□

Kim et al. [119] have also proved the following usual strong and strict converse duality theorems, we state these without proof.

Theorem 7.81 (Strong duality). *Under the generalized invexity conditions of Theorem 7.80, if (x^*, u^*) is an optimal solution for (P), then there exists $\lambda(t)$ and $\mu(t)$ such that (x^*, u^*, λ, μ) is optimal for (MWD), and the corresponding objective values are equal.*

□

Theorem 7.82 (Strict converse duality). *Let (x^*, u^*) be optimal for (P). Under the generalized invexity conditions of Theorem 7.80, if $(\bar{x}, \bar{u}, \bar{\lambda}, \bar{\mu})$ is optimal for (MWD), $\int_a^b f dt$, is quasi-invex with respect to η and ξ, and $\int_a^b (-\lambda^T (G - \dot{x}) - \mu^T R) dt$ is strictly pseudo-invex at $(\bar{x}, \bar{u})$ with respect to the same functions η and ξ, then $(x^*, u^*) = (\bar{x}, \bar{u})$, that is, $(\bar{x}, \bar{u})$ is optimal for (P).*

□

7.4 Multiobjective Control Problems

Throughout this section f_i, $i = 1, 2, \ldots, p$ be a function from $I \times R^n \times R^m$ into R, G be a function from $I \times R^n \times R^m$ into R^n and R be a function from $I \times R^n \times R^m$ into R^k. Consider a multiobjective control problem (MVP): Consider the following multiobjective control problem:

(VCP)

$$\text{Minimize} \int_a^b f(t, x(t), u(t)) dt$$
$$\text{Subject to} \quad x(a) = \alpha, \quad x(b) = \beta,$$

$$\dot{x}(t) = G(t, x(t), u(t)), \quad t \in I$$
$$R(t, x(t), u(t)) \geq 0, \quad t \in I.$$

Definition 7.83. *A feasible solution $(x^*(t), u^*(t))$ for (VCP) is said to be an efficient solution for (VCP) if for all feasible solutions $(x(t), u(t))$ for (VCP),*

$$\int_a^b f_i(t, x(t), u(t))dt \leq \int_a^b f_i(t, x^*(t), u^*(t))dt, \quad \forall i = \{1, 2, \ldots, p\}$$

$$\text{and} \quad \int_a^b f_j(t, x(t), u(t))dt < \int_a^b f_j(t, x^*(t), u^*(t))dt, \quad \text{for some } j.$$

Lee et al. [137] established the following sufficient optimality theorem for multiobjective case:

Theorem 7.84. *Suppose that (x^*, u^*) is feasible for (VCP) such that there exist $\tau^* \in \Lambda^+, \lambda^*(t)$, and $\mu^*(t)$ such that*

$$\sum_{i=1}^{p} \tau_i^* f_{ix}(t, x^*(t), u^*(t)) - G_x(t, x^*(t), u^*(t))\lambda^*(t)$$
$$- R_x(t, x^*(t), u^*(t))\mu^*(t) = \dot{\lambda}^*(t),$$
$$\sum_{i=1}^{p} \tau_i^* f_{iu}(t, x^*(t), u^*(t)) - G_u(t, x^*(t), u^*(t))\lambda^*(t)$$
$$- R_u(t, x^*(t), u^*(t))\mu^*(t) = 0,$$
$$\mu^*(t) R(t, x^*(t), u^*(t))\mu^*(t) = 0, \quad \text{and}$$
$$\mu^*(t) \geq 0,$$

hold throughout the interval I (except that at t corresponding to discontinuities of $\mu^*(t)$, () holds for right and left limits). If $\int_a^b f_i dt, \ i = 1, 2 \ldots, p$, are invex with respect to η and ξ, and $\int_a^b (-(\lambda^*)^T (G - \dot{x}) - (\mu^*)^T R)dt$, is invex with respect to the same functions η and ξ, then (x^*, u^*) is an efficient solution for (VCP).

Proof. Suppose that (x^*, u^*) is not an efficient solution of (VCP). Then there exists $(x, u) \neq (x^*, u^*)$ such that (x, u) is feasible for (VCP) and

$$\int_a^b f_i(t, x(t), u(t))dt \leq \int_a^b f_i(t, x^*(t), u^*(t))dt, \quad \forall\, i = \{1, 2, \ldots, p\}$$

$$\text{and} \quad \int_a^b f_j(t, x(t), u(t))dt < \int_a^b f_j(t, x^*(t), u^*(t))dt, \quad \text{for some } j.$$

Since $f_i dt, \ i = 1, 2, \ldots, p$ are invex with respect to η and ξ, we get

$$\int_a^b \{\eta^T f_{ix}(t,x^*(t),u^*(t)) + \xi^T f_{iu}(t,x^*(t),u^*(t))\}dt \leq 0, \quad \forall\, i = \{1,2,\ldots,p\}$$

and $$\int_a^b \{\eta^T f_{jx}(t,x^*(t),u^*(t)) + \xi^T f_{ju}(t,x^*(t),u^*(t))\}dt < 0, \quad \text{for some } j.$$

Since $\tau_i^* > 0$ for all i,

$$\int_a^b \sum_{i=1}^p \tau_i^* \eta^T f_{ix}(t,x^*(t),u^*(t))dt < \int_a^b \sum_{i=1}^p \tau_i^* \xi^T f_{iu}(t,x^*(t),u^*(t))dt.$$

Since $(\lambda^*)^T[G(t,x,u) - \dot{x}] = 0 = (\lambda^*)^T[G(t,x^*,u^*) - \dot{x}^*]$ and

$$(\lambda^*)^T R(t,x^*,u^*) = 0 \leq (\lambda^*)^T R(t,x,u),$$

we get

$$\int_a^b \left\{-(\lambda^*)^T(G(t,x^*,u^*) - \dot{x}^*) - (\mu^*(t))^T R(t,x^*,u^*)\right\} dt$$
$$\geq \int_a^b \left\{-(\lambda^*)^T(G(t,x,u) - \dot{x}^*] - (\mu^*(t))^T R(t,x,u)\right\} dt.$$

By the invexity of $\int_a^b (-(\lambda^*)^T(G - \dot{x}) - (\mu^*)^T R)dt$, we get

$$\int_a^b \{-\eta^T G_x(t,x^*,u^*)\lambda^*(t) + \frac{d\eta^T}{dt}\lambda^*(t) - \xi^T G_u(t,x^*,u^*)\lambda^*(t)$$
$$- \eta^T R_x(t,x^*,u^*)\mu^*(t) - \xi^T R_u(t,x^*,u^*)\}dt \leq 0. \tag{7.49}$$

Therefore, we get

$$\int_a^b \left\{\sum_{i=1}^p \tau_i^* \eta^T f_{ix}(t,x^*,u^*) + \sum_{i=1}^p \tau_i^* \xi^T f_{iu}(t,x^*,u^*)\right.$$
$$- \left(\eta^T G_x(t,x^*,u^*)\lambda^*(t) - \frac{d\eta^T}{dt}\lambda^*(t) + \xi^T G_u(t,x^*,u^*)\lambda^*(t)\right)$$
$$\left. - \left(\eta^T R_x(t,x^*,u^*)\mu^*(t) + \xi^T R_u(t,x^*,u^*)\right)\right\} dt < 0. \tag{7.50}$$

On the other hand, we get

$$\int_a^b \left\{\sum_{i=1}^p \tau_i^* \eta^T f_{ix}(t,x^*,u^*) + \sum_{i=1}^p \tau_i^* \xi^T f_{iu}(t,x^*,u^*)\right.$$
$$- \left(\eta^T G_x(t,x^*,u^*)\lambda^*(t) + \xi^T G_u(t,x^*,u^*)\lambda^*(t)\right)$$
$$\left. - \left(\eta^T R_x(t,x^*,u^*)\mu^*(t) + \xi^T R_u(t,x^*,u^*)\right) - \eta^T \frac{d}{dt}\lambda^*(t)\right\} dt = 0.$$

But,

$$\int_a^b \eta^T \frac{d}{dt}\lambda^*(t)dt = \eta^T\lambda^*(t)|_a^b - \int_a^b \frac{d\eta^T}{dt}\lambda^*(t)dt = -\int_a^b \frac{d\eta^T}{dt}\lambda^*(t)dt,$$

as fixed boundary conditions give $\eta = 0$ at a and b. From above we reach to a contradiction to (7.50). Hence (x^*, u^*) is optimal for (VCP).

For more results on optimality see, Gramatovici [68].

□

Lee et al. [137] defined the following Mond–Weir [180] type dual problem $(MWVD)$ for the primal problem (VCP):

MWVD

$$\underset{(x,u,\lambda,\mu)}{Maximize} \left(\int_a^b f_1(t,u,\dot{u})dt, \ldots, \int_a^b f_p(t,u,\dot{u})dt \right)$$

$$\text{Subject to } x(a) = \alpha, \ \ x(b) = \beta,$$

$$\sum_{i=1}^{p} \tau_i f_{ix}(t,x,u) - G_x(t,x,u)\lambda(t)R_x(t,x,u) = \dot{\lambda}(t),$$

$$\sum_{i=1}^{p} \tau_i f_{iu}(t,x,u) - G_u(t,x,u)\lambda(t)R_u(t,x,u) = 0,$$

$$\int_a^b \left\{ \lambda(t)^T[G(t,x,u) - \dot{x}] + \mu(t)^T R(t,x,u) \right\} dt \leq 0,$$

$$\mu(t) \geq 0, \ \ t \in I, \ \ \tau \in \Lambda^+ = \left\{ \tau \in R^p : \tau_i > 0 \text{ for all } i \text{ and } \sum_{i=1}^{p} \tau_i = 1 \right\},$$

where $\lambda : I \to R^n$ and $\mu : I \to R^k$. Lee et al. [137] established some duality theorems between the multiobjective control problem (VCP) and its Mond–Weir type dual problem $(MWVD)$.

Theorem 7.85 (Weak duality). *If $\int_a^b f_i dt$, $i = 1, \ldots, p$, are invex with respect to η and ξ, and $\int_a^b(-\lambda^T(G - \dot{x}) - \mu^T R)dt$ for any $\lambda(t) \in R^n$ and $\mu(t) \in R^m$ is invex with respect to the same functions η and ξ, then the following can not hold; for any feasible solution (x^*, u^*) of (VCP) and any feasible solution (x, u, λ, μ) of $(MWVD)$,*

$$\int_a^b f_i(t,x^*,u^*)dt \leq \int_a^b f_i(t,\bar{x},\bar{u})dt, \quad \forall i$$

$$\textit{and} \int_a^b f_j(t,x^*,u^*)dt < \int_a^b f_j(t,\bar{x},\bar{u})dt \quad \textit{for some } j.$$

Proof.

□

Theorem 7.86 (Strict converse duality). *Let (x^*, u^*) and $(x, u, \tau, \lambda, \mu)$ be feasible solutions of (VCP) and (MWVD), respectively such that*

$$\int_a^b \sum_{i=1}^{p} \tau_i f_i(t, x^*, u^*)dt < \int_a^b \sum_{i=1}^{p} \tau_i f_i(t, x, u)dt.()$$

If $\int_a^b f_i dt,\ i = 1, \ldots, p$ are strictly invex with respect to η and ξ, and $\int_a^b (-\lambda^T(G - \dot{x}) - \mu^T R)dt$ is invex with respect to the same functions η and ξ, then $(x^*, u^*) = (x, u)$.

Bhatia and Kumar [21] has given duality theorems for a similar problem to (MCP) under ρ-invexity, ρ-pseudo-invexity and ρ-quasi-invexity assumptions.

Zhian and Qingka [258] considered the same multiobjective control problem as given in this section above and defined the following two types of dual models:

(VCD1)

$$\text{Maximize } \left(\int_{t_0}^{t_f} \{f_1(t, x, u) + \mu(t)_\Sigma^T R_\Sigma(t, x, u)\}\, dt, \ldots, \right.$$
$$\left. \int_{t_0}^{t_f} \{f_p(t, x, u) + \mu(t)_\Sigma^T R_\Sigma(t, x, u)\}\, dt \right)$$

$$\text{Subject to } x(t_0) = x_0,\ \ x(t_f) = x_f,$$

$$\sum_{i=1}^{p} \tau_i f_{ix}(t, x, u) - G_x(t, x, u)\lambda(t)R_x(t, x, u) = \dot{\lambda}(t),$$

$$\sum_{i=1}^{p} \tau_i f_{iu}(t, x, u) - G_u(t, x, u)\lambda(t)R_u(t, x, u) = 0,$$

$$\int_{t_0}^{t_f} \lambda(t)^T G(t, x, u) - \dot{x}dt \geq 0$$

$$\int_{t_0}^{t_f} \mu(t)_{\Sigma'}^T R_{\Sigma'}(t, x, u)dt \geq 0,$$

$$\mu(t) \geq 0,\ \ t \in I,\ \ \tau \in \Lambda^+ = \left\{ \tau \in R^p : \tau_i > 0 \text{ for all } i \text{ and } \sum_{i=1}^{p} \tau_i = 1 \right\},$$

where $\lambda : I \to R^n$ and $\mu : I \to R^k$.

(VCD2)

$$\text{Maximize } \left(\int_{t_0}^{t_f} \{f_1(t, x, u) + \lambda(t)^T(G(t, x, u) - \dot{x}) + \mu(t)_\Sigma^T R_\Sigma(t, x, u)\}\, dt, \ldots, \right.$$

$$\left.\int_{t_0}^{t_f} \{f_p(t,x,u) + \lambda(t)^T(G(t,x,u) - \dot{x}) + \mu(t)_{\Sigma}^T R_{\Sigma}(t,x,u)\}\, dt\right)$$

$$\text{Subject to } x(t_0) = x_0, \quad x(t_f) = x_f,$$

$$\sum_{i=1}^{p} \tau_i f_{ix}(t,x,u) - G_x(t,x,u)\lambda(t) R_x(t,x,u) = \dot{\lambda}(t),$$

$$\sum_{i=1}^{p} \tau_i f_{iu}(t,x,u) - G_u(t,x,u)\lambda(t) R_u(t,x,u) = 0,$$

$$\int_{t_0}^{t_f} \mu(t)_{\Sigma'}^T R_{\Sigma'}(t,x,u) dt \geq 0,$$

$$\mu(t) \geq 0, \quad t \in I, \quad \tau \in \Lambda^+ = \left\{\tau \in R^p : \tau_i > 0 \text{ for all } i \text{ and } \sum_{i=1}^{p} \tau_i = 1\right\},$$

where $\lambda : I \to R^n$ and $\mu : I \to R^k$.

8

Invexity for Some Special Functions and Problems

8.1 Invexity of Quadratic Functions

There are few papers dealing with invexity of quadratic forms and functions; we know only the contributions of Smart [224], Mond and Smart [177] and Molho and Schaible [166]. The study of invex quadratic functions can improve optimality and duality results for that important class of problems formed by quadratic programming problems.

Consider the following quadratic program:
(QP)

$$\mathrm{Min}_x \left\{ \frac{1}{2} x^T M x + q^T x \right\}$$

$$\text{Subject to } Ax \leq b,$$

where M is a symmetric $n \times n$ matrix, A is an $m \times n$ matrix, $x \in R^n$, $q \in R^n$, $b \in R^m$.

The Wolfe dual of (QP) is the program:
(QD)

$$\mathrm{Max}_{x,u} \left\{ b^T u - \frac{1}{2} x^T M x \right\}$$

$$\text{Subject to } A^T u + Mx = -q$$

$$u \geq 0, \quad x \in R^n.$$

If M is nonsingular, then (QD) reduces to:
(QD)

$$\mathrm{Max}_u \left\{ d^T u + \frac{1}{2} u^T E u - \frac{1}{2} u^T A M^{-1} q \right\}$$

$$\text{Subject to } u \geq 0,$$

where $E = -AM^{-1}A^T$, $d = -b - AM^{-1}q = -b - A\bar{x}$ with $\bar{x} = M^{-1}q$.

From Sect. 5.4 we know that under assumption of invexity (with respect to a common η) of the objective function and the constraints of the primal problem, then weak duality holds:

$$\inf(QP) \geq \sup(QD).$$

The invexity assumption on the functions involved in (QD) allows also to establish a strong duality results, as the constraints of (QD) are linear and no constraint qualification is needed:

$$\inf(QP) = \sup(QD).$$

Here we follow the approach of Molho and Schaible [166]. Generalized convexity of quadratic functions on convex subsets of R^n has been studied extensively since the early works of Cottle, Ferland and Schaible (see, e.g., [10, 41, 218–220]). As said before, invexity of quadratic functions has not been studied so extensively.

Consider $f(x) = \frac{1}{2}x^T Mx + q^T x$, where M is a symmetric $n \times n$ matrix. Let C be an open set in R^n. We want to study the invexity of f on C. The following example shows that invexity of quadratic functions dos not trivially reduce to convexity. Consider the quadratic function

$$f(x) = \frac{1}{2}x^T Mx + q^T x,$$

where $M = \begin{bmatrix} -1 & 0 \\ 0 & 0 \end{bmatrix}$; q=$\begin{bmatrix} 1 \\ -1 \end{bmatrix}$;

$f(x)$ is not convex since M is not positive semidefinite. However it is invex, since it has no stationary points in R^n.

Now we prove that the absence of stationary points is a general feature of invex quadratic functions which are nonconvex, i.e., of "properly invex" quadratic functions. Let $S = \{x \in R^n : Mx + q = 0\}$, i.e., S is the set of stationary points of f in R^n.

Theorem 8.1. *Let $C \subseteq R^n$ be an open set and $f : C \to R$ be an invex quadratic function on C. Then either $C \cap S = \phi$ or f is convex on R^n.*

Proof. Suppose $C \cap S = \phi$. Let $x \in C \cap S = \phi$. Suppose f is not convex on R^n. Then M has at least one negative eigenvalue, say $\lambda_1 \leq 0$. Let x^1 be an associated normalized eigenvector. Since f is invex on C, $\bar{x}$ is a global minimizer of f on C. Now consider $x^i = \bar{x} + tx^1,\ t \neq 0$.

Since C is an open set, there exists $\bar{t} > 0$ such that for $|t| < \bar{t}$, we have $x^t \in C$. Moreover, it holds

$$\begin{aligned} f(x^T) &= \frac{1}{2}(x^t)^T Mx^t + q^T x^t \\ &= \frac{1}{2}(\bar{x} + tx^1)^T M(\bar{x} + tx^1) + q^T(\bar{x} + tx^1) \end{aligned}$$

$$= f(\bar{x}) + t(x^1)^T M\bar{x} + \frac{1}{2}\lambda_1 t^2 + tq^T x^1$$
$$= f(\bar{x}) + t((x^1)^T M\bar{x} + q) + \frac{1}{2}\lambda_1 t^2$$
$$= f(\bar{x}) + \frac{1}{2}\lambda_1 t^2, \quad \text{since } \bar{x} \in S.$$

As $\lambda_1 < 0$ we have $f(x^t) < f(\bar{x})$ for all $t \neq 0$ such that $|t| < \bar{t}$, contradicting the fact that $\bar{x}$ is a minimum point of f on C.

□

Consequently, if a given quadratic function is considered on an open set C which does not contain any stationary point, then the function is invex on C (this is true for any differentiable function and therefore also for any quadratic function, no mater how many negative eigenvalues M has). But if a quadratic function is considered on a set C which does contain at least one stationary point, then it can not be properly invex; in other words f is not invex whenever M has at least one negative eigenvalue. This is very different from the classical generalized convexity properties of quadratic functions. Let us recall what happens for pseudo convex quadratic functions, a proper subset of invex quadratic functions. If $C \cap S = \phi$, pseudo convexity can only occur if M has at most one simple negative eigenvalue and some other conditions are met (see Theorem 6.2 in [10]). On the other hand, if $C \cap S \neq \phi$, proper pseudo convexity is impossible on C, just like proper invexity is impossible in this case. Thus, the difference between pseudo convex and invex quadratic functions is very large, in the first case and does not exist in the second one.

Let us point out that $C \cap S = \phi$ for all sets $C \subseteq R^n$ if rank $(M, q) \neq$ rank (M). Therefore we have the following.

Corollary 8.2. *Any quadratic function* $f(x) = \frac{1}{2}x^T Mx + q^T x$ *with rank*(M, q) $\neq$ *rank*(M) *is invex on* R^n.

Since a proper invex quadratic function does not have a stationary point on an open set, the minimum (if it exists) can only be obtained at the boundary. The above characterization of invex quadratic functions may also be viewed as a special case of Theorem 2.3.1 due to Smart [224] and Mond and Smart [179]. Let us apply this result to the case of quadratic functions, i.e., $f(x) = \frac{1}{2}x^T Mx + q^T x$ (M symmetric matrix of order n, $q \in R^n$). The necessary and sufficient condition for invexity with respect to η, where $\eta(x,u) = \mu(x,u) + \Lambda(x,u)$, is

$$\forall x,\ u \in C\ Mu + q \in \Lambda^*(x,u)$$
$$\Rightarrow \frac{1}{2}x^T Mx + q^T x - \frac{1}{2}u^T Mu - q^T u - \mu(x,u)^T(Mu+q) \geq 0.$$

That is, for all $x,\ u \in C\ Mu + q \in \Lambda^*(x,u)$ implies

$$\frac{1}{2}x^T Mx - \mu(x,u)^T Mu - \frac{1}{2}u^T Mu + q^T(x - u - \mu(x,u)) \geq 0.$$

If $\mu(x,u) = x - u$, the condition becomes: for all $x, u \in C$ $Mu + q \in \Lambda^*(x,u)$, implies

$$(x-u)^T M(x-u) \geq 0.$$

Consider the case where Λ is independent of (x,u), i.e., $\Lambda(x,u) = \Lambda$.

For any given $u \in C$ for every $\nu \in R^n$ there exists $x \in C$ such that $x - u = \alpha\nu$ for some scalar $\alpha \neq 0$ since C is open. Therefore if there exists $u \in C$ such that $Mu + q \in \Lambda^*$ then M is positive semidefinite.

We analyze now the following special cases for Λ:

(a) Arbitrary invexity. If $\Lambda = R^n$, hence $\Lambda^* = \{0\}$, M is positive semidefinite, i.e., f is convex, whenever f has at least one stationary point in C and f is invex if it does not have any stationary point in C. This result was obtained in Theorem 8.1 using a different approach.
(b) Convexity. If $\Lambda^* = \{0\}$ hence $\Lambda = R^n$, (a) means that M is positive semidefinite, i.e., we obtain the usual necessary and sufficient conditions for the convexity of f on an open set.

We recall now the following results on twice continuously differentiable generalized convex functions which will be useful for comparison between invexity and pseudoconvexity.

Theorem 8.3. *Let f be a twice differentiable quasiconvex function on the open convex set $C \subseteq R^n$. Then $x^0 \in C, \nu \in R^n$ and $\nu^T \nabla f(x^0) = 0$ imply $\nu^T \nabla^2 f(x^0)\nu \geq 0$ (see [6, 10, 50]).*

Theorem 8.4. *Let f be a twice differentiable function on the open convex set $C \subseteq R^n$. Suppose that $x^0 \in C, \nu \in R^n$ and $\nu^T \nabla f(x^0) = 0$ imply $\nu^T \nabla^2 f(x^0)\nu \geq 0$ and $x^0 \in C$ imply $\nabla f(x^0) = 0$ is positive definite. Then f is pseudoconvex on C. (see [48, 50]).*

An immediate consequence of these results is the following corollary which sheds light on the relationships between invexity and pseudoconvexity.

Corollary 8.5. *Let f be an invex twice continuously differentiable function on the open convex set C such that $\nabla^2 f(x^0)$ is non-singular for all $x^0 \in C$ such that $\nabla f(x^0) = 0$. Then f is pseudoconvex if and only if $x^0 \in C, \nu \in R^n, \nu^T \nabla f(x^0) = 0$ imply $\nu^T \nabla^2 f(x^0)\nu \geq 0$.*

We now discuss some properties of locally invex C^2-functions. Let f be a C^2-functions in a neighbourhood of x^0. Then

$$\begin{aligned} f(x) = f(x^0) &+ (x - x^0)^T \nabla f(x^0) \\ &+ \frac{1}{2}(x - x^0)^T \nabla^2 f(x^0)(x - x^0) \\ &+ o(\|x - x^0\|)^2. \end{aligned}$$

Let $q(x)$ be the quadratic function approximating f in a neighbourhood of x^0, i.e.,

$$f(x) = f(x^0) + (x - x^0)^T \nabla f(x^0) + \frac{1}{2}(x - x^0)^T \nabla^2 f(x^0)(x - x^0).$$

Obviously

$$q(x) = f(x^0), \quad \nabla q(x) = \nabla f(x^0), \quad \nabla^2 q(x) = \nabla^2 f(x^0).$$

As a trivial consequence we obtain the following result.

Corollary 8.6. *Let f be a C^2-function in a neighbourhood of X^0. If f is invex in a neighbourhood of X^0, then q is invex in some neighbourhood of X^0.*

Consider a C^2-function f. Recalling the results established for invex quadratic functions, we see that if $\nabla f(x^0) \neq 0$ then there exists a neighbourhood of x^0 where f is invex. On the other hand, if $\nabla f(x^0) = 0$ then the approximating quadratic function is convex, while f itself may be locally nonconvex (see, e.g., p. 269 in [10]).

However, restricting ourselves to C^2-functions where $\nabla^2 f(x^0)$ is non-singular if $\nabla f(x^0) = 0$ a locally invex function is locally convex. In such a case, the notion of local invexity is scarcely significant: either it imposes no restriction on the behaviour of f or it coincides with local convexity. Hence the class of invex C^2-functions (with $\nabla^2 f(x^0)$ non-singular whenever $\nabla f(x^0) = 0$) is made up of functions that are convex in a neighbourhood of their stationary points.

8.2 Invexity in Fractional Functions and Fractional Programming Problems

The literature on fractional programming is abundant; see, e.g., the book of Stancu-Minasian [227] and Craven [42]. However, the papers on fractional programming under invexity assumptions are quite few. We may quote the seminal work of Craven [43], where the term "invex function" was proposed and the papers of Singh and Hanson [223], Khan and Hanson [117], Reddy and Mukherjee [208] and Craven and Mond [47]. Also Jeyakumar and Mond [105] have applied their definition of invex functions to the fractional programming problem.

The paper of Khan and Hanson [117] contains some errors and useless assumptions. The same holds for the paper of Reddy and Mukherjee [208]; so here we follow also the treatment of Craven and Mond [47]. Consider the nonlinear fractional programming problem

(FP)

$$\max \frac{f(x)}{g(x)}$$

$$\text{Subject to } h(x) \leq 0,$$

where f, g and h are all defined and differentiable on an open set $C \subseteq R^n$, $f, g : C \to R$, $h : C \to R^m$. We extend now this nonlinear fractional programming problem to the invex case; we establish sufficiency and duality theorems for an invex fractional programming problem.

We recall the notion of V-invex functions already given for vector-valued functions (see the comments after Theorem 5.14).

Definition 8.7. *A differentiable function $\varphi : R^n \to R$ is V-invex at the point y if there exist a positive weighting function $\beta(.,y)$ and a kernel function $\eta(.,.)$ such that for all x*

$$\varphi(x) - \varphi(y) \geq \beta(x,y)\nabla\varphi(y)\eta(x,y).$$

Obviously if $\beta(.,.) = 1$, we get the usual definition of invexity, which is always included in the above definition, by choosing, as kernel function $\widetilde{\eta} = \frac{\eta}{\beta}$. V-invex for a vector function means that each component is V-invex; the kernel function must be the same, but the weighting functions may differ.

Theorem 8.8. *If in (FP) we have $f(x) \geq 0$, $g(x) > 0$ with $f(x)$ and $-g(x)$ invex with the same kernel function η, then $\frac{f(x)}{g(x)}$ is invex at y, with kernel function $\eta(.,.)$ and weighting function $\frac{g(y)}{g(x)}$.*

Proof. Assume that $f(x) \geq 0$. Since also $g(x) > 0$ we have $\frac{f(y)}{g(y)} > 0$. Since $f(x)$ and $-g(x)$ are invex with the same kernel function η and $\frac{f(y)}{g(y)} \geq 0$ we have that $f(.) - \frac{f(y)}{g(y)}g(.)$ is also invex at y, with the same η. Thus

$$\begin{aligned} f(x) - \frac{f(y)}{g(y)}g(x) &= \left[f(x) - \frac{f(y)}{g(y)}g(x)\right] - \left[f(y) - \frac{f(y)}{g(y)}g(y)\right] \\ &\geq \left[\nabla f(y) - \frac{f(y)}{g(y)}\nabla g(y)\right]\eta(x,y). \end{aligned}$$

Hence

$$\frac{f(x)}{g(x)} - \frac{f(y)}{g(y)} \geq \left[\frac{g(y)}{g(x)}\right]\nabla\left[\frac{f(y)}{g(y)}\right]\eta(x,y).$$

□

Khan and Hanson [117] gave a similar result, but assuming, incorrectly, that $f(x) \leq 0$. If $x^0 \in C$ is a local minimum for problem (FP) and a constraint qualification is satisfied, then the following Kuhn–Tucker conditions hold for (FP):

$$\nabla\frac{f(x^0)}{g(x^0)} + {\lambda^0}^T\nabla h(x^0) = 0, \tag{8.1}$$

$${\lambda^0}^T h(x^0) = 0, \tag{8.2}$$

$$\lambda^0 \geq 0. \tag{8.3}$$

The following theorem establishes sufficient optimality conditions for (FP) under appropriate invexity and V-invexity assumptions.

Theorem 8.9. *Suppose x^0 is feasible for (FP) and that the Kuhn–Tucker conditions (8.1)–(8.3) are satisfied at x^0. Let $f(x) \geq 0$, $g(x) > 0$ where f and $-g$ are invex functions with respect to $\eta(x, x^0)$, and let h be V-invex at x^0 with respect to $\left(\frac{g(x^0)}{g(x)}\right) \eta(x, x^0)$. Then x^0 is a minimum for (FP).*

Proof. Let x be any feasible point for (FP). Then we have

$$\begin{aligned}
\frac{f(x)}{g(x)} - \frac{f(x^0)}{g(x^0)} &\geq \left[\frac{g(x^0)}{g(x)}\right] \nabla \left[\frac{f(x^0)}{g(x^0)}\right] \eta(x, x^0) \\
&= -\left[\frac{g(x^0)}{g(x)}\right] \lambda^{0^T} \nabla h(x^0) \eta(x, x^0) \\
&\geq -\left[\frac{g(x^0)}{g(x)}\right] \lambda^{0^T} \nabla h(x^0) \eta(x, x^0) \\
&\quad + \lambda^{0^T} h(x) - \lambda^{0^T} h(x^0) \\
&\geq 0.
\end{aligned}$$

Therefore, x^0 is a global minimum.

□

It should be noted that Jeyakumar and Mond [105] pointed out that the convex-concave fractional programming problem is not an invex problem. However, on the grounds of the previous results it turns out that (FP) is an invex programming problem under suitable assumptions of invexity of the objective function and of V-invexity of the constraints.

We turn now to duality for (FP). If f, $-g$ and h are convex, then $\frac{f(x)}{g(x)}$ is pseudoconvex (see, e.g., [143]) but not necessarily convex and the usual Wolfe duality results do not hold (see again [143]). This explains the various different duals for (FP) considered in the literature. We note however, that Hanson [83] has obtained duality results, with a Wolfe formulation for the dual, under invexity assumptions (see Sect. 5.4).

Khan and Hanson [117] established duality between (FP) and a slight variation of its Mond–Weir dual (see Sect. 5.4), namely the problem:

(FD)

$$\max \frac{f(u)}{g(u)}$$

$$\text{Subject to } \nabla \frac{f(u)}{g(u)} + \lambda^T \nabla h(u) = 0, \tag{8.4}$$

$$\lambda^T h(u) = 0, \tag{8.5}$$

$$\lambda \geq 0. \tag{8.6}$$

In the following two Theorems it is assumed that the functions $f(x) \geq 0$ and $f(x) \geq 0$ and that the functions $f(.)$ and $-g(.)$ are invex with respect to $\eta(x^0, u)$, hence $\frac{f(.)}{g(.)}$ (according to Theorem 8.8) is invex with respect to $\frac{g(u)}{g(x^0)}\eta(x^0, u)$. It is assumed, moreover that $\lambda^T h(.)$; $\lambda \geq 0$ is invex with respect to $\frac{g(u)}{g(x^0)}\eta(x^0, u)$.

Theorem 8.10 (Strong duality). *Assume that x^0 is minimal for (FP), then x^0 is maximal for (FD) and the optimal values of (FP) and (FD) are equal.*

Proof. Suppose that any vector u, λ satisfies the constraints of (FD). For x^0 to be maximal for (FD), we must show that

$$\frac{f(x^0)}{g(x^0)} - \frac{f(u)}{g(u)} \geq 0.$$

From the feasibility, we obtain

$$\begin{aligned}
\frac{f(x^0)}{g(x^0)} - \frac{f(u)}{g(u)} &\geq \frac{f(x^0)}{g(x^0)} - \frac{f(u)}{g(u)} - \lambda^T h(u) \\
&\geq \frac{f(x^0)}{g(x^0)}\eta(x^0, u)\nabla\frac{f(u)}{g(u)} - \lambda^T h(u) \\
&\quad \left(\text{since } \frac{f(x^0)}{g(x^0)} \text{ is invex with respect to } \frac{g(u)}{g(x^0)}\eta(x^0, u)^T\right) \\
&= -\frac{g(u)}{g(x^0)}\eta(x^0, u)^T \nabla\lambda^T h(u) - \lambda^T h(u) \quad \text{by (8.4)} \\
&\geq \lambda^T h(x^0), \\
&\quad \left(\text{since } \lambda^T h(.) \text{ is invex with respect to } \frac{g(u)}{g(x^0)}\eta(x^0, u)^T\right) \\
&\geq 0, \quad \text{being } h(x^0) \leq 0 \text{ and by (8.6).}
\end{aligned}$$

□

Theorem 8.11 (Converse duality). *If x^0 is maximal for (FD) under the assumptions of invexity already specified, then x^0 is minimal for (FP).*

Proof. Since x^0 is maximal for the dual problem (FD), there exists $\lambda^0 \geq 0$ such that the constraints are satisfied at (x^0, λ^0), i.e.,

$$\nabla\frac{f(x^0)}{g(x^0)} + \lambda^T h(x^0) = 0 \tag{8.7}$$

$$\lambda^{0^T} h(x^0) = 0 \tag{8.8}$$

$$\lambda^0 \geq 0. \tag{8.9}$$

For any $x \in C$ satisfying the constraints of (FP) we have

$$\frac{f(x)}{g(x)} - \frac{f(x^0)}{g(x^0)} \geq \frac{g(x^0)}{g(x)}\eta(x,x^0)\nabla\frac{f(x^0)}{g(x^0)}$$

$$\text{(by invexity of } \frac{f(.)}{g(.)} \text{ with respect to } \frac{g(x^0)}{g(x)}\eta(x,x^0))$$

$$-\frac{g(x^0)}{g(x)}\eta(x,x^0)^T\nabla\lambda^{0^T}h(x^0), \quad \text{by (8.7)}$$

$$\geq -\lambda^{0^T}h(x) - \lambda^0 h(x^0)$$

$$\text{(since } \lambda^{0^T}h(.) \text{ is invex with respect to } \frac{g(x^0)}{g(x)}\eta(x,x^0))$$

$$\geq 0,$$

being $h(x) \leq 0$ for any feasible $x \in C$, for problem (FP), and by (8.8) and (8.9).

□

8.3 Invexity in a Class of Nondifferentiable Problems

Mond [168] considered the following class of nondifferentiable mathematical programming problems:

(NDP)

$$\text{Min}F(x) = f(x) + (x^T Bx)^{\frac{1}{2}}$$

$$\text{Subject to } \; g(x) \geq 0,$$

where $f : R^n \to R$ and $g : R^n \to R^m$ are differentiable functions and B is an $n \times n$ symmetric positive semidefinite matrix. Assuming that f was convex and g was concave, a Wolfe-type dual problems was formulated and duality results proved. For other related studies and applications related to (NDP) see the bibliographical references of Mond [168].

Subsequently, Chandra et al. [30] weakened the convexity requirements for duality by giving a Mond–Weir type dual and assuming $f(x) + x^T Bw$ is pseudoconvex for all $w \in R^n$ and that $\lambda^T g(x)$ was quasiconvex for feasible multiplier vectors $\lambda \in R^m$.

A continuous version of (NDP) was studied by Chandra et al. [30] and a further weakening of the assumptions, both for (NDP) and for its continuous version, was considered by Mond and Smart [177], through the use of invexity and generalized invexity. In this section, we consider only the "static" version of the problem, i.e., (NDP) itself.

Two duals for (NDP) are to be formulated, making use of the necessary conditions for an optimal solution, due to Mond [168]. First, a regularity condition at feasible points is required. For each feasible $x^0 \in R^m$ define $I_0 = \{i : g_i(x^0) = 0\}$ and the following set

$$Z_0 = \{z : z^T \nabla g_i(x^0) \geq 0 \; \forall i \in I_0, \\ \frac{z^T \nabla f(x^0) + z^T B x^0}{(x^0 B x^0)^{\frac{1}{2}}} < 0, \text{ if } x^{0^T} B x^0 > 0, \\ z^T \nabla f(x^0) + (z^T B z)^{\frac{1}{2}} < 0 \text{ if } x^{0^T} B x^0 = 0\}.$$

Lemma 8.12. *If x^0 is an optimal solution for (NDP) and the corresponding set Z_0 is empty, then there exist $\lambda \in R^m$ and $w \in R^n$ such that*

$$\lambda^T g(x^0) = 0$$
$$\lambda^T \nabla g(x^0) = \nabla f(x^0) + Bw$$
$$w^T B w \leq 1$$
$$(x^{0^T} B x^0)^{\frac{1}{2}} = x^{0^T} B w$$
$$\lambda \geq 0.$$

The generalized Schwarz inequality will later be required:

$$x^{0^T} B w \leq (x^{0^T} B x^0)^{\frac{1}{2}} (w^T B w)^{\frac{1}{2}}, \quad \forall \, x, w \in R^n.$$

Mond [168] considered the following Wolfe-type dual problem:

$$\max G(u, \lambda) \equiv f(u) - \lambda^T g(u) + u^T [\lambda^T g(u) - \nabla f(u)]$$
$$\text{Subject to } \lambda^T \nabla g(u) = \nabla f(u) + Bw$$
$$w^T B w \leq 1$$
$$\lambda \geq 0,$$

where $u \in R^n$, $\lambda \in R^m$ and $w \in R^n$. Then it was assumed that f was convex and g was concave in order to prove weak, strong and converse duality.

This problem may be written equivalently as:
(NDD1)

$$\max \; G(u, \lambda, w) \equiv f(u) - \lambda^T g(u) + u^T B w$$
$$\text{Subject to } \lambda^T \nabla g(u) = \nabla f(u) + Bw, \tag{8.10}$$
$$w^T B w \leq 1, \tag{8.11}$$
$$\lambda \geq 0. \tag{8.12}$$

Theorem 8.13 (Weak duality). *Let $f(.) + .^T Bw$ and $-g$ be invex with respect to the same function η for all $w \in R^n$. Then* $\inf(NDP) \geq \sup(NDD1)$.

Proof. Let x be feasible for (NDP) and let (u, λ, w) be feasible for (NDD1). Then

$$
\begin{aligned}
F(x) - G(u, \lambda, w) &= f(x) + (x^T Bx)^{\frac{1}{2}} - f(u) - u^T Bw \\
&\geq \eta(x, u)^T [\nabla f(u) + Bw] - x^T Bw + (x^T Bx)^{\frac{1}{2}} \\
&\quad + \lambda^T g(u) \quad \text{(by invexity } f(.) + (.)^T Bw) \\
&= \eta(x, u)^T \lambda^T \nabla g(u) - x^T Bw + (x^T Bx)^{\frac{1}{2}} + \lambda^T g(u) \quad \text{by (8.10)} \\
&\geq \lambda^T g(x) - x^T Bw + (x^T Bx)^{\frac{1}{2}} \quad \text{by invexity of } g \text{ and (8.12)} \\
&\geq \lambda^T g(x) - x^T Bw + (x^T Bx)^{\frac{1}{2}} + (w^T Bw)^{\frac{1}{2}} \quad \text{by (8.11)} \\
&\geq 0.
\end{aligned}
$$

by feasibility, (8.11) and the generalized Schwarz inequality, as x and u, λ, w are arbitrary feasible solutions of (NDP) and (NDD1) respectively, then $\inf(NDP) \geq \sup(NDD1)$.

□

Strong duality also holds with the above invexity assumptions.

Theorem 8.14 (Strong duality). *Assume the invexity conditions of Theorem 8.13 are satisfied. If x^0 is an optimal solution for (NDP) and the corresponding set Z_0 is empty then there exist $\lambda^0 \in R^m$ and $w^0 \in R^n$, such that x^0, λ^0, w^0 is optimal for (NDD1), and the respective objective values are equal.*

Proof. Mond [168] shows the existence of λ^0 and w^0 such that x^0, λ^0, w^0 is feasible for (NDD1), and the objective values are equal. By Theorem 8.13, x^0, λ^0, w^0 is optimal for (NDD1).

□

Mond and Smart [177] require invexity of $f(.) + (.)^T Bw$ and not just f, since, in general, the sum of two invex functions is invex with respect to η only if they are both invex with respect to the same η. The term $x^T Bw$ is convex for each $w \in R^n$ and hence invex with $\eta(x, u) = x - u$. Mond and Smart [177] give a simple sufficient condition for $f(.) + (.)^T Bw$ to be invex, given that f is invex.

Theorem 8.15. *If f is invex with respect to η, with $\eta(x, u) = x - u + y(x, u)$ where $By(x, u) = 0$ then $f(.) + (.)^T Bw$ is also invex with respect to η.*

Proof. Let $x, u, w \in R^n$. Then

$$
\begin{aligned}
f(x) + x^T Bw &- f(u) - u^T Bw \\
&\geq (x - u + y(x, u))^T \nabla f(u) + (x - u)^T Bw \\
&\quad \text{(since } f \text{ is invex with respect to } \eta) \\
&\geq (x - u + y(x, u))^T \nabla f(u) + (x - u)^T Bw + y(x, u)^T Bw \\
&\geq (x - u + y(x, u))^T (\nabla f(u) + Bw).
\end{aligned}
$$

Therefore, $f(.) + (.)^T Bw$ is invex with respect to η.

□

Also the following example is taken from Mond and Smart [177].

Example 8.16. Consider (NDP) with $f : R^2 \to R$, $g : R^2 \to R$ and $B \in R^{2\times 2}$ defined by

$$f(x) = x_1^3 + x_1 - 10x_2^3 - x_2$$

and

$$g(x) = 2x_1 + x_2,$$

$$\begin{pmatrix} 4\,2 \\ 2\,1 \end{pmatrix}.$$

Now, $y(x,u) = 0$ if $y(x,u) = (z, -2z)^T$ for some $z \in R$ Then, $f(x) \geq f(u) \geq (x - u + y(x,u))^T \nabla f(u)$ is equivalent to

$$\begin{aligned} &x_1^3 + x_1 - 10x_2^3 - x_2 - u_1^3 - u_1 + 10u_2^3 + u_2 \\ &\quad \geq (x_1 - u_1)(3u_1^2 + 1) + (x_2 - u_2)(-30u_2^2 - 1) \\ &\quad\quad + (3u_1^2 + 1 + 60u_2^2 + 2)z. \end{aligned}$$

Thus, z can be chosen as

$$z = \frac{x_1^3 - 10x_2^3 + 2u_1^3 - 20u_2^3 - 3x_1u_1^2 + 30x_2u_2^2}{3u_1^2 + 60u_2^2 + 3},$$

also, as g is linear and $y(x,u)^T \nabla g(u) = 0$ for all $u \in R^2$, then g is also invex with respect to this η.

Chandra et al. [30] dealt with the following Mond–Weir type dual of (NDP):

(NDD2)

$$\max H(u,w) \equiv f(u) + u^T Bw$$

$$\text{Subject to } \lambda^T \nabla g(u) = \nabla f(u) + Bw \tag{8.13}$$

$$w^T Bw \leq 1 \tag{8.14}$$

$$\lambda \geq 0, \tag{8.15}$$

$$\lambda^T g(u) \leq 0. \tag{8.16}$$

There, weak duality was proved assuming that $f(.) + (.)^T Bw$ was pseudoconvex for all $w \in R^n$ and $\lambda^T g$ quasiconcave for any feasible multiplier λ.

Theorem 8.17 (Weak duality). *Let $f(.) + (.)^T Bw$ be pseudoinvex for all $w \in R^n$, and $-\lambda^T g$ quasiinvex for all feasible multiplier λ, both with respect to the same kernel η. Then $\inf(NDP) \geq \sup(NDD2)$.*

Proof. Let x be feasible for (NDP) and u, λ, w feasible for (NDD2). Then the feasibility of x, (8.15) and (8.16) imply that $\lambda^T g(x) - \lambda^T g(u) \geq 0$. Thus, $\eta(x,u)^T \nabla g(x) \geq 0$, since $-\lambda^T g$ is quasiinvex.

Equation (8.13) then gives $\eta(x,u)^T(\nabla f(x) + Bw) \geq 0$. Therefore,

$$f(x) + x^T Bw \geq f(u) + u^T Bw,$$

since $f(.) + (.)^T Bw$ is pseudoinvex with respect to η. But

$$\begin{aligned} f(x) + x^T Bw &\leq f(x) + (x^T Bx)^{\frac{1}{2}} (w^T Bw)^{\frac{1}{2}} \\ &\qquad \text{(by the Schwarz inequality)} \\ &\leq f(x) + (x^T Bx)^{\frac{1}{2}} \quad \text{(by (8.14))}. \end{aligned}$$

Therefore, $F(x) \geq H(u,w)$ for all x feasible for (NDP) and (u,λ,w) feasible for (NDD2), so $\inf(NDP) \geq \sup(NDD2)$.

□

Theorem 8.18 (Strong duality). *Assume the invexity conditions of Theorem 8.17 are satisfied. If x^0 is an optimal solution of (NDP), and the corresponding set Z_0 is empty, then there exist $\lambda^0 \in R^m$ and $w^0 \in R^n$ such that (x^0, λ^0, w^0) is optimal for (NDD2), and the respective objective values are equal.*

Proof. Chandra et al. [30] established the existence of λ^0 and w^0 such that (x^0, λ^0, w^0) is feasible for (NDD2), with the objective values equal. The optimality of (x^0, λ^0, w^0), then follows from Theorem 8.17.

□

Theorem 8.19 (Converse duality). *Let (x^*, λ^*, w^*) be optimal for either (NDD1) or (NDD2). Assume that $\nabla^2 f(u^*) - \nabla^2({\lambda^*}^T g(u^*))$ is non-singular. If either:*

1. *(x^*, λ^*, w^*) is optimal for (NDD1), and the invexity conditions of Theorem 8.13 are satisfied,*
2. *(x^*, λ^*, w^*) is optimal for (NDD2), the invexity conditions of Theorem 8.17 are satisfied, and ${\lambda^*}^T \nabla g(u^*) \neq 0$.*

Then, u^* is optimal for (NDP).

Proof. Mond [168] shows that if (x^*, λ^*, w^*) is optimal for (NDD1), then u^* is feasible for (NDP), and the objective values are equal. By Theorem 8.13, u^* is optimal for (NDP).

□

Chandra et al. [30] showed that if (x^*, λ^*, w^*) is optimal for (NDD2) and ${\lambda^*}^T \nabla g(u^*) \neq 0$ then u^* is feasible for (NDP) with the corresponding objective values equal. Optimality of u^*, follows from Theorem 8.17.

Mond et al. [173] extended the work of Mond [168] to multiobjective case:

(NDVP)

$$\text{Min}\left(f_1(x)+(x^TB_1x)^{\frac{1}{2}},\ldots,f_p(x)+(x^TB_px)^{\frac{1}{2}}\right)$$

$$\text{Subject to } \; g(x)\geq 0.$$

Mond et al. [173] formulated a Mond–Weir type dual to (NDVP) and established weak, strong and converse duality theorems under convexity hypothesis. These authors did not study optimality conditions for the (NDVP). We now state Kuhn–Tucker type necessary conditions.

Lemma 8.20 (Kuhn–Tucker type necessary condition). *Let x^* be an efficient solution of (NDVP). Then there exist $\tau\in R^n$, $\lambda\in R^m$ such that*

$$\sum_{i=1}^{p}\tau_i\left[f_i'(x^*)+B_iz_i\right]+\sum_{j\in I(x^*)}\lambda_jg_j'(x^*)=0, \tag{8.17}$$

$$\lambda_jg_j(x^*)=0, \tag{8.18}$$

$$z^TB_iz_i\leq 1,\quad i=1,\ldots,p, \tag{8.19}$$

$$({x^*}^TB_ix^*)^{\frac{1}{2}}={x^*}^TB_iz_i, \tag{8.20}$$

$$\tau>0,\;\; \lambda\geq 0,\;\; \sum_{i=1}^{p}\tau_i=1, \tag{8.21}$$

where $I(x^*)=\{j:g_j(x^*)=0\}\neq\phi$.

Mishra [153] studied the problem (NDVP) and established sufficient optimality conditions and duality results under a wider class of functions namely, V-invex functions (see Jeyakumar and Mond [105], see also this book).

Theorem 8.21 (Sufficient optimality condition). *Let x^* be an efficient solution of (NDVP) and let there exist scalars $\tau\in R^p$ and λ such that (8.17)–(8.21) hold. If $\left(\tau_1(f_1+.^TB_1z_1),\ldots,\tau_p(f_p+.^TB_pz_p)\right)$ is V-pseudo-invex and $\lambda_1g_1,\ldots,\lambda_mg_m$ is V-quasi-invex with respect to the same eta and for all piecewise smooth $z_i\in R^n$. Then x^* is an efficient solution for (NDVP).*

Proof. Let x be feasible for problem (NDVP). Then $x\in S,\ g(x)\leq 0$. Since $\lambda_jg_j(x^*)=0$ then

$$\sum_{j=1}^{m}\lambda_jg_j(x)\leq\sum_{j=1}^{m}\lambda_jg_j(x^*). \tag{8.22}$$

Since $\beta_j(x,x^*)>0,\ \forall\ j=1,\ldots,m$, we have

$$\sum_{j=1}^{m}\beta_j(x,x^*)\lambda_jg_j(x)\leq\sum_{j=1}^{m}\beta_j(x,x^*)\lambda_jg_j(x^*). \tag{8.23}$$

Then by V-pseudo-invexity of $\lambda_1 g_1, \ldots, \lambda_m g_m$, we get

$$\sum_{j=1}^{m} \lambda_j \nabla_x g_j(x^*) \leq 0.$$

Therefore, from (8.20), we have

$$\sum_{i=1}^{p} \tau_i \left[\nabla_x f_i(x^*) + B_i z_i\right] \geq 0. \tag{8.24}$$

From V-pseudo-invexity of $(\tau_1(f_1 + .^T B_1 z_1), \ldots, \tau_p(f_p + .^T B_p z_p))$, we have

$$\sum_{i=1}^{p} \alpha_i(x, x^*)\tau_i \left[f_i(x) + x^T B_i z\right] \geq \sum_{i=1}^{p} \alpha_i(x, x^*)\tau_i \left[f_i(x^*) + {x^*}^T B_i z\right].$$

That is

$$\alpha_i(x, x^*)\tau_i \left[f_i(x^*) + {x^*}^T B_i z\right] \leq \alpha_i(x, x^*)\tau_i \left[f_i(x) + x^T B_i z\right] \quad \forall i$$

$$\alpha_i(x, x^*)\tau_i \left[f_i(x^*) + {x^*}^T B_i z\right] < \alpha_j(x, x^*)\tau_j \left[f_j(x) + x^T B_j z\right]$$

for at least one j.

Since, $\alpha_i(x, x^*) > 0 \;\; \forall i$ and $\tau > 0$, we get

$$f_i(x^*) + {x^*}^T B_i z \leq f_i(x) + x^T B_i z \quad \forall i \quad \text{and}$$

$$f_i(x^*) + {x^*}^T B_i z < f_j(x) + x^T B_j z \quad \text{for at least one } j.$$

Thus, x^* is an efficient solution of (NDVP).

□

In relation to (NDVP) we associate the following dual nondifferentiable multiobjective maximization problem:

(NDVD)

$$\max \left(f_1(u) + (u^T B_1 z_1), \ldots, f_p(u) + (u^T B_p z_p)\right)$$

$$\text{Subject to } \sum_{i=1}^{p} \tau_i \left[\nabla f_i(u) + B_i z_i\right] + \sum_{j=1}^{m} \lambda_j \nabla_x g_j(u) = 0, \tag{8.25}$$

$$z^T B_i z_i \leq 1, \quad i = 1, \ldots, p, \tag{8.26}$$

$$\sum_{j=1}^{m} \lambda_j g_j(u) \geq 0, \tag{8.27}$$

$$\tau > 0, \;\; \lambda \geq 0, \;\; \sum_{i=1}^{p} \tau_i = 1. \tag{8.28}$$

Let H denote the set of feasible solutions for (NDVD).

Following, Mond et al. [173] to invexity, we obtained the duality results to V-invexity, following Mishra [153].

Theorem 8.22 (Weak duality). *Let $x \in K$ and $(u, \tau, \lambda, z_1, \dots, z_p) \in H$ and $\left(\tau_1(f_1 + .^T B_1 z_1), \dots, \tau_p(f_p + .^T B_p z_p)\right)$ is V-pseudo-invex and $\lambda_1 g_1, \dots, \lambda_m g_m$ is V-quasi-invex with respect to the same η and for all piecewise smooth $z_i \in R^n$. Then the following can not hold:*

$$f_i(x) + (x^T B_i x)^{\frac{1}{2}} \leq f_i(u) + u^T B_i z_i \quad \forall i = 1, \dots, p \quad \text{and}$$
$$f_{i_0}(x) + (x^T B_{i_0} x)^{\frac{1}{2}} \leq (f_{i_0}(u) + u^T B_{i_0} z_{i_0} \quad \text{for at least one } i_0.$$

Proof. Let x be feasible for (NDVP). Then $x \in S, g(x) \leq 0$. Since $\lambda_j g_j(x^*) = 0$, then

$$\lambda_j g_j(x) \leq \lambda_j g_j(u), \quad j = 1, \dots, m,$$

and

$$\lambda_{j_0} g_{j_0}(x) \leq \lambda_{j_0} g_{j_0}(u), \quad \text{for at least one } j_0 \in \{i, \dots, m\}.$$

Since $\beta_j(x, u) > 0, \ \forall j = 1, \dots, m$, we have

$$\sum_{j=1}^{m} \beta_j(x, u) \lambda_j g_j(x) \leq \sum_{j=1}^{m} \beta_j(x, u) \lambda_j g_j(u).$$

Then by V-quasi-invexity of $\lambda_1 g_1, \dots, \lambda_m g_m$, we get

$$\sum_{j=1}^{m} \lambda_j \nabla_x g_j(u) \eta(x, u) \leq 0,$$

and so from (8.25), we have

$$\sum_{i=1}^{p} \tau_i \left[\nabla_x f_i(u) + B_i z_i\right] \geq 0.$$

Thus, from V-pseudo-invexity of

$$\left(\tau_1(f_1 + .^T B_1 z_1), \dots, \tau_p(f_p + .^T B_p z_p)\right),$$

we have

$$\sum_{i=1}^{p} \alpha_i(x, u) \tau_i \left[f_i(x) + x^T B_i z\right] \geq \sum_{i=1}^{p} \alpha_i(x, u) \tau_i \left[f_i(u) + x^T B_i z\right].$$

But

$$\begin{aligned} x^T B_i z_i &\leq (x^T B_i x)^{\frac{1}{2}} (z_i^T B_i z_i)^{\frac{1}{2}} \quad \text{(by Schwarz inequality)} \qquad (8.29) \\ &\leq (x^T B_i x)^{\frac{1}{2}} \quad \text{by (8.26).} \end{aligned}$$

Now, from (8.21) and (8.29), we have

$$\alpha_i(x, u) \tau_i \left[(f_i(u) + (x^T B_i x)^{\frac{1}{2}})\right] \geq \alpha_i(x, u) \tau_i \left[(f_i(u) + x^T B_i z)\right].$$

That is

$$\alpha_i(x,u)\tau_i\left[(f_i(x)+(x^TB_ix)^{\frac{1}{2}})\right]\geq\alpha_i(x,u)\tau_i\left[(f_i(u)+u^TB_iz_i)\right]\quad\forall i,$$

and

$$\alpha_{i_0}(x,u)\tau_{i_0}\left[(f_{i_0}(x)+(x^TB_{i_0}x)^{\frac{1}{2}})\right]$$
$$\geq\alpha_{i_0}(x,u)\tau_{i_0}\left[(f_{i_0}(u)+u^TB_{i_0}z_{i_0}\right]\quad\text{for at least one } i_0.$$

Since, $\alpha_i(x,u)>0,\ \forall i$ and $\tau\geq 0$ we get

$$f_i(x)+(x^TB_ix)^{\frac{1}{2}}\geq f_i(u)+u^TB_iz_i\quad\forall i=1,\dots,p,$$

and

$$f_{i_0}(x)+(x^TB_{i_0}x)^{\frac{1}{2}}>(f_{i_0}(u)+u^TB_{i_0}z_{i_0}\quad\text{for at least one } i_0.$$

Thus, the following can not hold:

$$f_i(x)+(x^TB_ix)^{\frac{1}{2}}\leq f_i(u)+u^TB_iz_i\quad\forall i=1,\dots,p,$$

and

$$f_{i_0}(x)+(x^TB_{i_0}x)^{\frac{1}{2}}\leq(f_{i_0}(u)+u^TB_{i_0}z_{i_0}\quad\text{for at least one } i_0.$$

□

Remark 8.23. If we replace the V-invexity assumptions by invexity assumptions, we get:

Theorem 8.24 (Weak duality). *Let $x\in K$ and $(u,\tau,\lambda,z_1,\dots,z_p)\in H$ and $(\tau_1(f_1+.^TB_1z_1),\dots,\tau_p(f_p+.^TB_pz_p)$ is pseudo-invex and $\lambda_1g_1,\dots,\lambda_mg_m$ is quasi-invex with respect to the same η and for all piecewise smooth $z_i\in R^n$. Then the following can not hold:*

$$f_(x)+(x^TB_ix)^{\frac{1}{2}}\leq f_i(u)+u^TB_iz_i\quad\forall i=1,\dots,p,$$

and

$$f_{i_0}(x)+(x^TB_{i_0}x)^{\frac{1}{2}}\leq(f_{i_0}(u)+u^TB_{i_0}z_{i_0}\quad\textit{for at least one } i_0.$$

If we take the convexity assumptions, we get the weak duality obtained by Mond et al. [173]:

Theorem 8.25 (Weak duality). *Let $x\in K$ and $(u,\tau,\lambda,z_1,\dots,z_p)\in H$ and $(\tau_1(f_1+.^TB_1z_1),\dots,\tau_p(f_p+.^TB_pz_p))$ and $\lambda_1g_1,\dots,\lambda_mg_m$ are convex for all piecewise smooth $z_i\in R^n$. Then the following can not hold:*

$$f_i(x)+(x^TB_ix)^{\frac{1}{2}}\leq f_i(u)+u^TB_iz_i\quad\forall i=1,\dots,p,$$

and

$$f_{i_0}(x)+(x^TB_{i_0}x)^{\frac{1}{2}}\leq(f_{i_0}(u)+u^TB_{i_0}z_{i_0}\quad\textit{for at least one } i_0.$$

Theorem 8.26. *Let $x \in K$ and $(u, \tau, \lambda, z_1, \ldots, z_p) \in H$ and the V-pseudo-invexity and V-quasi-invexity conditions of Theorem 8.22 hold. If*

$$u^T B_i u = u^T B_i z_i, \quad \textit{for} \quad i = 1, \ldots, p, \tag{8.30}$$

and the objective values are equal, then x is properly efficient for (NDVP) and $(u, \tau, \lambda, z_1, \ldots, z_p)$ is properly efficient for (NDVD).

Proof. Suppose x is not an efficient solution for (NDVP), then there exists $x_0 \in K$ such that

$$f_i(x_0) + (x_0^T B_i x_0)^{\frac{1}{2}} \leq f_i(x) + (x^T B_i x)^{\frac{1}{2}} \quad \forall i = 1, \ldots, p,$$

and

$$f_{i_0}(x_0) + (x_0^T B_{i_0} x_0)^{\frac{1}{2}} < (f_{i_0}(x) + (x^T B_{i_0} x)^{\frac{1}{2}} \quad \text{for at least one } i_0.$$

Using (8.20), we get

$$f_i(x_0) + (x_0^T B_i x_0)^{\frac{1}{2}} \leq f_i(u) + u^T B_i z_i \quad \forall i = 1, \ldots, p,$$

$$f_{i_0}(x_0) + (x_0^T B_{i_0} x_0)^{\frac{1}{2}} \leq f_{i_0}(u) + u^T B_{i_0} z_{i_0} \quad \text{for at least one } i_0.$$

This is a contradiction of weak duality Theorem 8.22. Hence x is an efficient solution for (NDVP). Similarly it can be ensured that $(u, \tau, \lambda, z_1, \ldots, z_p)$ is an efficient solution of (NDVD).

Now suppose that x is not properly efficient of (NDVP). Therefore, for every positive function $M > 0$ there exists $x_0 \in X$ feasible for (NDVP) and an index i such that

$$f_i(x) + (x^T B_i x)^{\frac{1}{2}} - \left(f_i(x_0) + (x_0^T B_i x_0)^{\frac{1}{2}}\right)$$
$$> M\left(f_j(x_0) + (x_0^T B_j x_0)^{\frac{1}{2}} - f_j(x) + (x^T B_j x)^{\frac{1}{2}}\right)$$

for all j satisfying

$$f_j(x_0) + (x_0^T B_j x_0)^{\frac{1}{2}} > f_j(x) + (x^T B_j x)^{\frac{1}{2}},$$

whenever

$$f_i(x_0) + (x_0^T B_i x_0)^{\frac{1}{2}} < f_i(x) + (x^T B_i x)^{\frac{1}{2}}.$$

This means $f_i(x) + (x^T B_i x)^{\frac{1}{2}} - \left(f_i(x_0) + (x_0^T B_i x_0)^{\frac{1}{2}}\right)$ can be made arbitrarily large and hence for $\tau > 0$ the inequality

$$\sum_{i=1}^{p} \tau_i \left(f_i(x) + (x^T B_i x)^{\frac{1}{2}} - \left(f_i(x_0) + (x_0^T B_i x_0)^{\frac{1}{2}}\right)\right) > 0, \tag{8.31}$$

is obtained.

Now from feasibility conditions, we have

$$\lambda_j g_j(x_0) \leq \lambda_j g_j(u), \quad \forall j = 1, \ldots, m.$$

Since $\beta_j(x_0, u) > 0, \forall j = 1, \ldots, m$

$$\sum_{j=1}^{m} \beta_j(x_0, u)\lambda_j g_j(x_0) \leq \sum_{j=1}^{m} \beta_j(x_0, u)\lambda_j g_j(u).$$

Then by V-quasi-invexity of $\lambda_1 g_1, \ldots, \lambda_m g_m$, we get

$$\sum_{j=1}^{m} \lambda_j \nabla_x g_j(u)\eta(x, u) \leq 0$$

Therefore, from (8.25), we get

$$\sum_{i=1}^{p} \tau_i^0 f_i(u^0)\eta(\bar{x}, u^0) \geq 0.$$

Since $\tau \geq 0$, $\sum_{i=1}^{p} \tau_i = 1$, we have

$$\sum_{i=1}^{p} \tau_i \left(\nabla_x f_i(u) + B_i z_i\right) \eta(\bar{x}, u^0) \geq 0.$$

By using V-pseudo-invexity conditions, we have

$$\sum_{i=1}^{p} \alpha_i(x_0, u)\tau_i \left(f_i(x_0) + x_0^T B_i z_i\right) \geq \sum_{i=1}^{p} \alpha_i(x_0, u)\tau_i \left(f_i(u) + u^T B_i z_i\right).$$

Since $\alpha_i(x_0, u) > 0, \forall i = 1, \ldots, p$, we have

$$\sum_{i=1}^{p} \tau_i \left(f_i(x_0) + x_0^T B_i z_i\right) \geq \sum_{i=1}^{p} \tau_i \left(f_i(u) + u^T B_i z_i\right).$$

Since the objective values of (NDVP) and (NDVD) are equal, we have

$$\sum_{i=1}^{p} \tau_i \left(f_i(x) + (x^T B_i x)^{\frac{1}{2}}\right) \leq \sum_{i=1}^{p} \tau_i \left(f_i(x_0) + (x_0^T B_i x_0)^{\frac{1}{2}}\right).$$

This, yields

$$\sum_{i=1}^{p} \tau_i \left(f_i(x) + (x^T B_i x)^{\frac{1}{2}} - \sum_{i=1}^{p} \tau_i \left(f_i(x_0) + (x_0^T B_i x_0)^{\frac{1}{2}}\right)\right) \leq 0,$$

which is a contradiction to (8.31). Hence x is a properly efficient solution for (NDVP).

We now suppose that $(u, \tau, \lambda, z_1, \ldots, z_p)$ is not properly efficient solution for (NDVD). Therefore, for every positive function $M > 0$ there exists a feasible $(u_0, \tau_0, \lambda_0, z_1^0, \ldots, z_p^0)$ feasible for (NDVD) and an index i such that

$$f_i(u_0) + u_0^T B_i z_i^0 - \left(f_i(u) + u^T B_i z_i\right) > M\left(f_j(u) + u^T B_j z_j - f_j(u_0) + u_0^T B_j z_i^0\right)$$

for all j satisfying

$$f_j(u_0) + u_0^T B_j z_j^0 < f_j(u) + u^T B_j z_j,$$

whenever

$$f_i(u_0) + u_0^T B_i z_i^0 > f_i(u) + u^T B_i z_i.$$

This means $f_i(u_0) + (u_0^T B_i z_i^0 - f_i(u)) - u^T B_i z_i$ can be made arbitrarily large and hence for $\tau > 0$ the inequality

$$\sum_{i=1}^{p} \tau_i \left(f_i(u_0) + u_0^T B_i z_i^0 - f_i(u) - u^T B_i z_i\right) > 0, \tag{8.32}$$

is obtained.

Since x, $(u, \tau, \lambda, z_1, \ldots, z_p)$ feasible for (NDVP) and (NDVD), respectively, it follows as in first part

$$\sum_{i=1}^{p} \tau_i \left(f_i(u_0) + u_0^T B_i z_i^0 - f_i(u) - u^T B_i z_i\right) \leq 0,$$

which contradicts (8.32). Hence x, $(u, \tau, \lambda, z_1, \ldots, z_p)$ is properly efficient solution for (NDVD).

□

Theorem 8.27 (Strong duality). *Let x be a properly efficient solution for (NDVP) at which a suitable constraint qualification is satisfied. Let the V-pseudo-invexity and V-quasi-invexity conditions of Theorem 8.22 be satisfied. Then there exists $(\tau, \lambda, z_1, \ldots, z_p)$ such that $(x = u, \tau, \lambda, z_1, \ldots, z_p)$ is a properly efficient solution for (NDVD) and*

$$f_i(x) + (x^T B_i x)^{\frac{1}{2}} = f_i(u) + u^T B_i z_i \quad \forall i = 1, \ldots, p.$$

Proof. Since x is properly efficient solution for (NDVP) and a constraint qualification is satisfied at x, from the Kuhn–Tucker necessary condition Lemma 8.12, there exists $(\tau, \lambda, z_1, \ldots, z_p)$ such that $(x, \tau, \lambda, z_1, \ldots, z_p)$ is feasible for (NDVD). Since $(x^T B_i x)^{\frac{1}{2}} = x^T B_i z_i, \forall i = 1, \ldots, p$ the values of (NDVP) and (NDVD) are equal at x. By Theorem 8.24, $(x = u, \tau, \lambda, z_1, \ldots, z_p)$ is properly efficient solution of (NDVD).

□

Remark 8.28. If we replace V-invexity in Theorem 8.27 by invexity, we get the following strong duality:

Theorem 8.29 (Strong duality). *Let x be a properly efficient solution for (NDVP) at which a suitable constraint qualification is satisfied. Let the (V-pseudo-invexity and V-quasi-invexity conditions of Theorem 8.22 replaced by invexity. Then there exists $(\tau, \lambda, z_1, \ldots, z_p)$ such that $(x = u, \tau, \lambda, z_1, \ldots, z_p)$ is a properly efficient solution for (NDVD) and*

$$f_i(x) + (x^T B_i x)^{\frac{1}{2}} = f_i(u) + u^T B_i z_i \quad \forall i = 1, \ldots, p.$$

If we take the convexity assumptions, we get Theorem 2 (Strong duality) of Mond et al. [173].

Mangasarian [144] proposed the following second order dual model to (NDP):

(ND2MD)

$$\max f(u) - y^T g(u) + u^T Bw - \frac{1}{2} p^T \nabla^2 \left[f(u) - y^T g(u) \right] p$$

$$\begin{aligned} \text{Subject to } \nabla f(u) - \nabla y^T g(u) + Bw + \nabla^2 f(u)p - \nabla^2 y^T g(u)p &= 0 \\ w^T Bw \leq 1, \quad y \geq 0, \end{aligned}$$

where $u, w, p \in R^n$ and $y \in R^m$.

(ND2D)

$$\max f(u) + u^T Bw - \frac{1}{2} p^T \nabla^2 f(u)p$$

$$\begin{aligned} \text{Subject to } \nabla f(u) - \nabla y^T g(u) + Bw + \nabla^2 f(u)p - \nabla^2 y^T g(u)p &= 0 \\ y^T g(u) - \frac{1}{2} p^T \nabla^2 y^T g(u)p &\leq 1 \\ w^T Bw \leq 1, \quad y \geq 0, \end{aligned}$$

where $u, w, p \in R^n$ and $y \in R^m$.

These authors established duality theorems between (NDP) and (ND2MD) and (ND2D) under second-order convexity assumptions. Zhang [256] proposed Mangasarian type and Mond–Weir type higher order duals to (NDP) as follows:

(NDHMD)

$$\max f(u) + h(u,p) + (u+p)^T Bw - y^T g(u) - y^T k(u,p)p$$

$$\begin{aligned} \text{Subject to } \nabla_p h(u,p) + Bw &= \nabla_p (y^T k(u,p)) \\ w^T Bw \leq 1, \quad y \geq 0, \end{aligned}$$

where $u, w, p \in R^n$ and $y \in R^m$.

(NDHD)

$$\max f(u) + h(u,p) + u^T Bw - \frac{1}{2} p^T \nabla_p h(u,p)$$
$$\text{Subject to } \nabla_p h(u,p) + Bw = \nabla_p (y^T k(u,p))$$
$$y^T g(u) - y^T k(u,p) - p^T \nabla_p (y^T k(u,p)) \leq 0$$
$$w^T Bw \leq 1, \quad y \geq 0,$$

where $u, w, p \in R^n$ and $y \in R^m$.

Zhang [256] obtained duality results under higher-order invexity and generalized higher-order invexity assumptions between (NDP) and (NDHMD) and (NDHD). Later Mishra and Rueda [161] considered the (NDP), (NDHMD) and (NDHD) and established several duality theorems under generalized invexity assumptions. Furthermore, Mishra and Rueda [161] considered the following general Mond–Weir type higher order dual and obtained weak, strong and strict converse duality results.

(NDHGD)

$$\max f(u) + h(u,p) + u^T Bw - p^T \nabla_p h(u,p) - \sum_{i \in I_0} y_i g_i(u)$$
$$- \sum_{i \in I_0} y_i k_i(u,p) + p^T \nabla_p \sum_{i \in I_0} y_i k_i(u,p)$$
$$\text{Subject to } \nabla_p h(u,p) + Bw = \nabla_p (y^T k(u,p)) \sum_{i \in I_\alpha} y_i g_i(u)$$
$$+ \sum_{i \in I_\alpha} y_i k_i(u,p) - p^T \nabla_p \left(\sum_{i \in I_\alpha} y_i k_i(u,p) \right)$$
$$\leq 0, \quad \alpha = 1, \ldots, r$$
$$w^T Bw \leq 1, \quad y \geq 0,$$

where $u, w, p \in R^n$ and $y \in R^m$.

Throughout rest of this section we follow the following notations: if x and $y \in R^n$, then by $x \leqq y$ we mean $x_i \leqq y_i$ for all i, $x \leq y$ means $x_i \leqq y_i$ for all i and $x_j < y_j$ for at least one j, $1 \leq j \leq n$. By $x < y$ we mean $x_i < y_i$ for all i.

Mishra et al. [159] considered the following nondifferentiable multiobjective programming problem:

(VP)

$$\text{Min} \left(f_1(x) + s(x : C_1), \ldots, f_p(x) + s(x : C_p) \right)$$
$$\text{Subject to } g(x) \geqq 0, \quad x \in D$$

where f and g are twice differentiable functions from R^n to R^l and R^m, respectively; C_i for each $i \in L = \{1, \ldots, l\}$ is a compact convex set of R^n and D is an open subset of R^n.

Let $f : R^n \to R^l$ be twice differentiable and $\eta : R^n \times R^n \to R^n$.

Definition 8.30. *f is said to be higher order invex at u with respect to η and h if for all x*

$$f_i(x) - f_i(u) \geq \eta(x,u)^T \nabla_p h_i(u,p) - p^T \nabla_p h_i(u,p), \quad \forall i \in \{1,\ldots,l\}.$$

These authors extend the Definitions 2.2–2.6 of Aghezzaf and Hachimi [2] to the higher order context as follows:

Definition 8.31. *f is said to be higher order weak strictly pseudo invex at u with respect to η and h if for all x*

$$\begin{aligned} f_i(x) &\leq \{f_i(u) - h_i(u,p) + p^T \nabla_p h_i(u,p)\} \\ &\Rightarrow \eta(x,u)^T \nabla_p h_i(u,p) < 0, \quad \forall i \in \{1,2,\ldots,l\}. \end{aligned}$$

Definition 8.32. *f is said to be higher order strong pseudo invex at u with respect to η and h if for all x*

$$\begin{aligned} f_i(x) &\leq \{f_i(u) + h_i(u,p) - p^T \nabla_p h_i(u,p)\} \\ &\Rightarrow \eta(x,u)^T \nabla_p h_i(u,p) \leq 0, \quad \forall i \in \{1,\ldots,l\}. \end{aligned}$$

Definition 8.33. *f is said to be higher order weak quasi invex at u with respect to η and h if for all x*

$$\begin{aligned} f_i(x) &\leq \{f_i(u) + h_i(u,p) - p^T \nabla_p h_i(u,p)\} \\ &\Rightarrow \eta(x,u)^T \nabla_p h_i(u,p) \leqq 0, \quad \forall i \in \{1,\ldots,l\}. \end{aligned}$$

Definition 8.34. *f is said to be higher order weak pseudo invex at u with respect to η and h if for all x*

$$\begin{aligned} f_i(x) &< f_i(u) + h_i(u,p) - p^T \nabla_p h_i(u,p) \\ &\Rightarrow \eta(x,u)^T \nabla_p h_i(u,p) \leq 0, \quad \forall i \in \{1,2,\ldots,l\}. \end{aligned}$$

Definition 8.35. *f is said to be higher order strong quasi invex at u with respect to η and h if for all x*

$$\begin{aligned} f_i(x) &\leqq f_i(u) + h_i(u,p) - p^T \nabla_p h_i(u,p) \\ &\Rightarrow \eta(x,u)^T \nabla_p h_i(u,p) \leq 0, \quad \forall i \in \{1,\ldots,l\}. \end{aligned}$$

Mishra et al. [159] introduced the following higher order dual model in relation to (VP):

(NDVHD)

$$\begin{aligned} V - \text{Max}(&f_1(u) + h_1(u,p) + u^T w_1 - p^T \nabla_p h_1(u,p), \ldots, \\ &f_l(u) + h_l(u,p) + u^T w_l - p^T \nabla_p h_l(u,p)) \end{aligned}$$

$$\text{Subject to } \sum_{i=1}^{l} \lambda_i \left[\nabla_p h_i(u,p) + w_i\right] - \sum_{j=1}^{m} y_j \nabla_p k_j(u,p)$$

$$= \sum_{j=1}^{m} y_j \left[g_j(u) + k_j(u,p) - p^T \nabla_p k_j(u,p)\right] \leqq 0,$$

$$y \geqq 0, \quad \lambda \in \Lambda = \left\{\lambda \in R^l : \lambda \geq 0, \sum_{i=1}^{l} \lambda_i = 1\right\},$$

$$w_i \in C_i, \quad i = 1, \ldots, l,$$

where $h : R^n \times R^n \to R^l$ and $k : R^n \times R^n \to R^m$ are differentiable functions; $\nabla_p h_i(u,p)$ denotes the $n \times 1$ gradient of h_i with respect to p and $\nabla_p(y^T k_i(u,p))$ denotes the $n \times 1$ gradient of $y^T k$ with respect to p.

The problem (NDVHD) may be regarded as a multiple objective higher order nondifferentiable Mond–Weir type [181] vector dual to (VP).

Remark 8.36. If $h_i(u,p) = p^T \nabla f_i(u)$, $\forall i = \{1, \ldots, l\}$ and $k_j(u,p) = p^T \nabla g_j(u)$, $\forall j = \{1, \ldots, m\}$ (NDVHD) then becomes the Mond–Weir type vector dual of Tanino and Sawaragi [234] for (VP). If $h_i(u,p) = p^T \nabla f_i(u) + \frac{1}{2} p^T \nabla^2 f_i(u) p$, $\forall i = \{1, \ldots, l\}$ and $k_j(u,p) = p^T \nabla g_j(u) + \frac{1}{2} p^T \nabla^2 g_j(u) p$, $\forall j = \{1, \ldots, m\}$ then (NDVHD) becomes the second order nondifferentiable version of Mond–Weir type vector dual of Zhang [257] for (VP).

Mishra et al. [159] established the following higher order duality results under higher order generalized convexity assumptions:

Theorem 8.37 (Weak duality). *Let x be feasible for (VP), (u, λ, y, w, p) feasible for (NDVHD) and $\lambda > 0$. Assume that $f_i(.) + .^T w_i$ is higher order strong pseudoinvex with respect to $h_i(., p)$ and $-\sum_{j=1}^{m} y_j g_j(.)$ is higher order quasiinvex with respect to $-k_j(., p)$ then the following cannot hold:*

$$f_i(x) + S(x : C_i) \leqq f_i(u) + u^T w_i + h_i(u,p) - p^T \nabla_p h_i(u,p) \quad \forall i \in \{1, \ldots, l\}$$

$$f_{i_0}(x) + S(x : C_{i_0}) < f_{i_0}(u) + u^T w_{i_0} + h_{i_0}(u,p) - p^T \nabla_p h_{i_0}(u,p) \quad \textit{for some } i_0 \in \{1, \ldots, l\}.$$

Proof. Suppose contrary to the result of the theorem that

$$f_i(x) + S(x : C_i) \leqq f_i(u) + u^T w_i + h_i(u,p) - p^T \nabla_p h_i(u,p) \quad \forall i \in \{1, 2, \ldots, l\}$$

$$f_{i_0}(x) + S(x : C_{i_0}) < f_{i_0}(u) + u^T w_{i_0} + h_{i_0}(u,p) - p^T \nabla_p h_{i_0}(u,p) \quad \text{for some } i_0 \in \{1, 2, \ldots, l\}.$$

Since $x^T w_i \leq S(x : C_i)$, $\forall i = 1, 2, \ldots, l$ by higher order strong pseudoinvexity of $f_i(.) + .^T w_i$ with respect to $h_i(., p)$ and $\lambda > 0$, we get

$$\eta(x,u)^T \sum_{i=1}^{l} \lambda_i \left[\nabla_p h_i(u,p) + w_i\right] < 0.$$

Since x is feasible for (VP) and (u, λ, y, w, p) is feasible for (NDVHD), using higher order quasiinvexity of $-\sum_{j=1}^{m} y_j g_j(.)$ with respect to $-k_j(., p)$ we get

$$\eta(x,u)^T \sum_{j=1}^{m} y_j \left[\nabla_p k_j(u,p)\right] \geqq 0.$$

From the above two inequalities, we get

$$\sum_{i=1}^{l} \lambda_i \left[\nabla_p h_i(u,p) + w_i\right] - \sum_{j=1}^{m} y_j \left[\nabla_p k_j(u,p)\right] < 0.$$

which contradicts the first duality constraint of the dual problem (NDVHD). □

The proof of the following weak duality theorem is similar to the proof of Theorem 8.37, so we state it without proof.

Theorem 8.38 (Weak duality). *Let x be feasible for (VP), (u, λ, y, w, p) feasible for (NDVHD) and $\lambda > 0$. Assume that $f_i(.) + .^T w_i$ is higher order strictly pseudoinvex with respect to $h_i(., p)$ and $-\sum_{j=1}^{m} y_j g_j(.)$ is higher order quasi-invex with respect to $-k_j(., p)$ then the following cannot hold:*

$$f_i(x) + S(x : C_i) \leqq f_i(u) + u^T w_i + h_i(u,p) - p^T \nabla_p h_i(u,p) \quad \forall i \in \{1, 2, \ldots, l\}$$

$$f_{i_0}(x) + S(x : C_{i_0}) < f_{i_0}(u) + u^T w_{i_0} + h_{i_0}(u,p) - p^T \nabla_p h_{i_0}(u,p) \quad \text{for some } i_0 \in \{1, 2, \ldots, l\}.$$

Theorem 8.39 (Weak duality). *Let x be feasible for (VP), (u, λ, y, w, p) feasible for (NDVHD) and $\lambda > 0$. Assume that $f_i(.) + .^T w_i$ is higher order weak quasi-invex with respect to $h_i(., p)$ and $-\sum_{j=1}^{m} y_j g_j(.)$ is higher order strictly pseudoinvex with respect to $-k_j(., p)$ then the following cannot hold:*

$$f_i(x) + S(x : C_i) \leqq f_i(u) + u^T w_i + h_i(u,p) - p^T \nabla_p h_i(u,p) \quad \forall i \in \{1, 2, \ldots, l\}$$

$$f_{i_0}(x) + S(x : C_{i_0}) < f_{i_0}(u) + u^T w_{i_0} + h_{i_0}(u,p) - p^T \nabla_p h_{i_0}(u,p) \quad \text{for some } i_0 \in \{1, 2, \ldots, l\}.$$

Proof. Using the hypothesis given in this Theorem, we get

$$\eta(x,u)^T \sum_{i=1}^{l} \lambda_i \left[\nabla_p h_i(u,p) + w_i\right] < 0.$$

$$\eta(x,u)^T \sum_{j=1}^{m} y_j \left[\nabla_p k_j(u,p)\right] > 0.$$

From the above two inequalities, we get

$$\sum_{i=1}^{l} \lambda_i \left[\nabla_p h_i(u,p) + w_i\right] - \sum_{j=1}^{m} y_j \left[\nabla_p k_j(u,p)\right] < 0.$$

which contradicts the first duality constraint of the dual problem (NDVHD). □

Theorem 8.40 (Strong duality). *Let x_0 be an efficient solution for (VP) at which a Kuhn–Tucker constraint qualification is satisfied and let*

$$h(x_0,0)=0, \quad k(x_0,0)=0,$$
$$\nabla_p h(x_0,0)=\nabla f(x_0), \quad \nabla_p k(x_0,0)=\nabla g(x_0).$$

Then there exist $y \in R^m$ and $\lambda \in R^l$ such that $(x_0,\lambda,p=0)$ is feasible for (VP) and $(x_0,y,w,\lambda,p=0)$ is feasible for (NDVHD) and the corresponding values of (VP) and (NDVHD) are equal. If for all feasible $(x_0,y,w,\lambda,p=0)$ the assumptions of Theorem 8.37 or Theorem 8.38 are satisfied, then $(x_0,y,w,\lambda,p=0)$ is efficient for (NDVHD).

Proof. Since x_0 is an efficient solution and hence also a weak minimum for (VP) at which a Kuhn–Tucker constraint qualification is satisfied, then by Theorem 8.37 or Theorem 8.38 or Theorem 8.39, $(x_0,y,w,\lambda,p=0)$ must be an efficient solution for (NDVHD).

□

Mishra et al. [159] also considered the following general higher order nondifferentiable dual to (VP):

(NDHGVHD)

$$V-\mathrm{Max}\Bigg(f_1(u)+h_1(u,p)+u^T w_1-\sum_{i\in I_0} y_i g_i(u)-\sum_{i\in I_0} y_i k_i(u,p)$$
$$-p^T\left[\nabla_p h_1(u,p)-\nabla_p\left(\sum_{i\in I_0} y_i k_i(u,p)\right)\right],\ldots,$$
$$f_i(u)+h_i(u,p)+u^T w_i-\sum_{i\in I_0} y_i g_i(u)-\sum_{i\in I_0} y_i k_i(u,p)$$
$$-p^T\left[\nabla_p h_1(u,p)-\nabla_p\left(\sum_{i\in I_0} y_i k_i(u,p)\right)\right]\Bigg)$$

$$\text{Subject to } \sum_{i=1}^{l} \lambda_i \left[\nabla_p h_i(u,p) + w_i\right] - \sum_{j=1}^{m} y_j \nabla_p k_j(u,p) = 0$$

$$\sum_{j\in I_0} y_j \left[g_j(u) + k_j(u,p) - p^T \nabla_p k_j(u,p)\right] \leqq 0, \quad \alpha = 1,2,\ldots,r$$

$$y \geqq 0,\ \lambda \in \Lambda = \left\{\lambda \in R^l : \lambda \geq 0, \sum_{i=1}^{l} \lambda_i = 1\right\},$$

$$w_i \in C_i, \quad i = 1,2,\ldots,l,$$

where $I_\alpha \subseteq M = \{1,2,\ldots,m\}$, $\alpha = 0,1,2,\ldots,r$ with $I_\alpha \cap I_\beta = \phi$ if $\alpha \neq \beta$ and $\cup_{\alpha=0}^{r} I_\alpha = M$.

Remark 8.41. If $h_i(u,p) = p^T \nabla f_i(u)$, $\forall i = \{1,2,\ldots,l\}$ and $k_j(u,p) = p^T \nabla g_j(u)$, $\forall j = \{1,2,\ldots,m\}$ (NDHGVD) then becomes the Mond–Weir type vector dual of Tanino and Sawaragi [234] for (VP). If $h_i(u,p) = p^T \nabla f_i(u) + \frac{1}{2} p^T \nabla^2 f_i(u) p$, $\forall i = \{1,2,\ldots,l\}$ and $k_j(u,p) = p^T \nabla g_j(u) + \frac{1}{2} p^T \nabla^2 g_j(u) p$, $\forall j = \{1,2,\ldots,m\}$ then (NDHGVD) becomes the second order nondifferentiable version of Mond–Weir type vector dual of Zhang [257] for (VP).

The following duality theorems can be proved on the similar lines to that of the proofs of Theorems 8.37–8.40, respectively therefore we omit the proofs of following theorems.

Theorem 8.42 (Weak duality). *Let x be feasible for (VP), (u,λ,w,y,p) feasible for (NDHGVD) and $\lambda > 0$. Assume that $f_i(.) + .^T w_i - \sum_{i\in I_0} y_i g_i(.)$ is higher order strong pseudoinvex with respect to $h_i(.,p)$ and $-\sum_{j\in I_0} y_j g_j(.)$ is higher order quasi-invex with respect to $-k_j(.,p)$ then the following cannot hold:*

$$\begin{aligned} & f_i(x) + S(x : C_i) \\ & \quad \leqq f_i(u) + u^T w_i - \sum_{i\in I_0} y_i g_i(u) + h_i(u,p) - \sum_{i\in I_0} y_i k_i(u,p) \\ & \qquad - p^T \left(\nabla_p h_i(u,p) - \nabla_p \sum_{i\in I_0} y_i k_i(u,p) \right) \quad \forall i \in \{1,2,\ldots,l\} \end{aligned}$$

and

$$\begin{aligned} & f_{i_0}(x) + S(x : C_{i_0}) \\ & \quad < f_{i_0}(u) + u^T w_{i_0} - \sum_{i\in I_0} y_{i_0} g_{i_0}(u) + h_{i_0}(u,p) - \sum_{i\in I_0} y_{i_0} k_{i_0}(u,p) \\ & \qquad - p^T \left(\nabla_p h_{i_0}(u,p) - \nabla_p \sum_{i\in I_0} y_{i_0} k_{i_0}(u,p) \right) \end{aligned}$$

for at least one $i_0 \in \{1,2,\ldots,l\}$

Theorem 8.43 (Weak duality). *Let x be feasible for (VP), (u, λ, w, y, p) feasible for (NDHGVD) and $\lambda > 0$. Assume that $f_i(.) + .^T w_i - \sum_{i \in I_0} y_i g_i(.)$ is higher order weak strictly pseudoinvex with respect to $h_i(., p)$ and $-\sum_{j \in I_0} y_j g_j(.)$ is higher order quasi-invex with respect to $-k_j(., p)$ then the following cannot hold:*

$$
\begin{aligned}
&f_i(x) + S(x : C_i) \\
&\leqq f_i(u) + u^T w_i - \sum_{i \in I_0} y_i g_i(u) + h_i(u, p) - \sum_{i \in I_0} y_i k_i(u, p) \\
&\quad - p^T \left(\nabla_p h_i(u, p) - \nabla_p \sum_{i \in I_0} y_i k_i(u, p) \right) \quad \forall i \in \{1, 2, \ldots, l\}
\end{aligned}
$$

and

$$
\begin{aligned}
&f_{i_0}(x) + S(x : C_{i_0}) \\
&< f_{i_0}(u) + u^T w_{i_0} - \sum_{i \in I_0} y_{i_0} g_{i_0}(u) + h_{i_0}(u, p) - \sum_{i \in I_0} y_{i_0} k_{i_0}(u, p) \\
&\quad - p^T \left(\nabla_p h_{i_0}(u, p) - \nabla_p \sum_{i \in I_0} y_{i_0} k_{i_0}(u, p) \right) \\
&\quad \text{for at least one } i_0 \in \{1, 2, \ldots, l\}
\end{aligned}
$$

Theorem 8.44 (Weak duality). *Let x be feasible for (VP), (u, λ, w, y, p) feasible for (NDHGVD) and $\lambda > 0$. Assume that $f_i(.) + .^T w_i - \sum_{i \in I_0} y_i g_i(.)$ is higher order weak quasiinvex with respect to $h_i(., p)$ and $-\sum_{j \in I_0} y_j g_j(.)$ is higher order strictly pseudoinvex with respect to $-k_j(., p)$ then the following cannot hold:*

$$
\begin{aligned}
&f_i(x) + S(x : C_i) \\
&\leqq f_i(u) + u^T w_i - \sum_{i \in I_0} y_i g_i(u) + h_i(u, p) - \sum_{i \in I_0} y_i k_i(u, p) \\
&\quad - p^T \left(\nabla_p h_i(u, p) - \nabla_p \sum_{i \in I_0} y_i k_i(u, p) \right) \quad \forall i \in \{1, 2, \ldots, l\}
\end{aligned}
$$

and

$$
\begin{aligned}
&f_{i_0}(x) + S(x : C_{i_0}) \\
&< f_{i_0}(u) + u^T w_{i_0} - \sum_{i \in I_0} y_{i_0} g_{i_0}(u) + h_{i_0}(u, p) - \sum_{i \in I_0} y_{i_0} k_{i_0}(u, p) \\
&\quad - p^T \left(\nabla_p h_{i_0}(u, p) - \nabla_p \sum_{i \in I_0} y_{i_0} k_{i_0}(u, p) \right) \\
&\quad \text{at least one } i_0 \in \{1, 2, \ldots, l\}
\end{aligned}
$$

Theorem 8.45 (Strong duality). *Let x_0 be an efficient solution for (VP) at which a Kuhn–Tucker constraint qualification is satisfied and let*

$$h(x_0, 0) = 0, \quad k(x_0, 0) = 0,$$
$$\nabla_p h(x_0, 0) = \nabla f(x_0), \quad \nabla_p k(x_0, 0) = \nabla g(x_0).$$

Then there exist $y \in R^m$ and $\lambda \in R^l$ such that $(x_0, \lambda, p = 0)$ is feasible for (VP) and $(x_0, y, w, \lambda, p = 0)$ is feasible for (NDHGVD) and the corresponding values of (VP) and (NDHGVD) are equal. If for all feasible $(x_0, y, w, \lambda, p = 0)$ the assumptions of Theorem 8.42 or Theorem 8.43 or Theorem 8.44 are satisfied, then $(x_0, y, w, \lambda, p = 0)$ is efficient for (NDHGVD).

8.4 Nondifferentiable Symmetric Duality and Invexity

Following the earlier work of Dorn [59], Cottle [40] and Dantzig et al. [53] on symmetric duality, many researchers attempted to generalize the formulation and weaken the convexity-concavity assumptions required on $f(x, y)$. For results on symmetric duality one can see Problems (SP), (SD), (MWSP) and (MWSD) and discussions thereafter in Sect. 5.4 of Chap. 5.

In this section we discuss Nondifferentiable and fractional symmetric duality. Mond and Schechter [175] studied nondifferentiable symmetric duality as follows:

If F is a twice differentiable function from $R^n \times R^n$ to R, then $\nabla_x F$ and $\nabla_y F$ denote gradient (column) vectors of F with respect to x and y, respectively and $\nabla_{yy} F$ and $\nabla_{yx} F$ denote, respectively, the $m \times n$ and $n \times m$ matrices of second-order partial derivatives. Let C be a compact convex set in R_n. The support function of C is defined by

$$s(x\,|C) = \max\left\{x^T y : y \in C\right\}.$$

Yang et al. [251] presented the following pair of nondifferentiable fractional symmetric dual pair:

(FP)

$$\text{Min } \frac{f(x, y) + s(x : C) - y^T z}{g(x, y) + s(x : E - y^T r}$$

$$\text{Subject to } (\nabla_y f(x, y) - z)(g(x, y) - s : (x : E) + y^T r)$$
$$- (f(x, y) + s(x : C) - y^T z)(\nabla_y g(x, y) + r) \in Q*$$

$$y^T(\nabla_y f(x, y) - z)(g(x, y) - s : (x : E) + y^T r)$$
$$- (f(x, y) + s(x : C) - y^T z)(\nabla_y g(x, y) + r) \geq 0,$$

$$z \in D, \quad r \in F, \quad x \in P.$$

(FD)

$$\max \frac{f(u,v) + s(v:D) - u^T w}{g(u,v) + s(v:F - u^T t}$$

$$\text{Subject to } -(\nabla_u f(u,v) + w)(g(u,v) + s:(v:F) - u^T t) + (f(u,v) \\ - s(v:D) + u^T w)(\nabla_u g(u,v) - t) \in P*$$

$$u^T(\nabla_u f(u,v) + w)(g(u,v) + s:(v:F) - u^T t) \\ -(f(u,v) - s(v:D) + u^T w)(\nabla_u g(u,v) - t) \leq 0,$$

$$w \in C, \;\; t \in E, \;\; v \in Q,$$

where f and g are twice differentiable functions from $R^n \times R^m$ to R, C and E are compact convex sets in R^n, and D and F are compact convex sets in R^m, P and Q are two closed convex cones with nonempty interiors in R^n and R^m, respectively. It is assumed that in the feasible regions the denominator of the objective function is nonnegative and the numerator is positive.

Remark 8.46. If A is a positive semidefinite matrix, it can be easily verified that $(x^T Ax)^{\frac{1}{2}} = s(x\,|C)$, where $C = \{Ay : y^T Ay \leq 1\}$ and that this set C is compact convex.

(a) If $P = R^n_+$, $Q = R^m_+$, $g \equiv 1$ and $C = D = E = F = \{0\}$ we obtain the fractional symmetric dual pair studied by Chandra et al. [30].
(b) If in the feasible regions $g \equiv 1$, $E = F = \{0\}$, $(x^T Bx)^{\frac{1}{2}} = s(x\,|C)$, $C = \{By : y^T By \leq 1\}$, $(x^T Ax)^{\frac{1}{2}} = s(x\,|C)$ where $D = \{Ay : y^T Ay \leq 1\}$, $P = R^n_+$ and $Q = R^m_+$, then the above programs (FP) and (FD) become the symmetric dual nondifferentiable pair of problems studied by Chandra et al. [30].
(c) If in (FP) and (FD), $P = R^n_+$, $Q = R^m_+$, $g \equiv 1$, $E = F = \{0\}$ and we obtain symmetric dual pair of problems studied by Mond and Schechter [175].

Yang et al. [251] gave weak, strong and converse duality theorems for (FP) and (FD):

Theorem 8.47 (Weak duality). *Let (x, y, z, r) be feasible to (FP) and let (u, v, w, t) be feasible for (FD). Let*

$$(f(.,v) + (.)^T w)(g(u,v) + s(v:F) - u^T t) \\ -\left(f(u,v) - s(v:D) + u^T w\right)\left(g(.,v) + (.)^T t\right)$$

be pseudoinvex with respect to η_1 at u and

$$(f(x,.) + (.)^T z)(g(x,y) - s(x:E) + y^T r) \\ -\left(f(x,y) + s(x:C) - y^T z\right)\left(g(x,.) + (.)^T r\right)$$

be pseudoincave with respect to η_2 *at* y. *If* $\eta_1(x,u)+u \in P$ *and* $\eta_1(v,y)+y \in Q$ *then*

$$\frac{f(x,y)+s(x:C)-y^T z}{g(x,y)-s(x:E)+t^T r} \geq \frac{f(u,v)-s(v:D)-u^T w}{g(u,v)+s(v:F)-u^T t}.$$

Proof. For the proof, the reader is refer to Yang et al. [251].

□

Yang et al. [251] have also established the following Strong duality theorem:

Theorem 8.48 (Strong duality). *Let* $\bar{x},\bar{y},\bar{z},\bar{r}$ *be a solution to (FP). Suppose that all the conditions in Theorem 8.47 are fulfilled. Furthermore, assume that*

(a) $\nabla_{yy} f(\bar{x},\bar{y}) \left(g(\bar{x},\bar{y}) - s(\bar{x}:E) + \bar{y}^T\bar{r}\right)$
$- \left(f(\bar{x},\bar{y}) + s(\bar{x}:C) - \bar{y}^T\bar{z}\right) \nabla_{yy} g(\bar{x},\bar{y})$ *is positive or negative definite;*
(b) $(\nabla_y f(\bar{x},\bar{y}) - \bar{z}) \left(g(\bar{x},\bar{y}) - s(\bar{x}:E) + \bar{y}^T\bar{r}\right)$
$- \left(f(\bar{x},\bar{y}) + s(\bar{x}:C) - \bar{y}^T\bar{z}\right) (\nabla_y g(\bar{x},\bar{y}) + \bar{r}) \neq 0.$

Then there exist $\bar{w} \in R^n$, $\bar{t} \in R^m$ *such that* $\bar{x},\bar{y},\bar{w},\bar{t}$ *is a solution of (FD).*

Yang et al. [251] also stated the following converse duality theorem without proof:

Theorem 8.49 (Converse duality). *Let* $\bar{u},\bar{v},\bar{w},\bar{t}$ *be a solution of (FD). Suppose that all the conditions in Theorem 8.47 are fulfilled. Furthermore, assume that*

(a) $\nabla_{uu} f(\bar{u},\bar{v})(g(\bar{u},\bar{v}) - s(\bar{u}:E) + \bar{v}^T\bar{t}) - (f(\bar{u},\bar{v}) + s(\bar{u}:C) - \bar{v}^T\bar{w}) \nabla_{uu} g(\bar{u},\bar{v})$ *is positive or negative definite;*
(b) $(\nabla_v f(\bar{u},\bar{v}) - \bar{w}) \left(g(\bar{u},\bar{v}) - s(\bar{u}:E) + \bar{v}^T\bar{t}\right)$
$- \left(f(\bar{u},\bar{v}) + s(\bar{u}:C) - \bar{v}^T\bar{w}\right) (\nabla_v g(\bar{u},\bar{v}) + \bar{t}) \neq 0.$

Then there exist $\bar{z} \in R^n$, $\bar{r} \in R^n$ *such that* $\bar{u},\bar{v},\bar{z},\bar{r}$ *is a solution of (FP).*

Later Yang et al. [252] extended the above results to Multiobjective case by considering the following pair of problems:

(MFP)

$$\max \left(\frac{f_1(x,y)+s(x:C_1)-y^T z_1}{g_1(x,y)-s(x:E_1+y^T r_1)}, \ldots, \frac{f_k(x,y)+s(x:C_k)-y^T z_k}{g_k(x,y)-s(x:E_k+y^T r_k)} \right)$$

Subject to

$$\sum_{i=1}^{k} \lambda_i [\nabla_y f_i(x,y) - z_i - \frac{f_i(x,y)+s(x:C_i)-y^T z_i}{g_i(x,y)-s(x:E_i)+y^T r_i} (\nabla_y g_i(x,y) + r_i)] \leq 0,$$

$$y^T \sum_{i=1}^{k} \lambda_i [\nabla_y f_i(x,y) - z_i$$

$$- \frac{f_i(x,y) + s(x:C_i) - y^T z_i}{g_i(x,y) - s(x:E_i) + y^T r_i} (\nabla_y g_i(x,y) + r_i)] \leq 0,$$

$$z_i \in D_i, \;\; r_i \in F_i, \;\; 1 \leq i \leq k,$$
$$\lambda > 0, \;\; \lambda^T e = 1, \;\; x \geq 0,$$

(MFD)

$$\text{Min} \left(\frac{f_1(u,v) - s(v:D_1) + u^T w_1}{g_1(u,v) + s(v:F_1 - u^T t_1)}, \ldots, \frac{f_k(u,v) - s(v:D_k) + u^T w_k}{g_k(u,v) + s(v:F_k - u^T t_k)} \right)$$

Subject to

$$\sum_{i=1}^{k} \lambda_i [\nabla_u f_i(u,v) + w_i$$

$$- \frac{f_i(u,v) - s(v:D_i) + u^T w_i}{g_i(u,v) + s(v:F_i) - u^T t_i} (\nabla_u g_i(u,v) - t_i)] \geq 0,$$

$$u^T \sum_{i=1}^{k} \lambda_i [\nabla_u f_i(u,v) + w_i$$

$$- \frac{f_i(u,v) - s(v:D_i) + u^T w_i}{g_i(u,v) + s(v:F_i) - u^T t_i} (\nabla_u g_i(u,v) - t_i)] \geq 0,$$

$$w_i \in C_i, \;\; t_i \in E_i, \;\; 1 \leq i \leq k,$$

$$\lambda > 0, \;\; \lambda^T e = 1, \;\; v \geq 0,$$

where $e = (1, \ldots, 1)^T \in R^k$; f_i and g_i $i = (1, 2, \ldots, k)$ are twice differentiable functions from $R^n \times R^m$ to R. C_i, E_i, $i = (1, 2, \ldots, k)$ are compact convex sets in R^n and D_i, F_i, $i = (1, 2, \ldots, k)$ are compact convex sets in R^m. It is assumed that in the feasible regions the denominator of the objective function is nonnegative and the numerator is positive.

Remark 8.50. A frequently occurring example of a nondifferentiable support function is $(x^T Ax)^{\frac{1}{2}}$, where A, is a positive semidefinite matrix. It can be easily verified that $(x^T Ax)^{\frac{1}{2}} = s(x : C)$ where $C = \{Ay : y^T Ay \leq 1\}$, and that this set C is compact convex.

(a) If in the feasible regions $k \equiv 1$, $g_i \equiv 1$, $(x_i^T Bx)^{\frac{1}{2}} = s(x : C_i)$ where $C_i = \{B_i y : y^T B_i y \leq 1\}$, $(x_i^T Cx)^{\frac{1}{2}} = s(x : D_i)$ where $D_i = \{C_i y : y^T C_i y \leq 1\}$, $i = (1, \ldots, k)$ then (MFP) and (MFD) become a pair of symmetric dual nondifferentiable programs considered by Chandra et al. [30].

(b) If in ((MFP) and (MFD), $B_i = \{0\}$ and $C_i = \{0\}$, $i = (1, \ldots, k)$ and in the feasible regions $g_i \equiv 1$, $i = (1, \ldots, k)$, we obtain the symmetric dual Multiobjective programming problems studied by Weir and Mond [246].
(c) If in (MFP) and (MFD), $k = 1$, $g_i \equiv 1$, we obtain symmetric dual problems studied by Mond and Schechter [175].

Yang et al. [252] established usual weak and strong duality results for (MFP) and (MFD).

Kim et al. [124] introduced the following pair of symmetric dual multiobjective fractional variational problems:

(CMFP)

$$\mathrm{Min}\left(\frac{\int_a^b f_1(t,x(t),x^{\cdot}(t),y(t),y^{\cdot}(t))dt}{\int_a^b g_1(t,x(t),x^{\cdot}(t),y(t),y^{\cdot}(t))dt},\ldots,\frac{\int_a^b f_k(t,x(t),x^{\cdot}(t),y(t),y^{\cdot}(t))dt}{\int_a^b g_k(t,x(t),x^{\cdot}(t),y(t),y^{\cdot}(t))dt}\right)$$

Subject to $x(a) = 0 = x(b),\ \ y(a) = 0 = y(b)$

$x^{\cdot}(a) = 0 = x^{\cdot}(b),\ \ y^{\cdot}(a) = 0 = y^{\cdot}(b)$

$$\sum_{i=1}^k \tau_i \left\{[f_{iy} - Df_{iy^{\cdot}}]\, G_i(x,y) - [g_{iy} - Dg_{iy^{\cdot}}]\, F_i(x,y)\right\} \leqq 0,$$

$$\int_a^b y(t)^T \sum_{i=1}^k \tau_i \left\{[f_{iy} - Df_{iy^{\cdot}}]\, G_i(x,y) - [g_{iy} - Dg_{iy^{\cdot}}]\, F_i(x,y)\right\} dt \geqq 0,$$

$\tau > 0,\ \ \tau^T e = 1,\ \ t \in I,$

(CMFD)

$$\max\left(\frac{\int_a^b f_1(t,u(t),u^{\cdot}(t),v(t),v^{\cdot}(t))dt}{\int_a^b g_1(t,u(t),u^{\cdot}(t),v(t),v^{\cdot}(t))dt},\ldots,\frac{\int_a^b f_k(t,u(t),u^{\cdot}(t),v(t),v^{\cdot}(t))dt}{\int_a^b g_k(t,u(t),u^{\cdot}(t),v(t),v^{\cdot}(t))dt}\right) \tag{8.33}$$

Subject to $u(a) = 0 = u(b),\ \ v(a) = 0 = v(b)$

$u^{\cdot}(a) = 0 = u^{\cdot}(b),\ \ v^{\cdot}(a) = 0 = v^{\cdot}(b)$

$$\sum_{i=1}^k \tau_i \left\{[f_{iu} - Df_{iu^{\cdot}}]\, G_i(u,v) - [g_{iu} - Dg_{iu^{\cdot}}]\, F_i(u,v)\right\} \geqq 0,$$

$$\int_a^b u(t)^T \sum_{i=1}^k \tau_i \left\{[f_{iu} - Df_{iu^{\cdot}}]\, G_i(u,v) - [g_{iu} - Dg_{iu^{\cdot}}]\, F_i(u,v)\right\} dt \leqq 0,$$

$\tau > 0,\ \ \tau^T e = 1,\ \ t \in I,$

where $f_i : I \times R^n \times R^n \times R^m \times R^m \to R_+$ and $g_i : I \times R^n \times R^n \times R^m \times R^m \to R_+ \backslash \{0\}$ are continuously differentiable functions and $F_i(x,y) = \int_a^b f_i(t,x,x^\cdot,y,y^\cdot)dt$ and $G_i(x,y) = \int_a^b f_i(t,x,x^\cdot,y,y^\cdot)dt$.

Remark 8.51. If the time independence of programs (CMFP) and (CMFD) is removed and f and g are considered to have domain $R^n \times R^m$, we obtain the symmetric dual fractional pair given by Weir [243].

Kim et al. [124] established usual weak, strong and converse duality theorems as well as self-duality relations under invexity–incavity assumptions. Later Mishra et al. [162] studied nondifferentiable Multiobjective fractional symmetric duality under invexity–incavity assumptions, which extends and unifies earlier results in literature on symmetric duality. Mishra et al. [162] considered the problem of finding functions $x : [a,b] \to R^n$, and $y : [a,b] \to R^m$, where $(x^\cdot(t), y^\cdot(t))$ is piecewise smooth on $[a,b]$, to solve the following pair of symmetric dual multi-objective nondifferentiable fractional variational problems introduced as follows:

(MNFP)

$$\text{Min} \frac{\int_a^b \left(f(t,x(t),x^\cdot(t),y(t),y^\cdot(t)) + s(x(t)|C) - y(t)^T z\right) dt}{\int_a^b \left(g(t,x(t),x^\cdot(t),y(t),y^\cdot(t)) - s(x(t)|E) + y(t)^T r\right) dt}$$
$$= \frac{\int_a^b \left(f_1(t,x(t),x^\cdot(t),y(t),y^\cdot(t)) + s(x(t)|C_1) - y(t)^T z_1\right) dt}{\int_a^b \left(g_1(t,x(t),x^\cdot(t),y(t),y^\cdot(t)) - s(x(t)|E_1) + y(t)^T r_1\right) dt}, \ldots,$$
$$\frac{\int_a^b \left(f_k(t,x(t),x^\cdot(t),y(t),y^\cdot(t)) + s(x(t)|C_k) - y(t)^T z_k\right) dt}{\int_a^b \left(g_k(t,x(t),x^\cdot(t),y(t),y^\cdot(t)) - s(x(t)|E_k) + y(t)^T r_k\right) dt}$$

$$\text{Subject to } x(a) = 0 = x(b), \quad y(a) = 0 = y(b),$$
$$x^\cdot(a) = 0 = x^\cdot(b), \quad y^\cdot(a) = 0 = y^\cdot(b),$$
$$\sum_{i=1}^k \tau_i \{[f_{iy} - Df_{iy^\cdot} - z_i]\, G_i(x,y)$$
$$- [g_{iy} - Dg_{iy^\cdot} + r_i]\, F_i(x,y)\} \leqq 0,$$
$$\int_a^b y(t)^T \sum_{i=1}^k \tau_i \{[f_{iy} - Df_{iy^\cdot} - z_i]\, G_i(x,y)$$
$$\times [g_{iy} - Dg_{iy^\cdot} + r_i]\, F_i(x,y)\}\, dt \geqq 0,$$

$$\tau > 0, \quad \tau^T e = 1, \quad t \in I,$$
$$z_i \in D_i, \quad r_i \in H_i, \quad i = 1,2,\ldots,k.$$

(MNFD)

$$\max \frac{\int_a^b \left(f(t,u(t),u^{\cdot}(t),v(t),v^{\cdot}(t)) - s(v(t)|D) + u(t)^T w\right) dt}{\int_a^b \left(g(t,u(t),u^{\cdot}(t),v(t),v^{\cdot}(t)) + s(v(t)|H) - u(t)^T s\right) dt}$$
$$= \frac{\int_a^b \left(f_1(t,u(t),u^{\cdot}(t),v(t),v^{\cdot}(t)) - s(v(t)|D_1) + u(t)^T w_1\right) dt}{\int_a^b \left(g_1(t,u(t),u^{\cdot}(t),v(t),v^{\cdot}(t)) + s(v(t)|H_1) - u(t)^T s_1\right) dt}, \ldots,$$
$$\frac{\int_a^b \left(f_k(t,u(t),u^{\cdot}(t),v(t),v^{\cdot}(t)) - s(v(t)|D_k) + u(t)^T w_k\right) dt}{\int_a^b \left(g_k(t,u(t),u^{\cdot}(t),v(t),v^{\cdot}(t)) + s(v(t)|H_k) - u(t)^T s_k\right) dt}$$

$$\text{subject to } u(a) = 0 = u(b), \quad v(a) = 0 = v(b),$$
$$u^{\cdot}(a) = 0 = u^{\cdot}(b), \quad v^{\cdot}(a) = 0 = v^{\cdot}(b),$$

$$\sum_{i=1}^{k} \tau_i \left\{ [f_{iu} - Df_{iu^{\cdot}} + w_i]\, G_i^*(u,v) \right.$$
$$\left. - [g_{iu} - Dg_{iu^{\cdot}} - s_i]\, F_i^*(u,v) \right\} \geqq 0,$$

$$\int_a^b u(t)^T \sum_{i=1}^{k} \tau_i \left\{ [f_{iu} - Df_{iu^{\cdot}} + w_i]\, G_i^*(u,v) \right.$$
$$\left. \times [g_{iu} - Dg_{iu^{\cdot}} - s_i]\, F_i^*(u,v) \right\} \leqq 0,$$

$$\tau > 0, \quad \tau^T e = 1, \quad t \in I,$$
$$w_i \in C_i, \quad s_i \in E_i, \quad i = 1, 2, \ldots, k,$$

where $f_i : I \times R^n \times R^n \times R^m \times R^m \to R_+$ and $g_i : I \times R^n \times R^n \times R^m \times R^m \to R_+ \backslash \{0\}$ are continuously differentiable functions and

$$F_i(x,y) = \int_a^b \left\{ f_i(t,x,x^{\cdot},y,y^{\cdot}) + s(x(t)|C_i) - y(t)^T z_i \right\} dt;$$
$$G_i(x,y) = \int_a^b \left\{ g_i(t,x,x^{\cdot},y,y^{\cdot}) - s(x(t)|E_i) + y(t)^T r_i \right\} dt;$$
$$F_i^*(u,v) = \int_a^b \left\{ f_i(t,u,u^{\cdot},v,v^{\cdot}) - s(v(t)|D_i) + u(t)^T w_i \right\} dt;$$

and

$$G_i^*(u,v) = \int_a^b \left\{ g_i(t,u,u^{\cdot},v,v^{\cdot}) + s(v(t)|H_i) - u(t)^T s_i \right\} dt.$$

In the above problems (MNFP) and (MNFD), the numerators are nonnegative and denominators are positive; the differential operator D is given by

$$y = Dx \Leftrightarrow x(t) = \alpha + \int_a^t y(s)ds,$$

and $x(a) = \alpha$, $x(b) = \beta$ are given boundary values; thus $D = \frac{d}{dt}$ except at discontinuities. Let $f_x = f_x(t, x(t), x^{\cdot}(t), y(t), y^{\cdot}(t))$, $f_{x^{\cdot}} = f_{x^{\cdot}}(t, x(t), x^{\cdot}(t), y(t), y^{\cdot}(t))$, etc.

All the above statements for F_i, G_i, F_i^* and G_i^* will be assumed to hold for subsequent results. It is to be noted that

$$Df_{iy^{\cdot}} = f_{iy^{\cdot}t} + f_{iy^{\cdot}y}y^{\cdot} + f_{iy^{\cdot}y^{\cdot}}y^{\cdot\cdot} + f_{iy^{\cdot}x}x^{\cdot} + f_{iy^{\cdot}x^{\cdot}}x^{\cdot\cdot}$$

and consequently

$$\frac{\partial}{\partial y}Df_{iy^{\cdot}} = Df_{iy^{\cdot}y}, \quad \frac{\partial}{\partial y^{\cdot}}Df_{iy^{\cdot}} = Df_{iy^{\cdot}y^{\cdot}} + f_{iy^{\cdot}y}, \quad \frac{\partial}{\partial y^{\cdot\cdot}}Df_{iy^{\cdot}} = f_{iy^{\cdot}y^{\cdot}},$$
$$\frac{\partial}{\partial x}Df_{iy^{\cdot}} = Df_{iy^{\cdot}x}, \quad \frac{\partial}{\partial x^{\cdot}}Df_{iy^{\cdot}} = Df_{iy^{\cdot}x^{\cdot}} + f_{iy^{\cdot}x}, \quad \frac{\partial}{\partial x^{\cdot\cdot}}Df_{iy^{\cdot}} = f_{iy^{\cdot}x^{\cdot}}.$$

In order to simplify the notations we introduce

$$p_i = \frac{F_i(x,y)}{G_i(x,y)} = \frac{\int_a^b \left\{f_i(t,x,x^{\cdot},y,y^{\cdot}) + s(x(t)|C_i) - y(t)^T z_i\right\} dt}{\int_a^b \left\{g_i(t,x,x^{\cdot},y,y^{\cdot}) - s(x(t)|E_i) + y(t)^T r_i\right\} dt}$$

and

$$q_i = \frac{F_i^*(u,v)}{G_i^*(u,v)} = \frac{\int_a^b \left\{f_i(t,u,u^{\cdot},v,v^{\cdot}) - s(v(t)|D_i) + u(t)^T w_i\right\} dt}{\int_a^b \left\{g_i(t,u,u^{\cdot},v,v^{\cdot}) + s(v(t)|H_i) - u(t)^T s_i\right\} dt},$$

and express problems (MNFP) and (MNFD) equivalent as follows:

(EMSP)

$$\min p = (p_1, \ldots, p_k)^T$$

$$\text{subject to } x(a) = 0 = x(b), \quad y(a) = 0 = y(b),$$
$$x^{\cdot}(a) = 0 = x^{\cdot}(b), \quad y^{\cdot}(a) = 0 = y^{\cdot}(b),$$

$$\int_a^b \left\{f_i(t,x,x^{\cdot},y,y^{\cdot}) + s(x\,|C_i) - y^T z_i\right\} dt$$
$$- p_i \int_a^b \left\{g_i(t,x,x^{\cdot},y,y^{\cdot}) - s(x\,|E_i) + y^T r_i\right\} dt = 0;$$

$$\sum_{i=1}^k \tau_i \left\{[f_{iy} - Df_{iy^{\cdot}} - z_i] - p_i\,[g_{iy} - Dg_{iy^{\cdot}} + r_i]\right\} \leqq 0, \quad t \in I;$$

$$\int_a^b y(t)^T \sum_{i=1}^k \tau_i \left\{[f_{iy} - Df_{iy^{\cdot}} - z_i] - p_i\,[g_{iy} - Dg_{iy^{\cdot}} + r_i]\right\} \geqq 0, \quad t \in I;$$

$$\tau > 0, \quad \tau^T e = 1, \quad t \in I,$$
$$z_i \in D_i, \quad r_i \in H_i, \quad i = 1, 2, \ldots, k.$$

(EMSD)

$$\max q = (q_1, \ldots, q_k)^T$$

$$\text{Subject to } u(a) = 0 = u(b),\ \ v(a) = 0 = v(b),$$
$$u^{\cdot}(a) = 0 = u^{\cdot}(b),\ \ v^{\cdot}(a) = 0 = v^{\cdot}(b),$$

$$\int_a^b \{f_i(t,u,u^{\cdot},v,v^{\cdot}) - s(v\,|D_i) + u^T w_i\}\,dt$$
$$-\,q_i \int_a^b \{g_i(t,u,u^{\cdot},v,v^{\cdot}) + s(v\,|H_i) - u^T s_i\}\,dt = 0;$$

$$\sum_{i=1}^k \tau_i \{[f_{iu} - Df_{iu^{\cdot}} + w_i] - q_i\,[g_{iu} - Dg_{iu^{\cdot}} - s_i]\} \geqq 0, \quad t \in I;$$

$$\int_a^b u(t)^T \sum_{i=1}^k \tau_i \{[f_{iu} - Df_{iu^{\cdot}} + w_i] - q_i\,[g_{iu} - Dg_{iu^{\cdot}} - s_i]\} \leqq 0, \quad t \in I;$$

$$\tau > 0,\ \ \tau^T e = 1,\ \ t \in I,$$
$$w_i \in C_i,\ \ S_i \in E_i,\ \ i = 1, 2, \ldots, k.$$

In the above problems (EMSP) and (EMSD), it is to be noted that p and q are also nonnegative.

Remark 8.52. If the time dependence of programs (MNFP) and (MNFD) is removed and the functions involved are considered to have domain $R^n \times R^m$, we obtain the symmetric dual fractional pair given by

$$\text{(SNMFP)}\ \min\left[\frac{f_1(x,y) + s(x|C_1) - y^T z_1}{g_1(x,y) - s(x|E_1) + y^T r_1}, \ldots, \frac{f_k(x,y) + s(x|C_k) - y^T z_k}{g_k(x,y) - s(x|E_k) + y^T r_k}\right]$$

subject to

$$\sum_{i=1}^k \tau_i[\nabla_y f_i(x,y) - z_i - \frac{f_i(x,y) + s(x|C_i) - y^T z_i}{g_i(x,y) - s(x|E_i) + y^T r_i}\,(\nabla_y g_i(x,y) + r_i)] \le 0,$$

$$y^T \sum_{i=1}^k \tau_i[\nabla_y f_i(x,y) - z_i - \frac{f_i(x,y) + s(x|C_i) - y^T z_i}{g_i(x,y) - s(x|E_i) + y^T r_i}\,(\nabla_y g_i(x,y) + r_i)] \ge 0,$$

$$z_i \in D_i,\ \ r_i \in F_i,\ \ 1 \le i \le k,\ \ \tau > 0,\ \tau^T e = 1,\ \ x \ge 0.$$

$$\text{(SNMFD)}\quad \max \left[\frac{f_1(u,v) - s(v|D_1) + u^T w_1}{g_1(u,v) + s(v|F_1) - u^T t_1}, \ldots, \frac{f_k(u,v) - s(v|D_k) + u^T w_k}{g_k(u,v) + s(v|F_k) - u^T t_k}\right]$$

subject to

$$\sum_{i=1}^{k} \tau_i [\nabla_u f_i(u,v) + w_i - \frac{f_i(u,v) - s(v|D_i) + u^T w_i}{g_i(u,v) + s(v|F_i) - u^T t_i} (\nabla_u g_i(u,v) - t_i)] \geq 0,$$

$$u^T \sum_{i=1}^{k} \tau_i [\nabla_u f_i(u,v) + w_i - \frac{f_i(u,v) - s(v|D_i) + u^T w_i}{g_i(u,v) + s(v|F_i) - u^T t_i} (\nabla_u g_i(u,v) - t_i)] \leq 0,$$

$$w_i \in C_i,\ \ t_i \in E_i,\ \ 1 \leq i \leq k,\ \ \tau > 0,\ \ \tau^T e = 1,\ \ v \geq 0.$$

The pair of problems (SNMFP) and (SNMFD) obtained above is exactly the pair of problems (FP) and (FD) considered by Yang et al. [252], see also this section above.

If we set $k = 1$, and our problems are time independent, we get the following pair of problems:

(SNFP)

$$\min \left[\frac{f(x,y) + s(x|C) - y^T z}{g(x,y) - s(x|E) + y^T r}\right]$$

Subject to

$$\left[\nabla_y f(x,y) - z - \frac{f(x,y) + s(x|C) - y^T z}{g(x,y) - s(x|E) + y^T r} (\nabla_y g(x,y) + r)\right] \leq 0,$$

$$y^T \left[\nabla_y f(x,y) - z - \frac{f(x,y) + s(x|C) - y^T z}{g(x,y) - s(x|E) + y^T r} (\nabla_y g(x,y) + r)\right] \geq 0,$$

$$z \in D,\ \ r \in F,\ \ x \geq 0.$$

(SNFD)

$$\max \left[\frac{f(u,v) - s(v|D) + u^T w}{g(u,v) + s(v|F) - u^T t}\right]$$

Subject to

$$\left[\nabla_u f(u,v) + w - \frac{f(u,v) - s(v|D) + u^T w}{g(u,v) + s(v|F) - u^T t} \left(\nabla_u g(u,v) - t\right)\right] \geq 0,$$

$$u^T \left[\nabla_u f(u,v) + w - \frac{f(u,v) - s(v|D) + u^T w}{g(u,v) + s(v|F) - u^T t} \left(\nabla_u g(u,v) - t\right)\right] \leq 0,$$

$$w \in C, \quad t \in E, \quad v \geq 0.$$

The pair of problems (SNFP) and (SNFD) is exactly the pair of problems (FP) and (FD) considered by Yang et al. [252], see above in this section as well.

Mishra et al. [162] established usual duality results (for proofs, see, Mishra et al. [162]:

Theorem 8.53 (Weak duality). *Let $(x(t), y(t), p, \tau, z_1, z_2, \ldots, z_k, r_1, r_2, \ldots, r_k)$ be feasible for (EMSP) and let $(u(t), v(t), q, \tau, w_1, w_2, \ldots, w_k, s_1, s_2, \ldots, s_k)$ be feasible for (EMSD). Assume that $\int_a^b (f_i + .^T w_i)dt$ and $-\int_a^b (g_i - .^T s_i)dt$ are invex in x and $x^\cdot$ with respect to $\eta(x,u)$ and $-\int_a^b (f_i - .^T z_i)dt$ and $\int_a^b (g_i + .^T r_i)dt$ are invex in y and $y^\cdot$ with respect to $\xi(v,y)$ and $\eta(x,u) + u(t) \geqq 0$ and $\xi(v,y) + y(t) \geqq 0$, $\forall t \in I$, except possibly at corners of $(x^\cdot(t), y^\cdot(t))$ or $(u^\cdot(t), v^\cdot(t))$. Then one can not have $p \leq q$.*

Theorem 8.54 (Weak duality). *Let $(x(t), y(t), p, \tau, z_1, z_2, \ldots, z_k, r_1, r_2, \ldots, r_k)$ be feasible for (EMSP) and let $(u(t), v(t), q, \tau, w_1, w_2, \ldots, w_k, s_1, s_2, \ldots, s_k)$ be feasible for (EMSD). Assume that $\sum_{i=1}^k \tau_i \int_a^b (f_i + .^T w_i - q_i(g_i - .^T s_i))dt$ is pseudo invex in x and $x^\cdot$ with respect to $\eta(x,u)$ and $-\sum_{i=1}^k \tau_i \int_a^b (f_i - .^T z_i - p_i(g_i + .^T r_i))dt$ is pseudo invex in y and $y^\cdot$ with respect to $\xi(v,y)$, with $\eta(x,u) + u(t) \geqq 0$ and $\xi(v,y) + y(t) \geqq 0$, $\forall t \in I$, except possibly at corners of $(x^\cdot(t), y^\cdot(t))$ or $(u^\cdot(t), v^\cdot(t))$. Then one can not have $p \leq q$.*

The following strong duality Theorem 8.55 and converse duality Theorem 8.56 can be established on the lines of the proofs of Theorems 3.3 and 3.4 given by Kim et al. [124] in the light of the discussions given above in this section.

Theorem 8.55 (Strong duality). *Let $(\bar{x}(t), \bar{y}(t), \bar{p}, \bar{\tau}, \bar{z_1}, \bar{z_2}, \ldots, \bar{z_k}, \bar{r_1}, \bar{r_2}, \ldots, \bar{r_k})$ be a properly efficient solution for (EMSP) and fix $\tau = \bar{\tau}$ in (EMSD), and define*

$$\bar{p_i} = \frac{\int_a^b \{f_i(t, \bar{x}, \bar{x}^\cdot, \bar{y}, \bar{y}^\cdot) + s(\bar{x}(t)|C_i) - \bar{y}(t)^T \bar{z}_i\}\, dt}{\int_a^b \{g_i(t, \bar{x}, \bar{x}^\cdot, \bar{y}, \bar{y}^\cdot) - s(\bar{x}(t)|E_i) + \bar{y}(t)^T \bar{r}_i\}\, dt}, \quad i = 1, \ldots, k$$

Suppose that all the conditions in Theorem 8.53 or Theorem 8.54 are fulfilled. Furthermore, assume that

$$
\begin{aligned}
\text{(I)} \sum_{i=1}^{k} \bar{\tau}_i \int_a^b & \psi(t)^T [\{[(f_{iyy} - z_i) - \bar{p}_i(g_{iyy} + r_i)] \\
& - D[(f_{iy^{\cdot}y} - z_i) - \bar{p}_i(g_{iy^{\cdot}y} + r_i)\}] \\
& - D\{[(f_{iy^{\cdot}} - z_i - Df_{iy^{\cdot}y^{\cdot}}) - \bar{p}_i(g_{iyy^{\cdot}} + r_i - Dg_{iy^{\cdot}y^{\cdot}} - g_{iy^{\cdot}y})]\} \\
& + D^2\{-[(f_{iy^{\cdot}y^{\cdot}} - z_i) - \bar{p}_i(g_{iy^{\cdot}y^{\cdot}} + r^i)]\}]\psi(t)^T dt = 0
\end{aligned}
$$

implies that $\psi(t) = 0, \ \forall t \in I$, *and*

$$
\text{(II)} \left[\int_a^b \{(f_{1y} - z_1) - \bar{p}_1(g_{1y} + r_1)\} dt, \ldots, \int_a^b \{(f_{ky} - z_k) - \bar{p}_k(g_{ky} + r_k)\} dt \right]
$$

is linearly independent.
Then there exist $\bar{w}_i \in R^n, \ \bar{s}_i \in R^m, \ i = 1, 2, \ldots, k$ *such that* $(\bar{x}(t), \bar{y}(t), \bar{p}, \bar{\tau}, \bar{w}_1, \bar{w}_2, \ldots, \bar{w}_k, \bar{s}_1, \bar{s}_2, \ldots, \bar{s}_k)$ *is a properly efficient solution of (EMSD).*

Theorem 8.56 (Converse duality). *Let* $(\bar{x}(t), \bar{y}(t), \bar{p}, \bar{\tau}, \bar{z}_1, \bar{z}_2, \ldots, \bar{z}_k, \bar{r}_1, \bar{r}_2, \ldots, \bar{r}_k)$ *be a properly efficient solution for (EMSD) and fix* $\tau = \bar{\tau}$ *in (EMSP), and define* $\bar{p}_i$ *as in Theorem 8.55. Suppose that all the conditions in Theorem 8.53 or Theorem 8.54 are fulfilled. Furthermore, assume that (I) and (II) of Theorem 8.55 are satisfied. Then there exist* $\bar{w}_i \in R^n, \ \bar{s}_i \in R^m, \ i = 1, 2, \ldots, k$ *such that* $(\bar{x}(t), \bar{y}(t), \bar{p}, \bar{\tau}, \bar{w}_1, \bar{w}_2, \ldots, \bar{w}_k, \bar{s}_1, \bar{s}_2, \ldots, \bar{s}_k)$ *is a properly efficient solution of (EMSP).*

Following Kim et al. [124], we also present self-duality for (MNFP) and (MNFD) instead of for (EMSP) and (EMSD). Assume that $x(t)$ and $y(t)$ have the same dimension.

The function $f(t, x(t), x^{\cdot}(t), y(t), y^{\cdot}(t))$ is said to be skew-symmetric if

$$
f(t, x(t), x^{\cdot}(t), y(t), y^{\cdot}(t)) = -f(t, y(t), y^{\cdot}(t), x(t), x^{\cdot}(t))
$$

for all $x(t)$ and $y(t)$ in the domain of f and the function $g(t, x(t), x^{\cdot}(t), y(t), y^{\cdot}(t))$ will be called symmetric if

$$
g(t, x(t), x^{\cdot}(t), y(t), y^{\cdot}(t)) = g(t, y(t), y^{\cdot}(t), x(t), x^{\cdot}(t)).
$$

In order to establish the self-duality some conditions are required. We assume that $C = D, \ E = H$, and

$$
\begin{aligned}
g(t, u(t), u^{\cdot}(t), v(t), v^{\cdot}(t)) + s(v|E) - u(t)^T s \\
= g(t, v(t), v^{\cdot}(t), u(t), u^{\cdot}(t)) - s(u|E) + v(t)^T s.
\end{aligned}
$$

Theorem 8.57 (Self-duality). *If $f(t,x(t),x^{\cdot}(t),y(t),y^{\cdot}(t))$ is skew-symmetric and $g(t,x(t),x^{\cdot}(t),y(t),y^{\cdot}(t))$ is symmetric along with the assumptions $C = D$, $E = H$, and $g(t,u(t),u^{\cdot}(t),v(t),v^{\cdot}(t)) + s(v|E) - u(t)^T s = g(t,v(t),v^{\cdot}(t),u(t),u^{\cdot}(t)) - s(u|E) + v(t)^T s$, then (MNFP) and (MNFD) are self-dual. If (MNFP) and (MNFD) are dual problems, then with $(x^o(t),y^o(t),p^o,\tau^o,w^o,s^o)$ also $(y^o(t),x^o(t),p^o,\tau^o,w^o,s^o)$ are a joint optimal solution and the common optimal value is zero.*

Proof. As in Kim et al. [124], we have

$$\begin{aligned}
f_x(t,x(t),x^{\cdot}(t),y(t),y^{\cdot}(t)) &= -f_y(t,y(t),y^{\cdot}(t),x(t),x^{\cdot}(t))\\
f_y(t,x(t),x^{\cdot}(t),y(t),y^{\cdot}(t)) &= -f_x(t,y(t),y^{\cdot}(t),x(t),x^{\cdot}(t))\\
f_{x^{\cdot}}(t,x(t),x^{\cdot}(t),y(t),y^{\cdot}(t)) &= -f_{y^{\cdot}}(t,y(t),y^{\cdot}(t),x(t),x^{\cdot}(t))\\
f_{y^{\cdot}}(t,x(t),x^{\cdot}(t),y(t),y^{\cdot}(t)) &= -f_{x.}(t,y(t),y^{\cdot}(t),x(t),x^{\cdot}(t))
\end{aligned}$$

and with g symmetric, we have

$$\begin{aligned}
g_x(t,x(t),x^{\cdot}(t),y(t),y^{\cdot}(t)) &= g_y(t,y(t),y^{\cdot}(t),x(t),x^{\cdot}(t))\\
g_y(t,x(t),x^{\cdot}(t),y(t),y^{\cdot}(t)) &= g_x(t,y(t),y^{\cdot}(t),x(t),x^{\cdot}(t))\\
g_{x^{\cdot}}(t,x(t),x^{\cdot}(t),y(t),y^{\cdot}(t)) &= g_{y^{\cdot}}(t,y(t),y^{\cdot}(t),x(t),x^{\cdot}(t))\\
g_{y^{\cdot}}(t,x(t),x^{\cdot}(t),y(t),y^{\cdot}(t)) &= g_{x.}(t,y(t),y^{\cdot}(t),x(t),x^{\cdot}(t))
\end{aligned}$$

Expressing the dual problem (MNFD) as a minimization problem and making use of above relations and conditions given in the Theorem 8.57, we have

$$\min \frac{\int_a^b \{f(t,v(t),v^{\cdot}(t),u(t),u^{\cdot}(t)) + s(u(t)|C) - v(t)^T w\}\, dt}{\int_a^b \{g(t,v(t),v^{\cdot}(t),u(t),u^{\cdot}(t)) - s(u(t)|E) + v(t)^T s\}\, dt}$$

$$\begin{aligned}
&\text{subject to } u(a) = 0 = u(b), \;\; v(a) = 0 = v(b),\\
&u^{\cdot}(a) = 0 = u^{\cdot}(b), \;\; v^{\cdot}(a) = 0 = v^{\cdot}(b),\\
&\sum_{i=1}^{k} \tau_i \{[f_{iv} - Df_{iv^{\cdot}} - w_i]\, G_i(v,u) - [g_{iv} - Dg_{iv^{\cdot}} + s_i]\, F_i(v,u)\} \leqq 0,\\
&\int_a^b v(t)^T \sum_{i=1}^{k} \tau_i \{[f_{iv} - Df_{iv^{\cdot}} - w_i]\\
&\qquad\qquad \times G_i(v,u) - [g_{iv} - Dg_{iv^{\cdot}} + s_i]\, F_i(v,u)\} \geqq 0,\\
&\tau > 0, \;\; \tau^T e = 1, \;\; t \in I,\\
&w_i \in C_i, \;\; s_i \in E_i, \;\; i = 1,2,\ldots,k,
\end{aligned}$$

which is just the primal problem (MNFP).

Thus if $(x^o(t),y^o(t),p^o,\tau^o,w^o,s^o)$ is an optimal solution for (MNFD), then $(y^o(t),x^o(t),p^o,\tau^o,w^o,s^o)$ is an optimal solution for (MNFD).

Since f is skew-symmetric, g is symmetric, $C = D,\ E = H$, and

$$g(t, u(t), \dot{u}(t), v(t), \dot{v}(t)) + s(v|E) - u(t)^T s = g(t, v(t), \dot{v}(t), u(t), \dot{u}(t)) - s(u|E) + v(t)^T s,$$

we have

$$\frac{\int_a^b \{f(t, y^o(t), \dot{y}^o(t), x^o(t), \dot{x}^o(t)) + s(x^o(t)|C) - y^o(t)^T w^o\}\, dt}{\int_a^b \{g(t, y^o(t), \dot{y}^o(t), x^o(t), \dot{x}^o(t)) - s(x^o(t)|E) + y^o(t)^T s^o\}\, dt} = -\frac{\int_a^b \{f(t, x^o(t), \dot{x}^o(t), y^o(t), \dot{y}^o(t)) - s(y^o(t)|C) + x^o(t)^T w^o\}\, dt}{\int_a^b \{g(t, x^o(t), \dot{x}^o(t), y^o(t), \dot{y}^o(t)) + s(y^o(t)|E) - x^o(t)^T s^o\}\, dt}$$

Hence

$$\frac{\int_a^b \{f(t, x^o(t), \dot{x}^o(t), y^o(t), \dot{y}^o(t)) - s(y^o(t)|C) + x^o(t)^T w^o\}\, dt}{\int_a^b \{g(t, x^o(t), \dot{x}^o(t), y^o(t), \dot{y}^o(t)) + s(y^o(t)|E) - x^o(t)^T s^o\}\, dt} = \frac{\int_a^b \{f(t, y^o(t), \dot{y}^o(t), x^o(t), \dot{x}^o(t)) + s(x^o(t)|C) - y^o(t)^T w^o\}\, dt}{\int_a^b \{g(t, y^o(t), \dot{y}^o(t), x^o(t), \dot{x}^o(t)) - s(x^o(t)|E) + y^o(t)^T s^o\}\, dt} = -\frac{\int_a^b \{f(t, x^o(t), \dot{x}^o(t), y^o(t), \dot{y}^o(t)) - s(y^o(t)|C) + x^o(t)^T w^o\}\, dt}{\int_a^b \{g(t, x^o(t), \dot{x}^o(t), y^o(t), \dot{y}^o(t)) + s(y^o(t)|E) - x^o(t)^T s^o\}\, dt},$$

and so

$$\frac{\int_a^b \{f(t, x^o(t), \dot{x}^o(t), y^o(t), \dot{y}^o(t)) - s(y^o(t)|C) + x^o(t)^T w^o\}\, dt}{\int_a^b \{g(t, x^o(t), \dot{x}^o(t), y^o(t), \dot{y}^o(t)) + s(y^o(t)|E) - x^o(t)^T s^o\}\, dt} = \frac{\int_a^b \{f(t, y^o(t), \dot{y}^o(t), x^o(t), \dot{x}^o(t)) + s(x^o(t)|C) - y^o(t)^T w^o\}\, dt}{\int_a^b \{g(t, y^o(t), \dot{y}^o(t), x^o(t), \dot{x}^o(t)) - s(x^o(t)|E) + y^o(t)^T s^o\}\, dt} = 0.$$

References

1. Aghezzaf A, Hachimi M (2000) Generalized invexity and duality in multiobjective programming problems. Journal of Global Optimization 18: 91–101.
2. Aghezzaf A, Hachimi M (2001) Sufficient optimality conditions and duality in multiobjective optimization involving generalized convexity. Numerical Functional Analysis and Optimization 22(7–8): 775–788.
3. Ansari QH, Schaible S, Yao JC (1999) η-pseudolinearity. Riviste di Matematice per le Scienze Economiche e Sociali 22: 31–39.
4. Antczak T (2003) A new approach to multiobjective programming with a modified objective function. Journal of Global Optimization 27: 485–495.
5. Arrow KJ, Enthoven AC (1961) Quasi-concave programming. Econometrica 29: 779–800.
6. Avriel M (1972) R-convex functions. Mathematical Programming 2: 309–323.
7. Avriel M (1973) Solutions of certain nonlinear programs involving r-convex functions. Journal of Optimization Theory and Applications 11: 159–174.
8. Avriel M (1976) Nonlinear Programming: Analysis and Methods. Prentice-Hall, New Jersey.
9. Avriel M, Zang I (1980) Generalized arcwise-connected functions and characterizations of local–global minimum properties. Journal of Optimization Theory and Applications 32: 407–425.
10. Avriel M, Diewert WE, Schaible S, Zang I (1988) Generalized Concavity. Plenum, New York.
11. Balas E (1970) Minimax and duality for linear and nonlinear mixed integer programming. In: J. Abadie (ed.), Integer and Nonlinear Programming. North-Holland, Amsterdam.
12. Bazaraa MS, Shetty CM (1976) Foundations of Optimization. Springer, Berlin.
13. Bazaraa MS, Goode JJ, Shetty CM (1972) Constraint qualifications revisited. Management Science 18: 567–573.
14. Bector CR, Husain I (1992) Duality for multiobjective variational problems. Journal of Mathematical Analysis and Applications 166: 214–229.
15. Bector CR, Suneja SK, Gupta S (1992) Univex functions and univex nonlinear programming. In: Proceedings of the Administrative Sciences Association of Canada, 115–124.
16. Bector CR, Chandra S, Singh C (1993) A linearization approach to multiobjective programming duality. Journal of Mathematical Analysis and Application 175: 268–279.

17. Ben-Tal A (1977) On generalized means and generalized convex functions. Journal of Optimization Theory and Applications 21: 1–13.
18. Ben-Israel A, Mond B (1986) What is invexity? Journal of Australian Mathematical Society Series B 28: 1–9.
19. Berkovitz LD (1961) Variational methods in problems of control and programming. Journal of Mathematical Analysis and Applications 3: 145–169.
20. Bhatia D, Jain P (1994) Generalized (F,ρ)-convexity and duality for nonsmooth multiobjective programs. Optimization 31: 153–164.
21. Bhatia D, Kumar P (1995) Multiobjective control problem with generalized invexity. Journal of Mathematical Analysis and Applications 189: 676–692.
22. Bianchi M, Hadjisavvas N, Schaible S (2003) On pseudomonotone maps T for which $-T$ is also pseudomonotone. Journal of Convex Analysis 10: 149–168.
23. Bianchi M, Schaible S (2000) An extension of pseudolinear functions and variational inequality problems. Journal of Optimization Theory and Applications 104: 59–71.
24. Brandao AJV, Rojas-Medar MA, Silva GN (2000) Invex nonsmooth alternative theorems and applications. Optimization 48: 239–253 (Erratum in Optimization 51 (2002) 759).
25. Cambini A, Castagnoli E, Martein L, Mazzoleni P, Schaible S (1990) Generalized convexity and fractional programming with economic applications. In: Proceedings of the Third International Conference on Generalized Convexity, held at University of Pisa, Pisa, Italy, May 30 – June 1, 1988. Lecture Notes in Economics and Mathematical Systems, Vol. 345. Springer, Berlin.
26. Caprari E (2003) η-invex functions and (F,ρ)-convex functions: properties and equivalences. Optimization 52: 65–74.
27. Castagnoli E, Mazzoleni P (1989) Towards a unified type of concavity. In: C. Singh and B. K. Dass (eds.), Continuous-time, Fractional and Multiobjective Programming. Analytic Publishing, Delhi, 225–240.
28. Castellani M (2001) Nonsmooth invex functions and sufficient optimality conditions. Journal of Mathematical Analysis and Applications 255: 319–332.
29. Castellani M, Pappalardo M (1995) First order cone approximations and necessary optimality conditions. Optimization 35: 113–126.
30. Chandra S, Craven BD, Mond B (1985) Generalized concavity and duality with a square root term. Optimization 16: 653–662.
31. Chandra S, Craven BD, Huasain I (1985) A class of nondifferentiable continuous programming problem. Journal of Mathematical Analysis and Applications 107: 122–131.
32. Chandra S, Goyal A, Husain I (1998) On symmetric duality in mathematical programming with F-convexity. Optimization 43: 1–18.
33. Chandra S, Mond B, Smart I (1990) Constrained games and symmetric duality with pseudo-invexity. Opsearch 27: 14–30.
34. Chen X (2004) Minimax and symmetric duality for a class of multiobjective variational mixed integer programming problems. European Journal of Operational Research 154: 71–83.
35. Chen X, Yang J (2007) Symmetric duality for minimax multiobjective variational mixed integer programming problems with partial-invexity. European Journal of Operational Research 181(1): 76–85. doi: 10.1016/j.ejor.2006.04.045
36. Chew KL, Choo EU (1984) Pseudolinearity and efficiency. Mathematical Programming 28: 226–239.

37. Clarke FH (1976) A new approach to Lagrangian multipliers. Mathematics of Operations Research 1: 165–174.
38. Clarke FH (1983) Optimization and Nonsmooth Analysis. Wiley-Interscience, New York.
39. Coladas L, Wang S (1994) Optimality conditions for multiobjective and nonsmooth minimization in abstract spaces. Bulletin of the Australian Mathematical Society 50: 205–218.
40. Cottle RW (1963) Symmetric dual quadratic programs. Quarterly of Applied Mathematics 21: 237–243.
41. Cottle RW, Ferland JA (1972) Matrix-theoretic criteria for the quasi-convexity and pseudo-convexity of quadratic functions. Linear Algebra and Its Applications 5: 123–136.
42. Craven BD (1981) Duality for generalized convex fractional programs. In: S. Schaible and W. T. Ziemba (eds.), Generalized Concavity in Optimization and Economics. Academic, New York, 473–489.
43. Craven BD (1981) Invex functions and constrained local minima. Bulletin of the Australian Mathematical Society 24: 357–366.
44. Craven BD (1986) Nondifferentiable optimization by smooth approximations. Optimization 17: 3–17.
45. Craven BD, Glover BM (1985) Invex functions and duality. Journal of Australian Mathematical Society 24: 1–20.
46. Craven BD (1993) On continuous programming with generalized convexity. Asia-Pacific Journal of Operational Research 10: 219–232.
47. Craven BD, Mond B (1999) Fractional programming with invexity. In: A. Eberhard, R. Hill, D. Ralhp and B. M. Glover (eds.), Progress in Optimization from Australasia. Kluwer, Dordrecht, 79–89.
48. Crouzeix JP (1980) On second order conditions for quasiconvexity. Mathematical Programming 18: 349–352.
49. Crouzeix JP (1981) A duality framework in quasiconvex programming. In: S. Schaible and W. T. Ziemba (eds.), Convexity and Optimization in Economics. Academic, New York, 207–225.
50. Crouzeix JP, Ferland JA (1982) Criteria for quasiconvexity and pseudoconvexity: relationships and comparisons. Mathematical Programming 23: 193–205.
51. Crouzeix JP, Martinez-Legaz JE, Volle M (1998) Generalized convexity, generalized monotonicity. In: Proceedings of the Fifth International Symposium on Generalized Convexity, held at the Centre International de Rencontres Mathematiques (CIRM), Luminy-Marseille, France, June 17–21, 1996. Nonconvex Optimization and Its Applications, Vol. 27. Kluwer, Dordrecht.
52. Eberhard A, Hadjisavvas N, Luc DT (2002) Generalized convexity, generalized monotonicity and applications. In: Proceedings of the Seventh International Symposium on Generalized Convexity and Generalized Monotonicity, held at Hanoi, Vietnam, Aug. 27–31, 2001. Nonconvex Optimization and Its Applications, Vol. 77. Springer, Berlin.
53. Dantzig GB, Eisenberg E, Cottle RW (1965) Symmetric dual nonlinear programs. Pacific Journal of Mathematics 15: 809–812.
54. De Finetti B (1949) Sulle stratificazioni converse. Annali di Matematica Pura ed Applicata 30: 123–183.
55. Demyanov VF, Rubinov MA (1980) On quasidifferentiable functionals. Soviet Mathematics Doklady 21: 14–17.

56. Demyanov VF, Rubinov MA (1986) Quasidifferentiable Calculus. Optimization Software, New York.
57. Demyanov VF, Vasiliev LV (1985) Nondifferentiable Optimization. Optimization Software, New York.
58. Diewert WE, Avriel M, Zang I (1981) Nine kinds of quasi-convexity and concavity. Journal of Economic Theory 25: 397–420.
59. Dorn WS (1960) A symmetric dual theorem for quadratic programs. Journal of Operations Research Society Japan 2: 93–97.
60. Egudo RR, Hanson MA (1993) On sufficiency of Kuhn–Tucker conditions in nonsmooth multiobjective programming. Technical Report M-888, Florida State University.
61. Elster KH, Thierfelder J (1988) Abstract cone approximations and generalized differentiability in nonsmooth optimization. Optimization 19: 315–341.
62. Elster KH, Thierfelder J (1988) On cone approximations and generalized directional derivatives. In: F. H. Clarke, V. F. Demyanov and F. Giannessi (eds.), Nonsmooth Optimization and Related Topics. Plenum, New York, 133–154.
63. Fan K (1953) Minimax theorems. Proceedings of the National Academy of Sciences of the United States of America 39: 42–47.
64. Fan K, Glicksberg I, Hoffman AJ (1957) Systems of inequalities involving convex functions. Proceedings of American Mathematical Society 8: 617–622.
65. Fenchel W (1951) Convex cones, sets and functions. Mimeographed Lecture Notes. Princeton University, Princeton.
66. Friedrichs KD (1929) Verfahren der variations rechnung des minimum eines integral als das maximum eines anderen ausdruckes daeziestellen. Gottingen, Nachrichten.
67. Geoffrion AM (1968) Proper efficiency and the theory of vector optimization. Journal of Mathematical Analysis and Applications 22: 618–630.
68. Gramatovici S (2005) Optimality conditions in multiobjective control problems with generalized invexity. Analele Universitatii din Craiova. Seria Matematica-Informatica 32: 150–157.
69. Giorgi G (1990) A note on the relationships between convexity and invexity. Journal of Australian Mathematical Society Series B 32: 97–99.
70. Giorgi G, Guerraggio A (1998) Constraint qualifications in the invex case. Journal of Information and Optimization Sciences 19: 373–384.
71. Giorgi G, Guerraggio A (1996) Various types of nonsmooth invexity. Journal of Information and Optimization Sciences 17: 137–150.
72. Giorgi G, Guerraggio A (2000) Nonsmooth vector-valued invex functions and applications. Journal of Information and Optimization Sciences 21: 243–255.
73. Giorgi G, Guerraggio A (1998) The notion of invexity in vector optimization: smooth and nonsmooth case. In: J. P. Crouzeix, J. E. Martinez-Legaz and M. Volle (eds.), Nonconvex Optimization and Its Applications. Kluwer, Dordrecht, 389–405.
74. Giorgi G, Guerraggio A, Thierfelder J (2004) Mathematics of Optimization: Smooth and Nonsmooth Case. Elsevier, Amsterdam.
75. Giorgi G, Molho E (1992) Generalized invexity: relationships with generalized convexity and applications to optimality and duality conditions. In: P. Mazzoleni (ed.), Generalized Concavity for Economic Applications, Proceedings of the Workshop held in Pisa, April 2, 1992. Tecnoprint, Bologna, 53–70.

76. Gulati TR, Husain I, Ahmed A (1997) Multiobjective symmetric duality with invexity. Bulletin of the Australian Mathematical Society 56: 25–36.
77. Gulati TR, Islam AM (1994) Sufficiency and duality in multiobjective programming involving generalized F-convex functions. Journal of Mathematical Analysis and Applications 183: 181–195.
78. Hadjisavvas N, Schaible S (1993) On strong pseudomonotonicity and (semi) strict quasimonotonicity. Journal of Optimization Theory and Applications 79: 139–155.
79. Hadjisavvas N, Schaible S (1995) Errata corrige. Journal of Optimization Theory and Applications 85: 741–742.
80. Hadjisavvas N, Martinez-Legaz JE, Penot J-P (2001) Generalized convexity and generalized monotonicity. In: Proceedings of the Sixth International Symposium on Generalized Convexity/Monotonicity, held at Karlovassi-Samos, Greece, Aug. 30 – Sept. 3, 1999. Lecture Notes in Economics and Mathematical Systems, Vol. 502. Springer, Berlin.
81. Hadjisavvas N, Komlosi S, Schaible S (eds.) (2005) Handbook of Generalized Convexity and Generalized monotonicity. Springer, Berlin.
82. Hanson MA (1964) Bounds for functionally convex optimal control problems. Journal of Mathematical Analysis and Applications 8: 84–89.
83. Hanson MA (1981) On sufficiency of the Kuhn–Tucker conditions. Journal of Mathematical Analysis Applications 80: 545–550.
84. Hanson MA (1993) Second order invexity and duality in mathematical programming. Opsearch 30: 313–320.
85. Hanson MA, Pini R, Singh C (2001) Multiobjective programming under generalized type I invexity. Journal of Mathematical Analysis and Applications 261: 562–577.
86. Hanson MA, Mond B (1982) Further generalizations of convexity in mathematical programming. Journal of Information and Optimization Sciences 3: 25–32.
87. Hanson MA, Mond B (1987) Convex transformable programming problems and invexity. Journal of Information and Optimization Sciences 8: 201–207.
88. Hanson MA, Mond B (1987) Necessary and sufficient conditions in constrained optimization. Mathematical Programming 37: 51–58.
89. Hanson MA, Rueda NG (1989) A sufficient condition for invexity. Journal of Mathematical Analysis Applications 138: 193–198.
90. Hayashi M, Komiya H (1980) Perfect duality for convexlike programs. Journal of Optimization Theory and Applications 38: 179–189.
91. Hiriart-Urruty JB (1986) When is a point satisfying a global minimum of? American Mathematical Monthly 93: 556–558.
92. Hiriart-Urruty JB (1978) On optimality conditions in nondifferentiable programming. Mathematical Programming 14: 73–86.
93. Hiriart-Urruty JB (1979) Refinements of necessary optimality conditions in nondifferentiable programming I. Applied Mathematics and Optimization 5: 63–82.
94. Horst R (1984) On the convexification of nonlinear programming problems: an applications-oriented survey. European Journal of Operational Research 15: 382–392.
95. Huard P (1962) Dual programs. IBM Jornal of Research and Development 6: 137–139.

96. Ioffe AD (1986) Necessary and sufficient conditions for a local minimum I. A reduction theorem and first order conditions. SIAM Journal of Control and Optimization 17: 245–250.
97. Jahn J (1984) Scalarization in multiobjective optimization. Mathematical Programming 29: 203–219.
98. Jahn J, Krabs W (1988) Applications of multicriteria optimization in approximation theory. In: W. Stadler (ed.), Multicriteria Optimization in Engineering and in the Sciences. Plenum, New York, 49–75.
99. Jahn J, Sachs E (1986) Generalized quasiconvex mappings and vector optimization. SIAM Journal of Control and Optimization 24: 306–322.
100. Jeyakumar V (1985) Strong and weak invexity in mathematical programming. Methods of Operations Research 55: 109–125.
101. Jeyakumar V (1985) Convexlike alternative theorems and mathematical programming. Optimization 16: 643–652.
102. Jeyakumar V (1987) On optimality conditions in nonsmooth inequality constrained minimization. Numerecal Functional Analysis and Optimization 9: 535–546.
103. Jeyakumar V (1988) Equivalence of saddle-points and optima and duality for a class of nonsmooth non-convex problems. Journal of Mathematical Analysis and Applications 130: 334–343.
104. Jeyakumar V (1991) Composite nonsmooth programming with Gateaux differentiability. SIAM Journal on Optimization 1: 30–41.
105. Jeyakumar V, Mond B (1992) On generalized convex mathematical programming. Journal of Australian Mathematical Society Series B 34: 43–53.
106. Jeyakumar V, Yang XQ (1993) Convex composite multiobjective nonsmooth programming. Mathematical Programming 59: 325–343.
107. Jeyakumar V, Yang XQ (1995) On characterizing the solution sets of pseudolinear programs. Journal of Optimization Theory and Applications 87: 747–755.
108. Jourani A (1994) Constraint qualifications and Lagrange multipliers in nondifferentiable programming problems. Journal of Optimization Theory and Applications 81: 533–548.
109. Karamardian S (1967) Duality in mathematical programming. Journal of Mathematical Analysis and Applications 20: 344–358.
110. Karamardian S (1976) Complementarity over cones with monotone and pseudomonotone maps. Journal of Optimization Theory and Applications 18: 445–454.
111. Karamardian S, Schaible S (1990) Seven kinds of monotone maps. Journal of Optimization Theory and Applications 66: 37–46.
112. Karamardian S, Schaible S, Crouzeix JP (1993) Characterizations of generalized monotone maps. Journal of Optimization Theory and Applications 76: 399–413.
113. Kaul RN, Kaur S (1982) Generalizations of convex and related functions. European Journal of Operational Research 9: 369–377.
114. Kaul RN, Kaur S (1985) Optimality criteria in nonlinear programming involving nonconvex functions. Journal of Mathematical Analysis and Applications 105: 104–112.
115. Kaul RN, Suneja SK, Lalitha CS (1993) Duality in pseudolinear multiobjective fractional programming. Indian Journal of Pure and Applied Mathematics 24: 279–290.

116. Kaul RN, Suneja SK, Lalitha CS (1994) Generalized nonsmooth invexity. Journal of Information and Optimization Sciences 15: 1–17.
117. Khan ZA, Hanson MA (1997) On ratio invexity in mathematical programming. Journal of Mathematical Analysis and Applications 205: 330–336.
118. Kim DS (1988) Pseudo-invexity in mathematical programming. Atti dell'Accademia Peloritana dei Pericolanti. Classe di Scienze Fisiche, Matematiche e Naturali 66: 347–355.
119. Kim DS, Lee GM, Park JY, Son KH (1993) Control problems with generalized invexity. Mathematica Japonica 38: 263–269.
120. Kim DS, Kim MH (2005) Generalized type I invexity and duality in multiobjective variational problems. Journal of Mathematical Analysis and Applications 307: 533–554.
121. Kim DS, Lee GM (1993) Symmetric duality with pseudoinvexity in variational problems. Optimization 28: 9–16.
122. Kim DS, Lee WJ (1998) Symmetric duality for multiobjective variational problems with invexity. Journal of Mathematical Analysis and Applications 218: 34–48.
123. Kim DS, Yun YB, Lee WJ (1998) Multiobjective symmetric duality with cone constraints. European Journal of Operational Research 107: 686–691.
124. Kim DS, Lee WJ, Schaible S (2004) Symmetric duality for invex multiobjective fractional variational problems. Journal of Mathematical Analysis and Applications 289: 505–521.
125. Kim DS, Schaible S (2004) Optimality and duality for invex nonsmooth multiobjective programming problems. Optimization 53(2): 165–176.
126. Kim DS, Song YR (2001) Minimax and symmetric duality for nonlinear multiobjective mixed integer programming. European Journal of Operational Research 128: 435–446.
127. Klinger A, Mangasarian OL (1968) Logarithmic convexity and geometric programming. Journal of Mathematical Analysis and Applications 24: 388–408.
128. Komlosi S (1993) First and second order characterizations of pseudolinear functions. European Journal of Operations Research 67: 278–286.
129. Komlosi S, Rapcsak T, Schaible S (1994) Generalized convexity. In: Proceedings of the Fourth International Workshop on Generalized Convexity, held at Janus Pannonius University, Pecs, Hungary, Aug. 31 – Sept. 2, 1992. Lecture Notes in Economics and Mathematical Systems, Vol. 405. Springer, Berlin.
130. Konnov IV, Luc DT, Rubinov AM (2005) Generalized convexity and related topics. In: Proceedings of the Eighth International Conference on Generalized Convexity and Generalized Monotonicity, held at Varese, Italy, July 4–8, 2005. Lecture Notes Economics and Mathematical Systems, Vol. 583. Springer, Berlin.
131. Kortanek K, Evans JP (1967) Pseudoconcave programming and Lagrange regularity. Operations Research 15: 882–892.
132. Kruk S, Wolkowicz H (1999) Pseudolinear programming. SIAM Review 41: 795–805.
133. Kuhn HW, Tucker AW (1951) Nonlinear programming. In: Proceedings of the Second Berkeley Symposium on Mathematical Statistics and Probability, J. Neyman (ed.), University of California Press, Berkeley, 481–492.
134. Kuk H, Lee GM, Kim DS (1998) Nonsmooth multiobjective programs with $V-\rho$-invexity. Indian Journal of Pure and Applied Mathematics 29(4): 405–412.

135. Kumar V, Husain I, Chandra S (1995) Symmetric duality for minimax nonlinear mixed integer programming. European Journal of Operational Research 80: 425–430.
136. Lee GM (1994) Nonsmooth invexity in multiobjective programming. Journal of Information and Optimization Sciences 15: 127–136.
137. Lee GM, Kim DS, Lee BS (1996) Multiobjective control problems with invexity. Journal of Information and Optimization Sciences 17: 151–160.
138. Li ZM (1990) Generalized Kuhn–Tucker conditions of the vector extremum problem in the linear topological spaces. Journal of Systems Science Mathematical Science 10: 78–83.
139. Li ZF, Wang SY (1994) Lagrange multipliers and saddle points in multiobjective programming. Journal of Optimization Theory and Applications 83: 63–81.
140. Luc DT (1989) Theory of Vector Optimization. Springer, Berlin Heidelberg New York.
141. Mahajan DJ, Vartak MN (1977) Generalizations of some duality theorems in nonlinear programming. Mathematical Programming 1: 293–317.
142. Mangasarian OL (1965) Pseudo-convex functions. SIAM Journal on Control 3: 281–290.
143. Mangasarian OL (1969) Nonlinear Programming. McGraw-Hill, New York.
144. Mangasarian OL (1975) Second and higher-order duality in mathematical programming. Journal of Mathematical Analysis and Applications 51: 607–620.
145. Martin DH (1985) The essence of invexity. Journal of Optimization Theory and Applications 47: 65–76.
146. Martos B (1975) Nonlinear Programming. Theory and Methods. North Holland, Amsterdam.
147. Marusciac I (1982) On Fritz-John type optimality criterion in multiobjective optimization. L'Analyse Numericque et la theorie de l'Approximation 11: 109–114.
148. Mifflin R (1977) Semismooth and semiconvex functions in constrained optimization. SIAM Journal on Control and Optimization 15: 959–972.
149. Minami M (1983) Weak Pareto-optimal necessary conditions in a nondifferentiable multiobjective program on a Banach space. Journal of Optimization Theory and Applications 41: 451–461.
150. Minty GJ (1964) On the monotonicity of the gradient of a convex function. Pacific Journal of Mathematics 14: 243–247.
151. Mishra BK, Das C (1980) On minimax and symmetric duality for a nonlinear mixed integer programming problem. Opsearch 17: 1–11.
152. Mishra MS, Acharya D, Nanda S (1985) On a pair of nonlinear mixed integer programming problems. European Journal of Operational Research 19: 98–103.
153. Mishra SK (1995) V-Invex functions and their applications. Ph.D. Thesis, Institute of Technology, Banaras Hindu University, Varanasi, India.
154. Mishra SK (1997) On sufficiency and duality in nonsmooth multiobjective programming. Opsearch 34(4): 221–231.
155. Mishra SK (1998) On multiple-objective optimization with generalized univexity. Journal of Mathematical Analysis and Applications 224: 131–148.
156. Mishra SK (2001) On second order symmetric duality in mathematical programming. In: Manju Lata Agarwal and Kanwar Sen (eds.), Recent Developments in Operational Research. Narosa, New Delhi, 261–272.

157. Mishra SK, Lai KK (2007) Second order symmetric duality in multiobjective programming involving generalized cone-invex functions. European Journal of Operational Research 178: 20–26.
158. Mishra SK, Mukherjee RN (1996) On generalized convex multiobjective nonsmooth programming. Journal of Australian Mathematical Society Series B 38: 140–148.
159. Mishra SK, Wang SY, Lai KK (2004) Higher-order duality for a class of nondifferentiable multiobjective programming problems. International Journal of Pure and Applied Mathematics 11(2): 221–232.
160. Mishra SK, Rueda NG (2000) Higher-order generalized invexity and duality in mathematical programming. Journal of Mathematical Analysis and Applications 247: 173–182.
161. Mishra SK, Rueda NG (2002) Higher-order generalized invexity and duality in nondifferentiable mathematical programming. Journal of Mathematical Analysis and Applications 272: 496–506.
162. Mishra SK, Wang SY, Lai KK (2007) Symmetric duality for a class of nondifferentiable multiobjective fractional variational problems. Journal of Mathematical Analysis and Applications 333: 1093–1110.
163. Mishra SK, Wang SY, Lai KK (2008) V-Invex Functions and Vector Optimization. Optimization and Its Applications, Vol. 14. Springer, New York.
164. Mititelu S (2004) Invex functions. Revue Roumaine de Mathématique pures et appliquées 49: 529–544.
165. Mohan SR, Neogy SK (1994) On invex sets and preinvex functions. Journal of Mathematical Analysis and Applications 189: 901–908.
166. Molho E, Schaible S (1996) Invexity of quadratic functions and restricted local invexity. Journal of Information and Optimization Sciences 17: 127–136.
167. Mond B (1965) A symmetric dual theorem for nonlinear programs. Quarterly of Applied Mathematics 23: 265–269.
168. Mond B (1974) Nondifferentiable mathematical programming problems. Journal of Mathematical Analysis and Applications 46: 169–174.
169. Mond B (1983) Generalized convexity in mathematical programming. Bulletin of the Australian Mathematical Society 27: 185–202.
170. Mond B, Hanson MA (1967) Duality for variational problems. Journal of Mathematical Analysis and Applications 18: 355–364.
171. Mond B, Hanson MA (1968) Duality for control problems. SIAM Journal on Control 6: 114–120.
172. Mond B, Husain I (1989) Sufficient optimality criteria and duality for variational problems with generalized invexity. Journal of Australian Mathematical Society Series B 31: 108–121.
173. Mond B, Husain I, Durga Prasad MV (1991) Duality for a class of nondifferentiable multiobjective programs. Utilitas Mathematica 39: 3–19.
174. Mond B, Chandra S, Husain I (1988) Duality for variational problems with invexity. Journal of Mathematical Analysis and Applications 134: 322–328.
175. Mond B, Schechter M (1996) Nondifferentiable symmetric duality. Bulletin of the Australian Mathematical Society 53: 177–188.
176. Mond B, Smart I (1988) Duality and sufficiency in control problems with invexity. Journal of Mathematical Analysis and Applications 136: 325–333.
177. Mond B, Smart I (1989) Duality with invexity for a class of nondifferentiable static and continuous programming problems. Journal of Mathematical Analysis and Applications 141: 373–388.

178. Mond B, Smart I (1991) On invexity and other forms of generalized convexity. Management Science Recherche Operesionalle, Proceedings of the Annual Conference in the Administrative Sciences Associaton of Canada, Management Sciences Division 12: 16–25.
179. Mond B, Smart I (1991) The role of invexity in optimization. In: Proceedings of the International Workshop in Analysis and Applications, C. V. Stanojeric and O. Hadzic (eds.), Novi Sad, 177–196.
180. Mond B, Weir T (1981) Generalized concavity and duality. In: S. Schaible and W. T. Ziemba (eds.), Generalized Concavity in Optimization and Economics. Academic, New York, 263–279.
181. Mond B, Weir T (1981–1983) Generalized convexity and higher order duality. Journal of Mathematical Sciences 16–18: 74–94.
182. Mond B, Zhang J (1998) Higher order invexity and duality in mathematical programming. In: JP Crouzeix et al. (eds.), Generalized Convexity, Generalized Monotonicity: Recent Results. Kluwer, Dordrecht, 357–372.
183. Mukherjee RN, Mishra SK (1995) Generalized invexity and duality in multiple objective variational problems. Journal of Mathematical Analysis and Applications 195: 307–322.
184. Mukherjee RN, Mishra SK (1994) Sufficient optimality criteria and duality for multiobjective variational problems with V-invexity. Indian Journal of Pure and Applied Mathematics 25: 801–813.
185. Mukherjee RN, Rao CP (1997) Generalized F-convexity and its classification. Optimization 40: 335–341.
186. Naccache PM (1978) Connectedness of the set of nondominated outcomes in multicriteria optimization. Journal of Optimization Theory and Applications 25: 459–467.
187. Nahak C, Nanda S (1996) Duality for multiobjective variational problems with invexity. Optimization 36: 235–248.
188. Nahak C, Nanda S (2000) Symmetric duality with pseudoinvexity in variational problems. European Journal of Operational Research 122: 145–150.
189. Nanda S (1988) Invex generalizations of some duality results. Opsearch 25: 105–111.
190. Nanda S, Das LN (1994) Pseudo-invexity and symmetric duality in nonlinear programming. Optimization 28: 267–273.
191. Ortega IH, Rheinboldt WC (1970) Iterative Solutions of Nonlinear Equations in Several Variables. Academic, New York.
192. Osuna-Gomez R, Rufian-Lizana A, Ruiz-Canales P (1998) Invex functions and generalized convexity in multiobjective programming. Journal of Optimization Theory and Applications 98: 651–661.
193. Osuna-Gomez R, Beato-Moreno A, Rifian-Lizana A (1999) Generalized convexity in multiobjective programming. Journal of Mathematical Analysis and Application 233: 205–220.
194. Pareto V (1896) Course d'economie politique. Rouge, Lausane.
195. Parida J, Sahoo M, Kumar A (1989) A Variational-like inequality problem. Bulletin of the Australian Mathematical Society 39: 225–231.
196. Parida J, Sen A (1987) A variational-like inequality for multifunctions with applications. Journal of Optimization Theory and Applications 124: 73–81.
197. Passy U, Prisman EZ (1985) A convex-like duality scheme for quasi-convex programs. Mathematical Programming 32: 278–300.

198. Peng JW (2006) Criteria for generalized invex monotonicities without condition C. European Journal of Operational Research 170: 667–671.
199. Peterson DW (1973) A review of constraint qualifications in finite-dimensional spaces. SIAM Review 15: 639–654.
200. Phu HX (2004) Is invexity weaker than convexity? Vietnam Journal of Mathematics 32: 87–94.
201. Pini R (1991) Invexity and generalized convexity. Optimization 21: 513–525.
202. Pini, R, Singh C (1997) A survey of recent [1985–1995] advances in generalized convexity with applications to duality theory and optimality conditions. Optimization 39: 311–360.
203. Ponstein J (1967) Seven kinds of convexity. SIAM Review 9: 115–119.
204. Preda V (1992) On efficiency and duality for multiobjective programs. Journal of Mathematical Analysis and Applications 166: 365–377.
205. Pshenichnyi BN (1971) Necessary Conditions for an Extremum. Marcel Dekker, New York.
206. Rao CR, Mitra SK (1971) Generalized Inverse of Matrices and Its Applications. Wiley, New York.
207. Rapcsak T (1991) On pseudolinear functions. European Journal of Operations Research 50: 353–360.
208. Reddy LV, Mukherjee RN (1999) Some results on mathematical programming with generalized ratio invexity. Journal of Mathematical Analysis and Applications 240: 299–310.
209. Reiland TW (1990) Nonsmooth invexity. Bulletin of the Australian Mathematical Society 42: 437–446.
210. Reiland TW (1989) Generalized invexity for nonsmooth vector-valued mappings. Numerical Functional Analysis and Optimization 10: 1191–1202.
211. Rockafellar RT (1970) Convex Analysis. Princeton University Press, Princeton.
212. Rockafellar RT (1980) Generalized directional derivatives and subgradients of nonconvex functions. Canadian Journal of Mathematics 32: 257–280.
213. Rueda NG (1989) Generalized convexity in nonlinear programming. Journal of Information and Optimization Sciences 10: 395–400.
214. Rueda NG, Hanson MA (1988) Optimality criteria in mathematical programming involving generalized invexity. Journal of Mathematical Analysis and Applications 130: 375–385.
215. Rueda NG, Hanson MA, Singh C (1995) Optimality and duality with generalized convexity. Journal of Optimization Theory and Applications 86(2): 491–500.
216. Ruiz-Garzon G, Osuna-Gomez R, Rufian-Lizana A (2003) Generalized invex monotonicity. Euroepan Journal of Operational Research 144: 501–512.
217. Sawaragi Y, Nakayama H, Tanino T (1985) Theory of Multiobjective Optimization. Academic, New York.
218. Schaible S (1981) Generalized convexity of quadratic functions. In: S. Schaible and W. T. Ziemba (eds.), Generalized Convexity in Optimization and Economics. Academic, New York, 183–197.
219. Schaible S (1981) Quasiconvex, pseudoconvex and strictly pseudoconvex quadratic functions. Journal of Optimization Theory and Applications 35: 303–338.
220. Schaible S (1973) Quasiconcave, strictly quasiconcave and pseudoconcave functions. Methods of Operations Research 17: 308–316.

221. Schaible S, Ziemba WT (1981) Generalized convexity in optimization and economics. In: Proceedings of the First International Conference on Generalized Convexity, held at University of British Columbia, Vancouver, Canada, Aug. 4–15, 1980. Academic, New York.
222. Singh C, Dass BK (1989) Continuous-time fractional and multiobective programming. In: Proceedings of the second International Conference on generalized convexity, held at St. Lawrence University, Canton, New York, July 29 – Aug. 1, 1986. Analytic Publishing, Delhi.
223. Singh C, Hanson MA (1991) Multiobjective fractional programming duality theory. Naval Research Logistics Quarterly 38: 925–933.
224. Smart I (1990) Invex functions and their application to mathematical programming. Ph.D. Thesis, Department of Mathematics, La Trobe University, Australia.
225. Smart I (1996) On the continuity of the kernel of invex functions. Journal of Mathematical Analysis and Applications 197: 548–557.
226. Smart I, Mond B (1990) Symmetric duality with invexity in variational problems. Journal of Mathematical Analysis and Applications 152: 536–545.
227. Stancu-Minasian IM (1997) Fractional Programming: Theory, Methods and Applications. Kluwer, Dordrecht.
228. Stoxer J, Witzgall C (1970) Convexity and Optimization in Finite Dimensions I. Springer, Berlin.
229. Swartz C (1987) Pshenichnyi's theorem for vector minimization. Journal of Optimization Theory and Applications 53: 309–317.
230. Suneja SK, Lalitha CS, Khurana S (2003) Second order symmetric duality in multiobjective programming. European Journal of Operational Research 144: 492–500.
231. Tamura K, Arai S (1982) On proper and improper efficient solutions of optimal problems with multicriteria. Journal of Optimization Theory and Applications 38: 191–205.
232. Tanaka Y (1990) Note on generalized convex functions. Journal of Optimization Theory and Applications 66: 345–349.
233. Tanaka Y, Fukusima M, Ibaraki I (1989) On generalized pseudo-convex functions. Journal of Mathematical Analysis and Applications 144: 342–355.
234. Tanino T, Sawaragi Y (1979) Duality theory in multiobjective programming. Journal of Optimization Theory and Applications 27: 509–529.
235. Thompson WA, Parke DW (1973) Some properties of generalized concave functions. Operations Research 22: 305–313.
236. Udriste C, Ferrara M, Opris D (2004) Economic Geometric Dynamics. Geometry Balkan, Bucharest.
237. Valentine FA (1937) The problem of Lagrange with differential inequalities as added side conditions. Contributions to the Calculus of Variations (1933–1937). University of Chicago Press, Chicago.
238. Vartak MN, Gupta I (1989) Generalized convexity in multiobjective programming. Indian Journal of Pure and Applied Mathematics 20: 10–39.
239. Vial JP (1983) Strong and weak convexity of sets and functions. Mathematics of Operations Research 8: 231–259.
240. Ward DE (1987) Isotone tangent cones and nonsmooth optimization. Optimization 18: 769–783.
241. Ward DE (1988) The quantificational tangent cones. Canadian Journal of Mathematics 40: 666–694.

242. Weir T (1988) A note on invex functions and duality in multiple-objective optimization. Opsearch 25: 98–104.
243. Weir T (1991) Symmetric dual multiobjective fractional programming. Journal of the Australian Mathematical Society Series A 50: 67–74.
244. Weir T, Jeyakumar V (1988) A class of nonconvex functions and mathematical programming. Bulletin of the Australian Mathematical Society 38: 177–189.
245. Weir T, Mond B (1988) Pre-invex functions in multiple objective optimization. Journal of Mathematical Analysis and Applications 136: 29–38.
246. Weir T, Mond B (1988) Symmetric and self duality in multiple objective programming. Asia Pacific Journal of Operational Research 5: 124–133.
247. Wolfe P (1961) A duality theorem for nonlinear programming. Quarterly of Applied Mathematics 19: 239–244.
248. Yang XM, Yang XQ, Teo KL (2003) Generalized invexity and invariant monotonicity. Journal of Optimization Theory and Applications 117: 607–625.
249. Yang XM, Yang XQ, Teo KL (2005) Criteria for generalized invex monotonicities. European Journal of Operational Research 164: 115–119.
250. Yang XM, Li D (2001) On properties of preinvex functions. Journal of Mathematical Analysis and Applications 256: 229–241.
251. Yang XM, Teo KL, Yang XQ (2002) Symmetric duality for a class of nonlinear fractional programming problems. Journal of Mathematical Analysis and Applications 271: 7–15.
252. Yang XM, Wang SY, Deng XT (2002) Symmetric duality for a class of multiobjective fractional programming problems. Journal of Mathematical Analysis and Applications 274: 279–295.
253. Ye YL (1991) D-invexity and optimality conditions. Journal of Mathematical Analysis and Applications 162: 242–249.
254. Yen ND, Sach PH (1993) On locally lipschitz vector-valued invex functions. Bulletin of the Australian Mathematical Society 47: 259–271.
255. Zang I, Choo EU, Avriel M (1977) On functions whose stationary points are global minima. Journal of Optimization Theory and Applications 22: 195–208.
256. Zhang J (1998) Generalized convexity and higher order duality for mathematical programming problems. Ph.D. thesis, La Trobe University, Australia.
257. Zhang J (1999) Higher order convexity and duality in multiobjective programming problems. In: A. Eberhard, R. Hill, D. Ralph and B. M. Glover (eds.), Progress in Optimization, Contributions from Australasia. Applied Optimization, Vol. 30. Kluwer, Dordrecht, 101–116.
258. Zhian L, Qingka Y (2001) Duality for a class of multiobjective control problems with generalized invexity. Journal of Mathematical Analysis and Applications 256: 446–461.

Index

MIX
Papier aus verantwortungsvollen Quellen
Paper from responsible sources
FSC® C105338
www.fsc.org

If you have any concerns about our products,
you can contact us on
ProductSafety@springernature.com

In case Publisher is established outside the EU,
the EU authorized representative is:
Springer Nature Customer Service Center GmbH
Europaplatz 3, 69115 Heidelberg, Germany

Printed by Libri Plureos GmbH
in Hamburg, Germany